MW00339442

Numerical Methods for Unconstrained Optimization and Nonlinear Equations

SIAM's Classics in Applied Mathematics series consists of books that were previously allowed to go out of print. These books are republished by SIAM as a professional service because they continue to be important resources for mathematical scientists.

Editor-in-Chief
Gene H. Golub, *Stanford University*

Editorial Board
Richard A. Brualdi, *University of Wisconsin-Madison*
Herbert B. Keller, *California Institute of Technology*
Ingram Olkin, *Stanford University*
Robert E. O'Malley, Jr., *University of Washington*

Classics in Applied Mathematics

C. C. Lin and L. A. Segel, *Mathematics Applied to Deterministic Problems in the Natural Sciences*

Johan G. F. Belinfante and Bernard Kolman, *A Survey of Lie Groups and Lie Algebras with Applications and Computational Methods*

James M. Ortega, *Numerical Analysis: A Second Course*

Anthony V. Fiacco and Garth P. McCormick, *Nonlinear Programming: Sequential Unconstrained Minimization Techniques*

F. H. Clarke, *Optimization and Nonsmooth Analysis*

George F. Carrier and Carl E. Pearson, *Ordinary Differential Equations*

Leo Breiman, *Probability*

R. Bellman and G. M. Wing, *An Introduction to Invariant Imbedding*

Abraham Berman and Robert J. Plemmons, *Nonnegative Matrices in the Mathematical Sciences*

Olvi L. Mangasarian, *Nonlinear Programming*

*Carl Friedrich Gauss, *Theory of the Combination of Observations Least Subject to Errors: Part One, Part Two, Supplement.* Translated by G. W. Stewart

Richard Bellman, *Introduction to Matrix Analysis*

U. M. Ascher, R. M. M. Mattheij, and R. D. Russell, *Numerical Solution of Boundary Value Problems for Ordinary Differential Equations*

K. E. Brenan, S. L. Campbell, and L. R. Petzold, *Numerical Solution of Initial-Value Problems in Differential-Algebraic Equations*

Charles L. Lawson and Richard J. Hanson, *Solving Least Squares Problems*

J. E. Dennis, Jr. and Robert B. Schnabel, *Numerical Methods for Unconstrained Optimization and Nonlinear Equations*

*First time in print.

Numerical Methods for Unconstrained Optimization and Nonlinear Equations

J. E. Dennis, Jr.
Rice University
Houston, Texas

Robert B. Schnabel
University of Colorado
Boulder, Colorado

siam.

Society for Industrial and Applied Mathematics
Philadelphia

Copyright © 1996 by the Society for Industrial and Applied Mathematics.

This SIAM edition is an unabridged, corrected republication of the work first published by Prentice-Hall, Inc., Englewood Cliffs, NJ, 1983.

10 9 8 7 6 5 4 3 2 1

All rights reserved. Printed in the United States of America. No part of this book may be reproduced, stored, or transmitted in any manner without the written permission of the publisher. For information, write to the Society for Industrial and Applied Mathematics, 3600 University City Science Center, Philadelphia, PA 19104-2688.

Library of Congress Cataloging-in-Publication Data

Dennis, J. E. (John E.) , 1939-
    Numerical methods for unconstrained optimization and nonlinear
equations / J.E. Dennis, Jr., Robert B. Schnabel .
        p. cm. -- (Classics in applied mathematics ; 16)
    Originally published : Englewood Cliffs, N.J. : Prentice-Hall,
c 1983.
    Includes bibliographical references and indexes.
    ISBN 0-89871-364-1 (pbk)
    1. Mathematical optimization. 2. Equations--Numerical solutions.
I. Schnabel, Robert B. II. Title III. Series.
QA402.5.D44   1996   95-51776

The royalties from the sales of this book are being placed in a fund to help students attend SIAM meetings and other SIAM related activities. This fund is administered by SIAM and qualified individuals are encouraged to write directly to SIAM for guidelines.

siam. is a registered trademark.

To Catherine, Heidi, and Cory

Contents

Preface to the Classics Edition

We are delighted that SIAM is republishing our original 1983 book after what many in the optimization field have regarded as "premature termination" by the previous publisher. At 12 years of age, the book may be a little young to be a "classic," but since its publication it has been well received in the numerical computation community. We are very glad that it will continue to be available for use in teaching, research, and applications.

We set out to write this book in the late 1970s because we felt that the basic techniques for solving small to medium-sized nonlinear equations and unconstrained optimization problems had matured and converged to the point where they would remain relatively stable. Fortunately, the intervening years have confirmed this belief. The material that constitutes most of this book—the discussion of Newton-based methods, globally convergent line search and trust region methods, and secant (quasi-Newton) methods for nonlinear equations, unconstrained optimization, and nonlinear least squares—continues to represent the basis for algorithms and analysis in this field. On the teaching side, a course centered around Chapters 4 to 9 forms a basic, in-depth introduction to the solution of nonlinear equations and unconstrained optimization problems. For researchers or users of optimization software, these chapters give the foundations of methods and software for solving small to medium-sized problems of these types.

We have not revised the 1983 book, aside from correcting all the typographical errors that we know of. (In this regard, we especially thank Dr. Oleg Burdakov who, in the process of translating the book for the Russian edition published by Mir in 1988, found numerous typographical errors.) A main reason for not revising the book at this time is that it would have delayed its republication substantially. A second reason is that there appear to be relatively few places where the book needs updating. But inevitably there are some. In our opinion, the main developments in the solution of small to medium-sized unconstrained optimization and nonlinear equations problems since the publication of this book, which a current treatment should include, are

1. improved algorithms and analysis for trust region methods for unconstrained optimization in the case when the Hessian matrix is indefinite [1, 2] and

2. improved global convergence analysis for secant (quasi-Newton) methods [3].

A third, more recent development is the field of automatic (or computational) differentiation [4]. Although it is not yet fully mature, it is clear that this development is increasing the availability of analytic gradients and Jacobians and therefore reducing the cases where finite difference approximations to these derivatives are needed. A fourth, more minor but still significant development is a new, more stable modified Cholesky factorization method [5, 6]. Far more progress has been made in the solution of large nonlinear equations and unconstrained optimization problems. This includes the development or improvement of conjugate gradient, truncated-Newton, Krylov-subspace, and limited-memory methods. Treating these fully would go beyond the scope of this book even if it were revised, and fortunately some excellent new references are emerging, including [7]. Another important topic that is related to but not within the scope of this book is that of new derivative-free methods for unconstrained optimization [8].

The appendix of this book has had an impact on software in this field. The IMSL library created their unconstrained optimization code from this appendix, and the UNCMIN software [9] created in conjunction with this appendix has been and continues to be a widely used package for solving unconstrained optimization problems. This software also has been included in a number of software packages and other books. The UNCMIN software continues to be available from the second author (bobby@cs.colorado.edu).

Finally, one of the most important developments in our lives since 1983 has been the emergence of a new generation: a granddaughter for one of us, a daughter and son for the other. This new edition is dedicated to them in recognition of the immense joy they have brought to our lives and with all our hopes and wishes for the lives that lay ahead for them.

[1] J. J. Moré and D. C. Sorensen, *Computing a trust region step,* SIAM J. Sci. Statist. Comput., 4 (1983), pp. 553–572.

[2] G. A. Shultz, R. B. Schnabel, and R. H. Byrd, *A family of trust region based algorithms for unconstrained minimization with strong global convergence properties,* SIAM J. Numer. Anal., 22 (1985), pp. 47–67.

[3] R. H. Byrd, J. Nocedal, and Y. Yuan, *Global convergence of a class of quasi-Newton methods on convex problems,* SIAM J. Numer. Anal., 24 (1987), pp. 1171–1189.

[4] A. Griewank and G. F. Corliss, eds., *Automatic Differentiation of Algorithms: Theory, Implementation, and Application,* Society for Industrial and Applied Mathematics, Philadelphia, PA, 1991.

[5] R. B. Schnabel and E. Eskow, *A new modified Cholesky factorization,* SIAM J. Sci. Statist. Comput., 11 (1990), pp. 1136–1158.

[6] E. Eskow and R. B. Schnabel, *Software for a new modified Cholesky factorization,* ACM Trans. Math. Software, 17 (1991), pp. 306–312.

[7] C. T. Kelley, *Iterative Methods for Linear and Nonlinear Equations,* Society for Industrial and Applied Mathematics, Philadelphia, PA, 1995.

[8] J. E. Dennis, Jr. and V. Torczon, *Direct search methods on parallel computers,* SIAM J. Optim., 1 (1991), pp. 448–474.

[9] R. B. Schnabel, J. E. Koontz, and B. E. Weiss, *A modular system of algorithms for unconstrained minimization,* ACM Trans. Math. Software, 11 (1985), pp. 419–440.

Preface

This book offers a careful introduction, at a low level of mathematical and computational sophistication, to the numerical solution of problems in unconstrained optimization and systems of nonlinear equations. We have written it, beginning in 1977, because we feel that the algorithms and theory for small-to-medium-size problems in this field have reached a mature state, and that a comprehensive reference will be useful. The book is suitable for graduate or upper-level undergraduate courses, but also for self-study by scientists, engineers, and others who have a practical interest in such problems.

The minimal background required for this book would be calculus and linear algebra. The reader should have been at least exposed to multivariable calculus, but the necessary information is surveyed thoroughly in Chapter 4. Numerical linear algebra or an elementary numerical methods course would be helpful; the material we use is covered briefly in Section 1.3 and Chapter 3.

The algorithms covered here are all based on Newton's method. They are often called **Newton-like**, but we prefer the term **quasi-Newton**. Unfortunately, this term is used by specialists for the subclass of these methods covered in our Chapters 8 and 9. Because this subclass consists of sensible multidimensional generalizations of the secant method, we prefer to call them secant methods. Particular secant methods are usually known by the proper names of their discoverers, and we have included these servings of alphabet soup, but we have tried to suggest other descriptive names commensurate with their place in the overall scheme of our presentation.

The heart of the book is the material on computational methods for multidimensional unconstrained optimization and nonlinear equation problems covered in Chapters 5 through 9. Chapter 1 is introductory and will be more useful for students in pure mathematics and computer science than for readers with some experience in scientific applications. Chapter 2, which covers the one-dimensional version of our problems, is an overview of our approach to the subject and is essential motivation. Chapter 3 can be omitted by readers who have studied numerical linear algebra, and Chapter 4 can be omitted by those who have a good background in multivariable calculus. Chapter 10 gives a fairly complete treatment of algorithms for nonlinear least squares, an important type of unconstrained optimization problem that, owing to its special structure, is solved by special methods. It draws heavily on the chapters that precede it. Chapter 11 indicates some research directions in which the field is headed; portions of it are more difficult than the preceding material.

We have used the book for undergraduate and graduate courses. At the lower level, Chapters 1 through 9 make a solid, useful course; at the graduate level the whole book can be covered. With Chapters 1, 3, and 4 as remedial reading, the course takes about one quarter. The remainder of a semester is easily filled with these chapters or other material we omitted.

The most important omitted material consists of methods not related to Newton's method for solving unconstrained minimization and nonlinear equation problems. Most of them are important only in special cases. The Nelder-Meade simplex algorithm [see, e.g., Avriel (1976)], an effective algorithm for problems with less than five variables, can be covered in an hour. Conjugate direction methods [see, e.g., Gill, Murray, and Wright (1981)] properly belong in a numerical linear algebra course, but because of their low storage requirements they are useful for optimization problems with very large numbers of variables. They can be covered usefully in two hours and completely in two weeks.

The omission we struggled most with is that of the Brown-Brent methods. These methods are conceptually elegant and startlingly effective for partly linear problems with good starting points. In their current form they are not competitive for general-purpose use, but unlike the simplex or conjugate-direction algorithms, they would not be covered elsewhere. This omission can be remedied in one or two lectures, if proofs are left out [see, e.g., Dennis (1977)]. The final important omission is that of the continuation or homotopy-based methods, which enjoyed a revival during the seventies. These elegant ideas can be effective as a last resort for the very hardest problems but are not yet competitive for most problems. The excellent survey by Allgower and Georg (1980) requires at least two weeks.

We have provided many exercises; many of them further develop ideas that are alluded to briefly in the text. The large appendix (by Schnabel) is intended to provide both a mechanism for class projects and an important reference for readers who wish to understand the details of the algorithms and perhaps to develop their own versions. The reader is encouraged to read the preface to the appendix at an early stage.

Several problems of terminology and notation were particularly troublesome. We have already mentioned the confusion over the terms "quasi-Newton" and "secant methods." In addition, we use the term "unconstrained optimization" in the title but "unconstrained minimization" in the text, since technically we consider only minimization. For maximization, turn the problems upside-down. The important term "global" has several interpretations, and we try to explain ours clearly in Section 1.1. Finally, a major notational problem was how to differentiate between the ith component of an n-vector x, a *scalar* usually denoted by x_i, and the ith iteration in a sequence of such x's, a *vector* also usually denoted x_i. After several false starts, we decided to allow this conflicting notation, since the intended meaning is always clear from the context; in fact, the notation is rarely used in both ways in any single section of the text.

We wanted to keep this book as short and inexpensive as possible without slighting the exposition. Thus, we have edited some proofs and topics in a merciless fashion. We have tried to use a notion of rigor consistent with good taste but subservient to insight, and to include proofs that give insight while omitting those that merely substantiate results. We expect more criticism for omissions than for inclusions, but as every teacher knows, the most difficult but important part in planning a course is deciding what to leave out.

We sincerely thank Idalia Cuellar, Arlene Hunter, and Dolores Pendel for typing the numerous drafts, and our students for their specific identification of unclear passages. David Gay, Virginia Klema, Homer Walker, Pete Stewart, and Layne Watson used drafts of the book in courses at MIT, Lawrence Livermore Laboratory, University of Houston, University of New Mexico, University of Maryland, and VPI, and made helpful suggestions. Trond Steihaug and Mike Todd read and commented helpfully on portions of the text.

Rice University *J. E. Dennis, Jr.*

University of Colorado at Boulder *Robert B. Schnabel*

Before we begin, a program note

The first four chapters of this book contain the background material and motivation for the study of multivariable nonlinear problems. In Chapter 1 we introduce the problems we will be considering. Chapter 2 then develops some algorithms for nonlinear problems in just one variable. By developing these algorithms in a way that introduces the basic philosophy of all the nonlinear algorithms to be considered in this book, we hope to provide an accessible and solid foundation for the study of multivariable nonlinear problems. Chapters 3 and 4 contain the background material in numerical linear algebra and multivariable calculus required to extend our consideration to problems in more than one variable.

Introduction

This book discusses the methods, algorithms, and analysis involved in the computational solution of three important nonlinear problems: solving systems of nonlinear equations, unconstrained minimization of a nonlinear functional, and parameter selection by nonlinear least squares. Section 1.1 introduces these problems and the assumptions we will make about them. Section 1.2 gives some examples of nonlinear problems and discusses some typical characteristics of problems encountered in practice; the reader already familiar with the problem area may wish to skip it. Section 1.3 summarizes the features of finite-precision computer arithmetic that the reader will need to know in order to understand the computer-dependent considerations of the algorithms in the text.

1.1 PROBLEMS TO BE CONSIDERED

This book discusses three nonlinear problems in real variables that arise often in practice. They are mathematically equivalent under fairly reasonable hypotheses, but we will not treat them all with the same algorithm. Instead we will show how the best current algorithms seek to exploit the structure of each problem.

The simultaneous nonlinear equations problem (henceforth called "nonlinear equations") is the most basic of the three and has the least exploitable structure. It is

$$\text{Given} \quad F: \mathbb{R}^n \longrightarrow \mathbb{R}^n,$$

$$\text{find} \quad x_* \in \mathbb{R}^n \quad \text{for which} \quad F(x_*) = 0 \in \mathbb{R}^n, \tag{1.1.1}$$

where $\mathbb{R}^n$ denotes n-dimensional Euclidean space. Of course, (1.1.1) is just the standard way of denoting a system of n nonlinear equations in n unknowns, with the convention that right-hand side of each equation is zero. An example is

$$F(x_1, x_2) = \begin{pmatrix} x_1^2 + x_2^3 + 7 \\ x_1 + x_2 + 1 \end{pmatrix},$$

which has $F(x_*) = 0$ for $x_* = (1, -2)^T$.

Certainly the x_* that solves (1.1.1) would be a minimizer of

$$\sum_{i=1}^{n} (f_i(x))^2,$$

where $f_i(x)$ denotes the ith component function of F. This is a special case of the *unconstrained minimization* problem

$$\text{Given} \quad f: \mathbb{R}^n \longrightarrow \mathbb{R}$$

$$\text{find } x_* \in \mathbb{R}^n \quad \text{for which} \quad f(x_*) \le f(x) \text{ for every } x \in \mathbb{R}^n, \tag{1.1.2}$$

which is the second problem we will consider. Usually (1.1.2) is abbreviated to

$$\min_{x \in \mathbb{R}^n} f: \mathbb{R}^n \longrightarrow \mathbb{R}. \tag{1.1.3}$$

An example is

$$\min_{x \in \mathbb{R}^n} f(x_1, x_2, x_3) = (x_1 - 3)^2 + (x_2 + 5)^4 + (x_3 - 8)^2,$$

which has the solution $x_* = (3, -5, 8)^T$.

In some applications, one is interested in solving a constrained version of (1.1.3),

$$\min_{x \in \Omega \subset \mathbb{R}^n} f: \mathbb{R}^n \longrightarrow \mathbb{R}, \tag{1.1.4}$$

where Ω is a closed connected region. If the solution to (1.1.4) lies in the interior of Ω, then (1.1.4) can still be viewed as an unconstrained minimization problem. However, if x_* is a boundary point of Ω, then the minimization of f over Ω becomes a constrained minimization problem. We will not consider the constrained problem because less is known about how it should be solved, and there is plenty to occupy us in considering unconstrained problems. Furthermore, the techniques for solving unconstrained problems are the foundation for constrained-problem algorithms. In fact, many attempts to solve constrained problems boil down to either solving a related unconstrained minimization problem whose solution $\hat{x}$ is at least very near the solution x_* of the constrained problem, or to finding a nonlinear system of equations whose simultaneous solution is the same x_*. Finally, a large percentage of the problems that we have met in practice are either unconstrained or else constrained in a very trivial way—for example, every component of x might have to be nonnegative.

The third problem that we consider is also a special case of unconstrained minimization, but owing to its importance and its special structure it is a research area all by itself. This is the *nonlinear least-squares* problem:

$$\text{Given} \quad R: \mathbb{R}^n \longrightarrow \mathbb{R}^m, \quad m \geq n,$$

$$\text{find} \quad x_* \in \mathbb{R}^n \quad \text{for which} \quad \sum_{i=1}^m (r_i(x))^2 \text{ is minimized,} \qquad (1.1.5)$$

where $r_i(x)$ denotes the ith component function of R. Problem (1.1.5) is most frequently met within the context of curve fitting, but it can arise whenever a nonlinear system has more nonlinear requirements than degrees of freedom.

We are concerned exclusively with the very common case when the nonlinear functions F, f, or R are at least once, twice, or twice continuously differentiable, respectively. We do not necessarily assume that the derivatives are analytically available, only that the functions are sufficiently smooth. For further comments on the typical size and other characteristics of nonlinear problems being solved today, see Section 1.2.

The typical scenario in the numerical solution of a nonlinear problem is that the user is asked to provide a subroutine to evaluate the problem function(s), and a starting point x_0 that is a crude approximation to the solution x_*. If they are readily available, the user is asked to provide first and perhaps second derivatives. Our emphasis in this book is on the most common difficulties encountered in solving problems in this framework: (1) what to do if the starting guess x_0 is not close to the solution x_* ("global method") and how to combine this effectively with a method that is used in the vicinity of the answer ("local method"); (2) what to do if analytic derivatives are not available; and (3) the construction of algorithms that will be efficient if evaluation of the problem function(s) is expensive. (It often is, sometimes dramatically so.) We discuss the basic methods and supply details of the algorithms that are currently considered the best ones for solving such problems. We also give the analysis that we believe is relevant to understanding these methods and extending or improving upon them in the future. In particular, we try to identify and emphasize the ideas and techniques that have evolved as the central ones in this field. We feel that the field has jelled to a point where these techniques are identifiable, and while some improvement is still likely, one no longer expects new algorithms to result in quantum jumps over the best being used today.

The techniques for solving the nonlinear equations and unconstrained minimization problems are closely related. Most of the book is concerned with these two problems. The nonlinear least-squares problem is just a special case of unconstrained minimization, but one can modify unconstrained minimization techniques to take special advantage of the structure of the nonlinear least-squares problem and produce better algorithms for it. Thus Chapter 10 is really an extensive worked-out example that illustrates how to apply and extend the preceding portion of the book.

One problem that we do not address in this book is finding the "global minimizer" of a nonlinear functional—that is, the absolute lowest point of $f(x)$ in the case when there are many distinct local minimizers, solutions to (1.1.2) in open connected regions of $\mathbb{R}^n$. This is a very difficult problem that is not nearly as extensively studied or as successfully solved as the problems we consider; two collections of papers on the subject are Dixon and Szegö (1975, 1978). Throughout this book we will use the word "global," as in "global method" or "globally convergent algorithm" to denote a method that is designed to converge to a *local* minimizer of a nonlinear functional or *some* solution of a system of nonlinear equations, *from almost any starting point*. It might be appropriate to call such methods *local* or *locally convergent*, but these descriptions are already reserved by tradition for another usage. Any method that is guaranteed to converge from every starting point is probably too inefficient for general use [see Allgower and Georg (1980)].

1.2 CHARACTERISTICS OF "REAL-WORLD" PROBLEMS

In this section we attempt to provide some feeling for nonlinear problems encountered in practice. First we give three real examples of nonlinear problems and some considerations involved in setting them up as numerical problems. Then we make some remarks on the size, expense, and other characteristics of nonlinear problems encountered in general.

One difficulty with discussing sample problems is that the background and algebraic description of problems in this field is rarely simple. Although this makes consulting work interesting, it is of no help in the introductory chapter of a numerical analysis book. Therefore we will simplify our examples when possible.

The simplest nonlinear problems are those in one variable. For example, a scientist may wish to determine the molecular configuration of a certain compound. The researcher derives an equation $f(x)$ giving the potential energy of a possible configuration as a function of the tangent x of the angle between its two components. Then, since nature will cause the molecule to assume the configuration with the minimum potential energy, it is desirable to find the x for which $f(x)$ is minimized. This is a minimization problem in the single variable x. It is likely to be highly nonlinear, owing to the physics of the function f. It truly is unconstrained, since x can take any real value. Since the problem has only one variable, it should be easy to solve by the techniques of Chapter 2. However, we have seen related problems where f was a function of between 20 and 100 variables, and although they were not *difficult* to solve, the evaluations of f cost between \$5 and \$100 each, and so they were *expensive* to solve.

A second common class of nonlinear problems is the choice of some best one of a family of curves to fit data provided by some experiment or from some sample population. Figure 1.2.1 illustrates an instance of this problem that we encountered: 20 pieces of solar spectroscopy data y_i taken at wavelengths t_i were provided by a satellite, and the underlying theory implied that any m such pieces of data $(t_1, y_1), \ldots, (t_m, y_m)$, could be fitted by a bell-shaped curve. In practice, however, there was experimental error in the points, as shown in the figure. In order to draw conclusions from the data, one wants to find the bell-shaped curve that comes "closest" to the m points. Since the general equation for a bell-shaped curve is

$$y(x_1, x_2, x_3, x_4, t) = x_1 + x_2 e^{-(t+x_3)^2/x_4},$$

this means choosing x_1, x_2, x_3, and x_4 to minimize some aggregate measure of the discrepancies (residuals) between the data points and the curve; they are given by

$$r_i(x) \triangleq y(x_1, x_2, x_3, x_4, t_i) - y_i.$$

The most commonly used aggregate measure is the sum of the squares of the r_i's, leading to determination of the bell-shaped curve by solution of the nonlinear least-squares problem,

$$\min_{x \in \mathbb{R}^4} f(x) \triangleq \sum_{i=1}^m r_i(x)^2 = \sum_{i=1}^m [x_1 + x_2 e^{-(t_i+x_3)^2/x_4} - y_i]^2. \quad (1.2.1)$$

Some comments are in order. First, the reason problem (1.2.1) is called a *nonlinear* least-squares problem is that the residual functions $r_i(x)$ are nonlinear functions of some of the variables x_1, x_2, x_3, x_4. Actually r_i is linear in x_1 and x_2, and some recent methods take advantage of this (see Chapter 10). Second, there are functions other than the sum of squares that could be chosen to measure the aggregate distance of the data points from the bell-shaped

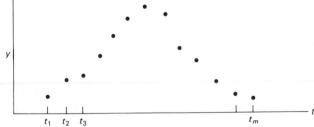

Figure 1.2.1 Data points to be fitted with a bell-shaped curve.

curve. Two obvious choices are

$$f_1(x) = \sum_{i=1}^{m} |r_i(x)|$$

and

$$f_\infty(x) = \max_{1 \leq i \leq m} |r_i(x)|.$$

The reasons one usually chooses to minimize $f(x)$ rather than $f_1(x)$ or $f_\infty(x)$ are sometimes statistical and sometimes that the resultant optimization problem is far more mathematically tractable, since the least-squares function is continuously differentiable and the other two are not. In practice, most data-fitting problems are solved using least squares. Often f is modified by introducing "weights" on the residuals, but this is not important to our discussion here.

As a final example, we give a version of a problem encountered in studying nuclear fusion reactors. A nuclear fusion reactor would be shaped like a doughnut, with some hot plasma inside (see Figure 1.2.2). An illustrative simplification of the actual problem is that we were asked to find the combination of the inner radius (r), width of the doughnut (w), and temperature of the plasma (t) that would lead to the lowest cost per unit of energy. Scientists had determined that the cost per unit of energy was modeled by

$$f(r, w, t) = 10^6 \cdot t^2 \left[c_1 \left(r^2 + \frac{2 \cdot r \cdot w}{3} + \frac{w^2}{6} \right) \right.$$

$$\left. + c_2 \left(c_4 + \frac{c_4^2}{r^2 - c_4^2} \right) \right] + c_3 \left(\frac{1}{2} + \frac{r}{w} \right) (10^3 t)^{-1.46}, \quad (1.2.2)$$

where c_1, c_2, c_3, c_4 are constants. Thus the nonlinear problem was to minimize f as a function of r, w, and t.

There were, however, other important aspects to this problem. The first was that, unlike the variables in the previous examples, r, w, and t could not assume arbitrary real values. For example, r and w could not be negative.

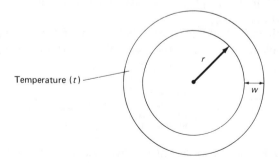

Figure 1.2.2   Nuclear fusion reactor.

Therefore, this was a constrained minimization problem. Altogether there were five simple linear constraints in the three variables.

It is important to emphasize that a constrained problem must be treated as such only if the presence of the constraints is expected to affect the solution, in the sense that the solution to the constrained problem is expected not to be a minimizer of the same function without the constraints. In the nuclear reactor problem, the presence of the constraints usually did make a difference, and so the problem was solved by constrained techniques. However, many problems with simple constraints, such as bounds on the variables, can be solved by unconstrained algorithms, because the constraints are satisfied by the unconstrained minimizer.

Notice that we said the constraints in the nuclear reactor problem *usually* made a difference. This is because we were actually asked to solve 625 instances of the problem, using different values for the constants c_1, c_2, c_3, and c_4. These constant values depended on factors, such as the cost of electricity, that would be constant at the time the reactor was running, but unknown until then. It was necessary to run different instances of the problem in order to see how the optimal characteristics of the reactor were affected by changes in these factors. Often in practical applications one wants to solve many related instances of a particular problem; this makes the efficiency of the algorithm more important. It also makes one willing to experiment with various algorithms initially, to evaluate them on the particular class of problems.

Finally, equation (1.2.2) was only the simple model of the nuclear fusion reactor. In the next portion of the study, the function giving the cost per unit of energy was not an analytic formula like (1.2.2); rather it was the output from a model of the reactor involving partial differential equations. There were also five more parameters (see Figure 1.2.3). The minimization of this sort of function is very common in nonlinear optimization, and it has some important influences on our algorithm development. First, a function like this is probably accurate to only a few places, so it wouldn't make sense to ask for many places of accuracy in the solution. Second, while the function f may be many times continuously differentiable, its derivatives usually are not obtainable. This is one reason why derivative approximation becomes so important. And finally, evaluation of f may be quite expensive, further stimulating the desire for efficient algorithms.

The problems above give some indication of typical characteristics of nonlinear problems. The first is their size. While certainly there are problems

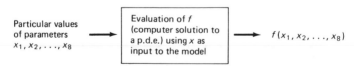

Figure 1.2.3 Function evaluation in refined model of the nuclear reactor problem.

that have more variables than those discussed above, most of the ones we see have relatively few variables, say 2 to 30. The state of the art is such that we hope to be able to solve most of the small problems, say those with from 2 to 15 variables, but even 2-variable problems can be difficult. Intermediate problems in this field are those with from 15 to 50 variables; current algorithms will solve many of these. Problems with 50 or more variables are large problems in this field; unless they are only mildly nonlinear, or there is a good starting guess, we don't have a good chance of solving them economically. These size estimates are very volatile and depend less on the algorithms than on the availability of fast storage and other aspects of the computing environment.

A second issue is the availability of derivatives. Frequently we deal with problems where the nonlinear function is itself the result of a computer simulation, or is given by a long and messy algebraic formula, and so it is often the case that analytic derivatives are not readily available although the function is several times continuously differentiable. Therefore it is important to have algorithms that work effectively in the absence of analytic derivatives. In fact, if a computer-subroutine library includes the option of approximating derivatives, users rarely will provide them analytically—who can blame them?

Third, as indicated above, many nonlinear problems are quite expensive to solve, either because an expensive nonlinear function is evaluated repeatedly or because the task is to solve many related problems. We have heard of a 50-variable problem in petroleum engineering where each function evaluation costs 100 hours of IBM 3033 time. Efficiency, in terms of algorithm running time and function and derivative evaluations, is an important concern in developing nonlinear algorithms.

Fourth, in many applications the user expects only a few digits of accuracy in the answer. This is primarily due to the approximate nature of the other parts of the problem: the function itself, other parameters in the model, the data. On the other hand, users often ask for more digits than they need. Although it is reasonable to want extra accuracy, just to be reasonably sure that convergence has been attained, the point is that the accuracy required is rarely near the computer's precision.

A fifth point, not illustrated above, is that many real problems are poorly scaled, meaning that the sizes of the variables differ greatly. For example, one variable may always be in the range 10^6 to 10^7 and another in the range 1 to 10. In our experience, this happens surprisingly often. However, most work in this field has not paid attention to the problem of scaling. In this book we try to point out where ignoring the affects of scaling can degrade the performance of nonlinear algorithms, and we attempt to rectify these deficiencies in our algorithms.

Finally, in this book we discuss only those nonlinear problems where the unknowns can have any real value, as opposed to those where some variables must be integers. All our examples had this form, but the reader may wonder if this is a realistic restriction in general. The answer is that there certainly are nonlinear problems where some variables must be integers because they represent things like people, trucks, or large widgits. However, this restriction makes the problems so much more difficult to solve—because all continuity is lost—that often we can best solve them by regarding the discrete variables as continuous and then rounding the solution values to integers as necessary. The theory does not guarantee this approach to solve the corresponding integer problem, but in practice it often produces reasonable answers. Exceptions are problems where some discrete variables are constrained to take only a few values such as 0, 1, or 2. In this case, discrete methods must be used. [See, e.g., Beale (1977), Garfinkel and Nemhauser (1972).]

1.3 FINITE-PRECISION ARITHMETIC AND MEASUREMENT OF ERROR

Some features of our computer algorithms, such as tests for convergence, depend on how accurately real numbers are represented on the computer. On occasion, arithmetical coding also is influenced by an understanding of computer arithmetic. Therefore, we need to describe briefly finite-precision arithmetic, which is the computer version of real arithmetic. For more information, see Wilkinson (1963).

In scientific notation, the number 51.75 is written $+0.5175 \times 10^{+2}$. Computers represent real numbers in the same manner, using a sign (+ in our example), a base (10), an exponent (+2), and a mantissa (0.5175). The representation is made unique by specifying that $1/\text{base} \leq \text{mantissa} < 1$—that is, the first digit to the right of the "decimal" point is nonzero. The length of the mantissa, called the *precision* of the representation, is especially important to numerical computation. The representation of a real number on a computer is called its *floating-point* representation; we will denote the floating-point representation of x by $\text{fl}(x)$.

On CDC machines the base is 2, and the mantissa has 48 places. Since $2^{48} \cong 10^{14.4}$, this means that we can accurately store up to 14 decimal digits. The exponent can range from -976 to $+1070$, so that the smallest and largest numbers are about 10^{-294} and 10^{322}. On IBM machines the base is 16; the mantissa has 6 places in single precision and 14 in double precision, which corresponds to about 7 and 16 decimal digits, respectively. The exponent can range from -64 to $+63$, so that the smallest and largest numbers are about 10^{-77} and 10^{76}.

The implications of storing real numbers to only a finite precision are important, but they can be summarized simply. First, since not every real number can be represented exactly on the computer, one can at best expect a solution to be as accurate as the computer precision. Second, depending on the computer and the compiler, the result of each intermediate arithmetic operation is either truncated or rounded to the accuracy of the machine. Thus the inaccuracy due to finite precision may accumulate and further diminish the accuracy of the results. Such errors are called *round-off errors*. Although the effects of round-off can be rather subtle, there are really just three fundamental situations in which it can unduly harm computational accuracy. The first is the addition of a sequence of numbers, especially if the numbers are decreasing in absolute value; the right-hand parts of the smaller numbers are lost, owing to the finite representation of intermediate results. (For an example, see Exercise 4.) The second is the taking of the difference of two almost identical numbers; much precision is lost because the leading left-hand digits of the difference are zero. (For an example, see Exercise 5.) The third is the solution of nearly singular systems of linear equations, which is discussed in Chapter 3. This situation is actually a consequence of the first two, but it is so basic and important that we prefer to think of it as a third fundamental problem. If one is alert to these three situations in writing and using computer programs, one can understand and avoid many of the problems associated with the use of finite-precision arithmetic.

A consequence of the use of finite-precision arithmetic, and even more, of the iterative nature of our algorithms, is that we do not get exact answers to most nonlinear problems. Therefore we often need to measure how close a number x is to another number y. The concept we will use most often is the *relative error* in y as an approximation to a nonzero x,

$$\frac{|x - y|}{|x|}.$$

This is preferable, unless $x = 0$, to the use of *absolute error*,

$$|x - y|,$$

because the latter measure is dependent on the scale of x and y but the former is not (see Exercise 6).

A common notation in the measurement of error and discussion of algorithms will be useful to us. Given two sequences of positive real numbers α_i, β_i, $i = 1, 2, 3, \ldots$, we write $\alpha_i = O(\beta_i)$ (read "α_i is big-oh of β_i") if there exists some positive constant c such that for all positive integers i, except perhaps some finite subset, $\alpha_i \leq c \cdot \beta_i$. This notation is used to indicate that the magnitude of each α_i is of the same order as the corresponding β_i, or possibly smaller. For further information see Aho, Hopcroft, and Ullman [1974].

Another effect of finite-precision arithmetic is that certain aspects of our algorithms, such as stopping criteria, will depend on the machine precision. It is important, therefore, to characterize machine precision in such a way that discussions and computer programs can be reasonably independent of any particular machine. The concept commonly used is *machine epsilon*, abbreviated *macheps*; it is defined as the smallest positive number τ such that $1 + \tau > 1$ on the computer in question (see Exercise 7). For example, on the CDC machine, since there are 48 base-2 places, macheps $= 2^{-47}$ with truncating arithmetic, or 2^{-48} with rounding. The quantity, macheps, is quite useful when we discuss computer numbers. For example, we can easily show that the relative error in the computer representation $fl(x)$ of any real nonzero number x is less than macheps; conversely, the computer representation of any real number x will be in the range $(x(1 - \text{macheps}), x(1 + \text{macheps}))$. Similarly, two numbers x and y agree in the leftmost half of their digits approximately when

$$\frac{|x - y|}{|x|} \leq \sqrt{\text{macheps}}.$$

This test is quite common in our algorithms.

Another way to view macheps is as a key to the difficult task of deciding when a finite-precision number could just as well be zero in a certain context. We are used to thinking of 0 as that unique solution to $x + 0 = x$ for every real number x. In finite precision, the additive identity role of 0 is played by an interval 0_x which contains 0 and is approximately equal to $(-\text{macheps} \cdot x, +\text{macheps} \cdot x)$. It is common that in the course of a computation we will generate finite-precision numbers x and y of different enough magnitude so that $fl(x + y) = fl(x)$. This means that y is zero in the context, and sometimes, as in numerical linear algebra algorithms, it is useful to monitor the computation and actually set y to zero.

Finally, any computer user should be aware of overflow and underflow, the conditions that occur when a computation generates a nonzero number whose exponent is respectively larger than, or smaller than, the extremes allowed on the machine. For example, we encounter an underflow condition when we reciprocate 10^{322} on a CDC machine, and we encounter an overflow condition when we reciprocate 10^{-77} on an IBM machine.

In the case of an overflow, almost any machine will terminate the run with an error message. In the case of an underflow, there is often either a compiler option to terminate, or one to substitute zero for the offending expression. The latter choice is reasonable sometimes, but not always (see Exercise 8). Fortunately, when one is using well-written linear algebra routines, the algorithms discussed in this book are not usually prone to overflow or underflow. One routine, discussed in Section 3.1, that does require care is computing Euclidean norm of a vector,

$$\|v\|_2 = \sqrt{v_1^2 + v_2^2 + \cdots + v_n^2}, \qquad v \in \mathbb{R}^n.$$

1.4 EXERCISES

1. Rephrase as a simultaneous nonlinear equation problem in standard form: Find $(x_1, x_2)^T$ such that

$$x_1^2 = x_2^2 + 3x_2 + 2,$$

$$x_1^3 + 2 = x_2^4.$$

2. A laboratory experiment measures a function f at 20 distinct points in time t (between 0 and 50). It is known that $f(t)$ is a sine wave, but its amplitude, frequency, and displacement in both the f and t directions are unknown. What numerical problem would you set up to determine these characteristics from your experimental data?

3. An economist has a complex computer model of the economy which, given the unemployment rate, rate of growth in the GNP, and the number of housing starts in the past year, estimates the inflation rate. The task is to determine what combination of these three factors will lead to the lowest inflation rate. You are to set up and solve this problem numerically.

   (a) What type of numerical problem might this turn into? How would you handle the variable "number of housing starts"?

   (b) What are some questions you would ask the economist in an attempt to make the problem as numerically tractable as possible (for example, concerning continuity, derivatives, constraints)?

   (c) Is the problem likely to be expensive to solve? Why?

4. Pretend you have a computer with base 10 and precision 4 that truncates after each arithmetic operation; for example, the sum of $24.57 + 128.3 = 152.87$ becomes 152.8. What are the results when 128.3, 24.57, 3.163, and 0.4825 are added in ascending order and in descending order in this machine? How do these compare with the correct ("infinite-precision") result? What does this show you about adding sequences of numbers on the computer?

5. Assume you have the same computer as in Exercise 4, and you perform the computation $(\frac{1}{3} - 0.3300)/0.3300$. How many correct digits of the real answer do you get? What does this show you about subtracting almost identical numbers on the computer?

6. What are the relative and absolute errors of the answer obtained by the computer in Exercise 5 as compared to the correct answer? What if the problem is changed to $((\frac{100}{3}) - 33)/33$? What does this show about the usefulness of relative versus absolute error?

7. Write a program to calculate machine epsilon on your computer or hand calculator. You may assume that macheps will be a power of 2, so that your algorithm can look like

$$\text{EPS} \quad := 1$$

$$\text{WHILE} \quad 1 + \text{EPS} > 1 \text{ DO}$$

$$\text{EPS} \quad := \text{EPS}/2.$$

Keep a counter that enables you to know at the end of your program which power of 2 macheps is. Print out this value and the decimal value of macheps. (*Note:* The value of macheps will vary by a factor of 2 depending on whether rounding or truncating arithmetic is used. Why?)

For further information on the computer evaluation of machine environment parameters, see Ford (1978).

8. In each of the following calculations an underflow will occur (on an IBM machine). In which cases is it reasonable to substitute zero for the quantity that underflows? Why?

   (a) $a := \sqrt{b^2 + c^2}$, with $b = 1$, $c = 10^{-50}$
   (b) $a := \sqrt{b^2 + c^2}$, with $b = c = 10^{-50}$
   (c) $v := (w * x)/(y * z)$, with $w = 10^{-30}$, $x = 10^{-60}$, $y = 10^{-40}$, $z = 10^{-50}$

2

Nonlinear Problems
in One Variable

We begin our study of the solution of nonlinear problems by discussing problems in just one variable: finding the solution of one nonlinear equation in one unknown, and finding the minimum of a function of one variable. The reason for studying one-variable problems separately is that they allow us to see those principles for constructing good local, global, and derivative-approximating algorithms that will also be the basis of our algorithms for multivariable problems, without requiring knowledge of linear algebra or multivariable calculus. The algorithms for multivariable problems will be more complex than those in this chapter, but an understanding of the basic approach here should help in the multivariable case.

Some references that consider the problems of this chapter in detail are Avriel (1976), Brent (1973), Conte and de Boor (1980), and Dahlquist, Björck, and Anderson (1974).

2.1 WHAT IS NOT POSSIBLE

Consider the problem of finding the real roots of each of the following three nonlinear equations in one unknown:

$$f_1(x) = x^4 - 12x^3 + 47x^2 - 60x,$$
$$f_2(x) = x^4 - 12x^3 + 47x^2 - 60x + 24,$$
$$f_3(x) = x^4 - 12x^3 + 47x^2 - 60x + 24.1.$$

(see Figure 2.1.1). It would be wonderful if we had a general-purpose computer routine that would tell us: "The roots of $f_1(x)$ are $x = 0, 3, 4,$ and 5; the real roots of $f_2(x)$ are $x = 1$ and $x \cong 0.888$; $f_3(x)$ has no real roots."

It is unlikely that there will ever be such a routine. In general, the questions of existence and uniqueness—does a given problem have a solution, and is it unique?—are beyond the capabilities one can expect of algorithms that solve nonlinear problems. In fact, we must readily admit that for any computer algorithm there exist nonlinear functions (infinitely continuously differentiable, if you wish) perverse enough to defeat the algorithm. Therefore, all a user can be guaranteed from any algorithm applied to a nonlinear problem is the answer, "An approximate solution to the problem is _____," or, "No approximate solution to the problem was found in the alloted time." In many cases, however, the supplier of a nonlinear problem knows from practical considerations that it has a solution, and either that the solution is unique or that a solution in a particular region is desired. Thus the inability to determine the existence or uniqueness of solutions is usually not the primary concern in practice.

It is also apparent that one will be able to find only approximate solutions to most nonlinear problems. This is due not only to the finite precision of our computers, but also to the classical result of Galois that for some polynomials of degree $n \geq 5$, no closed-form solutions can be found using integers and the operations $+$, $-$, $\times$, $\div$, exponentiation, and second through nth roots. Therefore, we will develop methods that try to find one approximate solution of a nonlinear problem.

2.2 NEWTON'S METHOD FOR SOLVING ONE EQUATION IN ONE UNKNOWN

Our consideration of finding a root of one equation in one unknown begins with Newton's method, which is the prototype of the algorithms we will generate. Suppose we wish to calculate the square root of 3 to a reasonable number of places. This can be viewed as finding an approximate root x_* of the func-

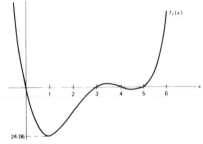

Figure 2.1.1   The equation $f_1(x) = x^4 - 12x^3 + 47x^2 - 60x$

tion $f(x) = x^2 - 3$ (see Figure 2.2.1). If our initial or current estimate of the answer is $x_c = 2$, we can get a better estimate x_+ by drawing the line that is tangent to $f(x)$ at $(2, f(2)) = (2, 1)$, and finding the point x_+ where this line crosses the x axis. Since

$$x_+ = x_c - \Delta x,$$

and

$$f'(x_c) = \frac{\Delta y}{\Delta x} = \frac{f(x_c)}{\Delta x},$$

we have that

$$f'(x_c)\Delta x = \Delta y = f(x_c)$$

or

$$x_+ = x_c - \frac{f(x_c)}{f'(x_c)} \qquad (2.2.1)$$

which gives

$$x_+ = 2 - \tfrac{1}{4} = 1.75.$$

The logical thing to do next is to apply the same process from the new current estimate $x_c = 1.75$. Using (2.2.1) gives $x_+ = 1.75 - (0.0625/3.5) = 1.732\tfrac{1}{7}$, which already has four correct digits of $\sqrt{3}$. One more iteration gives $x_+ \cong 1.7320508$, which has eight correct digits.

The method we have just developed is called the *Newton-Raphson method* or *Newton's method*. It is important to our understanding to take a more abstract view of what we have done. At each iteration we have constructed a *local model* of our function $f(x)$ and solved for the root of the model. In

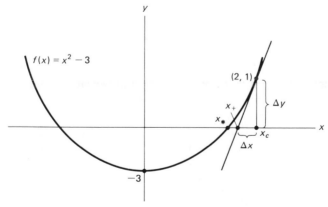

Figure 2.2.1   An iteration of Newton's method on $f(x) = x^2 - 3$ (not to scale)

the present case, our model

$$M_c(x) = f(x_c) + f'(x_c)(x - x_c) \qquad (2.2.2)$$

is just the unique line with function value $f(x_c)$ and slope $f'(x_c)$ at the point x_c. [We use capital M to be consistent with the multidimensional case and to differentiate from minimization problems where our model is denoted by $m_c(x)$.] It is easy to verify that $M_c(x)$ crosses the x axis at the point x_+ defined by (2.2.1).

Pedagogical tradition calls for us to say that we have obtained Newton's method by writing $f(x)$ as its Taylor series approximation around the current estimate x_c,

$$f(x) = f(x_c) + f'(x_c)(x - x_c) + \frac{f''(x_c)(x - x_c)^2}{2!} + \cdots$$

$$= \sum_{i=0}^{\infty} \frac{f^i(x_c)(x - x_c)^i}{i!}, \qquad (2.2.3)$$

and then approximating $f(x)$ by the affine* portion of this series, which naturally is given also by (2.2.2). Again the root is given by (2.2.1). There are several reasons why we prefer a different approach. It is unappealing and unnecessary to make assumptions about derivatives of any higher order than those actually used in the iteration. Furthermore, when we consider multivariable problems, higher-order derivatives become so complicated that they are harder to understand than any of the algorithms we will derive.

Instead, Newton's method comes simply and naturally from Newton's theorem,

$$f(x) = f(x_c) + \int_{x_c}^{x} f'(z)\, dz.$$

It seems reasonable to approximate the indefinite integral by

$$\int_{x_c}^{x} f'(z)\, dz \cong f'(x_c)(x - x_c)$$

and once more obtain the affine approximation to $f(x)$ given by (2.2.2). This type of derivation will be helpful to us in multivariable problems, where geometrical derivations become less manageable.

Newton's method is typical of methods for solving nonlinear problems; it is an iterative process that generates a sequence of points that we hope come increasingly close to a solution. The obvious question is, "Will it work?" The

---

* We will refer to (2.2.2) as an *affine* model, although colloquially it is often called a *linear* model. The reason is that an affine model corresponds to an affine subspace through $(x, f(x))$, a line that does not necessarily pass through the origin, whereas a linear subspace must pass through the origin.

answer is a qualified "Yes." Notice that if $f(x)$ were linear, Newton's method would find its root in one iteration. Now let us see what it will do for the general square-root problem;

$$\text{given } \alpha > 0, \qquad \text{find } x \text{ such that } f(x) = x^2 - \alpha = 0,$$

starting from a current guess $x_c \neq 0$. Since

$$x_+ = x_c - \frac{f(x_c)}{f'(x_c)} = x_c - \frac{x_c^2 - \alpha}{2x_c} = \frac{x_c}{2} + \frac{\alpha}{2x_c},$$

one has

$$x_+ - \sqrt{\alpha} = \frac{x_c}{2} + \frac{\alpha}{2x_c} - \sqrt{\alpha} = \frac{(x_c - \sqrt{\alpha})^2}{2x_c}, \qquad (2.2.4a)$$

or, using relative error, one has

$$\frac{x_+ - \sqrt{\alpha}}{\sqrt{\alpha}} = \left(\frac{x_c - \sqrt{\alpha}}{\sqrt{\alpha}}\right)^2 \cdot \left(\frac{\sqrt{\alpha}}{2x_c}\right). \qquad (2.2.4b)$$

Thus as long as the initial error $|x_c - \sqrt{\alpha}|$ is less than $|2x_c|$, the new error $|x_+ - \sqrt{\alpha}|$ will be smaller than the old error $|x_c - \sqrt{\alpha}|$, and eventually each new error will be much smaller than the previous error. This agrees with our experience for finding the square root of 3 in the example that began this section.

The pattern of decrease in error given by (2.2.4) is typical of Newton's method. The error at each iteration will be approximately the square of the previous error, so that, if the initial guess is good enough, the error will decrease and eventually decrease rapidly. This pattern is known as local q-quadratic convergence. Before deriving the general convergence theorem for Newton's method, we need to discuss rates of convergence.

2.3 CONVERGENCE OF SEQUENCES OF REAL NUMBERS

Given an iterative method that produces a sequence of points $x_1, x_2, \ldots$, from a starting guess x_0, we will want to know if the iterates converge to a solution x_*, and if so, how quickly. If we assume that we know what it means to write

$$\lim_{k \to \infty} a_k = 0$$

for a real sequence $\{a_k\}$, then the following definition characterizes the properties we will need.

Definition 2.3.1   Let $x_* \in \mathbb{R}$, $x_k \in \mathbb{R}$; $k = 0, 1, 2, \ldots$. Then the sequence

$\{x_k\} = \{x_0, x_1, x_2, \dots\}$ is said to *converge* to x_* if

$$\lim_{k \to \infty} |x_k - x_*| = 0.$$

If in addition, there exists a constant $c \in [0, 1)$ and an integer $\hat{k} \geq 0$ such that for all $k \geq \hat{k}$,

$$|x_{k+1} - x_*| \leq c|x_k - x_*| \tag{2.3.1}$$

then $\{x_k\}$ is said to be *q-linearly convergent* to x_*. If for some sequence $\{c_k\}$ that converges to 0,

$$|x_{k+1} - x_*| \leq c_k|x_k - x_*|, \tag{2.3.2}$$

then $\{x_k\}$ is said to *converge q-superlinearly* to x_*. If there exist constants $p > 1, c \geq 0$, and $\hat{k} \geq 0$ such that $\{x_k\}$ converges to x_* and for all $k \geq \hat{k}$,

$$|x_{k+1} - x_*| \leq c|x_k - x_*|^p, \tag{2.3.3}$$

then $\{x_k\}$ is said to converge to x_* with *q-order at least* p. If $p = 2$ or 3, the convergence is said to be *q-quadratic* or *q-cubic*, respectively.

If $\{x_k\}$ converges to x_* and, in place of (2.3.2),

$$|x_{k+j} - x_*| \leq c_k|x_k - x_*|$$

for some fixed integer j, then $\{x_k\}$ is said to be *j-step q-superlinearly convergent* to x_*. If $\{x_k\}$ converges to x_* and, in place of (2.3.3), for $k > \hat{k}$,

$$|x_{k+j} - x_*| \leq c|x_k - x_*|^p$$

for some fixed integer j, then $\{x_k\}$ is said to have *j-step q-order convergence* of order *at least* p.

An example of a q-linearly convergent sequence is

$$x_0 = 2, \quad x_1 = \tfrac{3}{2}, \quad x_2 = \tfrac{5}{4}, \quad x_3 = \tfrac{9}{8}, \quad \dots, \quad x_i = 1 + 2^{-i}, \quad \dots.$$

This sequence converges to $x_* = 1$ with $c = \tfrac{1}{2}$; on a CDC machine it will take 48 iterations until $fl(x_k) = 1$. An example of a q-quadratically convergent sequence is

$$x_0 = \tfrac{3}{2}, \quad x_1 = \tfrac{5}{4}, \quad x_2 = \tfrac{17}{16}, \quad x_3 = \tfrac{257}{256}, \quad \dots, \quad x_k = 1 + 2^{-2^k}, \quad \dots,$$

which converges to $x_* = 1$ with $c = 1$; on a CDC machine, $fl(x_6)$ will equal 1. In practice, q-linear convergence can be fairly slow, whereas q-quadratic or q-superlinear convergence is eventually quite fast. However, actual behavior also depends upon the constants c in (2.3.1–2.3.3); for example, q-linear convergence with $c = 0.001$ is probably quite satisfactory, but with $c = 0.9$ it is not. (For further examples see Exercises 2 and 3). It is worth emphasizing that the utility of q-superlinear convergence is directly related to how many iterations are needed for c_k to become small.

The prefix "q" stands for quotient and is used to differentiate from "r" (root) orders of convergence. R-order* is a weaker type of convergence rate; all that is said of the errors $|x_k - x_*|$, of a sequence with r-order p, is that they are bounded above by another sequence of q-order p. A definitive reference is Ortega and Rheinboldt [1970]. An iterative method that will converge to the correct answer at a certain rate, provided it is started close enough to the correct answer, is said to be *locally convergent* at that rate. In this book we will be interested mainly in methods that are locally q-superlinearly or q-quadratically convergent and for which this behavior is apparent in practice.

2.4 CONVERGENCE OF NEWTON'S METHOD

We now show that, for most problems, Newton's method will converge q-quadratically to the root of one nonlinear equation in one unknown, provided it is given a good enough starting guess. However, it may not converge at all from a poor start, so that we need to incorporate the global methods of Section 2.5. The local convergence proof for Newton's method hinges on an estimate of the errors in the sequence of affine models $M_c(x)$ as approximations to $f(x)$. Since we obtained the approximations by using $f'(x_c)(x - x_c)$ to approximate

$$\int_{x_c}^{x} f'(z) \, dz$$

we are going to need to make some smoothness assumptions on f' in order to estimate the error in the approximation, which is

$$f(x) - M_c(x) = \int_{x_c}^{x} [f'(z) - f'(x_c)] \, dz.$$

First we define the notion of Lipschitz continuity.

Definition 2.4.1   A function g is Lipschitz continuous with constant γ in a set X, written $g \in \text{Lip}_\gamma(X)$, if for every $x, y \in X$,

$$|g(x) - g(y)| \leq \gamma |x - y|.$$

In order to prove the convergence of Newton's method, we first prove a simple lemma showing that if $f'(x)$ is Lipschitz continuous, then we can obtain a bound on how close the affine approximation $f(x) + f'(x)(y - x)$ is to $f(y)$.

---

* We will capitalize the prefix letters R and Q when they begin a sentence, but not otherwise.

LEMMA 2.4.2 For an open interval D, let $f: D \to \mathbb{R}$ and let $f' \in \text{Lip}_\gamma(D)$. Then for any $x, y \in D$,

$$| f(y) - f(x) - f'(x)(y - x) | \le \frac{\gamma(y - x)^2}{2}. \qquad (2.4.1)$$

Proof. From basic calculus, $f(y) - f(x) = \int_x^y f'(z) \, dz$, or equivalently,

$$f(y) - f(x) - f'(x)(y - x) = \int_x^y [f'(z) - f'(x)] \, dz. \qquad (2.4.2)$$

Making the change of variables

$$z = x + t(y - x), \qquad dz = dt(y - x),$$

(2.4.2) becomes

$$f(y) - f(x) - f'(x)(y - x) = \int_0^1 [f'(x + t(y - x)) - f'(x)](y - x) \, dt,$$

and so by the triangle inequality applied to the integral and the Lipschitz continuity of f',

$$| f(y) - f(x) - f'(x)(y - x) | \le | y - x | \int_0^1 \gamma | t(y - x) | \, dt$$
$$= \gamma | y - x |^2 / 2. \qquad \square$$

Note that (2.4.1) closely resembles the error bound given by the Taylor series with remainder, with the Lipschitz constant γ taking the place of a bound on $| f''(\xi) |$ for $\xi \in D$. The main advantage of using Lipschitz continuity is that we do not need to discuss this next higher derivative. This is especially convenient in multiple dimensions.

We are now ready to state and prove a fundamental theorem of numerical mathematics. We will prove the most useful form of the result and leave the more general ones as exercises (see Exercises 13-14.)

THEOREM 2.4.3 Let $f: D \to \mathbb{R}$, for an open interval D, and let $f' \in \text{Lip}_\gamma(D)$. Assume that for some $\rho > 0$, $| f'(x) | \ge \rho$ for every $x \in D$. If $f(x) = 0$ has a solution $x_* \in D$, then there is some $\eta > 0$ such that: if $| x_0 - x_* | < \eta$, then the sequence $\{x_k\}$ generated by

$$x_{k+1} = x_k - \frac{f(x_k)}{f'(x_k)}, \qquad k = 0, 1, 2, \ldots$$

exists and converges to x_*. Furthermore, for $k = 0, 1, \ldots$,

$$| x_{k+1} - x_* | \le \frac{\gamma}{2\rho} | x_k - x_* |^2. \qquad (2.4.3)$$

Proof. Let $\tau \in (0, 1)$, let $\hat{\eta}$ be the radius of the largest open interval around x_* that is contained in D, and define $\eta = \min\{\hat{\eta}, \tau(2\rho/\gamma)\}$. We will show by induction that for $k = 0, 1, 2, \ldots$, (2.4.3) holds, and

$$|x_{k+1} - x_*| \le \tau |x_k - x_*| < \eta.$$

The proof simply shows at each iteration that the new error $|x_{k+1} - x_*|$ is bounded by a constant times the error the affine model makes in approximating f at x_*, which from Lemma 2.4.2 is $O(|x_k - x_*|^2)$. For $k = 0$,

$$x_1 - x_* = x_0 - x_* - \frac{f(x_0)}{f'(x_0)} = x_0 - x_* - \frac{f(x_0) - f(x_*)}{f'(x_0)}$$

$$= \frac{1}{f'(x_0)} [f(x_*) - f(x_0) - f'(x_0)(x_* - x_0)].$$

The term in brackets is $f(x_*) - M_0(x_*)$, the error at x_* in the local affine model at $x_c = x_0$. Thus from Lemma 2.4.2,

$$|x_1 - x_*| \le \frac{\gamma}{2|f'(x_0)|} |x_0 - x_*|^2,$$

and by the assumptions on $f'(x)$

$$|x_1 - x_*| \le \frac{\gamma}{2\rho} |x_0 - x_*|^2.$$

Since $|x_0 - x_*| \le \eta \le \tau \cdot 2\rho/\gamma$, we have $|x_1 - x_*| \le \tau |x_0 - x_*| < \eta$. The proof of the induction step then proceeds identically. $\qquad\square$

The condition in Theorem 2.4.3 that $f'(x)$ have a nonzero lower bound in D simply means that $f'(x_*)$ must be nonzero for Newton's method to converge quadratically. Indeed, if $f'(x_*) = 0$, then x_* is a multiple root, and Newton's method converges only linearly (see Exercise 12). To appreciate the difference, we give below sample iterations of Newton's method applied to $f_1(x) = x^2 - 1$ and $f_2(x) = x^2 - 2x + 1$, both starting from $x_0 = 2$. Notice how much more slowly Newton's method converges on $f_2(x)$ because $f_2'(x_*) = 0$.

EXAMPLE 2.4.4 Newton's Method Applied to Two Quadratics (CDC, Single Precision)

$f_1(x) = x^2 - 1$		$f_2(x) = x^2 - 2x + 1$
2	x_0	2
1.25	x_1	1.5
1.025	x_2	1.25
1.0003048780488	x_3	1.125
1.0000000464611	x_4	1.0625
1.0	x_5	1.03125

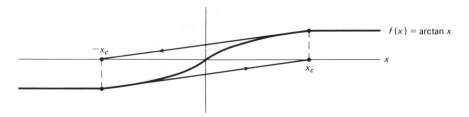

Figure 2.4.1   Newton's method applied to $f(x) = \arctan(x)$

It is also informative to examine the constant $\gamma/2\rho$ involved in the q-quadratic convergence relation (2.4.3). The numerator γ, a Lipschitz constant for f' on D, can be considered a measure of the nonlinearity of f. However, γ is a scale-dependent measure; multiplying f or changing the units of x by a constant will scale f' by that constant without making the function more or less nonlinear. A partially scale-free measure of nonlinearity is the relative rate of change in $f'(x)$, which is obtained by dividing γ by $f'(x)$. Thus, since ρ is a lower bound on $f'(x)$ for $x \in D$, γ/ρ is an upper bound on the relative nonlinearity of $f(x)$, and Theorem 2.4.3 says that the smaller this measure of relative nonlinearity, the faster Newton's method will converge. If f is linear, then $\gamma = 0$ and $x_1 = x_*$.

Theorem 2.4.3 guarantees the convergence of Newton's method only from a good starting point x_0, and indeed it is easy to see that Newton's method may not converge at all if $|x_0 - x_*|$ is large. For example, consider the function $f(x) = \arctan x$ (see Figure 2.4.1). For some $x_c \in [1.39, 1.40]$, if $x_0 = x_c$, then Newton's method will produce the cycle $x_1 = -x_c$, $x_2 = x_c$, $x_3 = -x_c, \ldots$. If $|x_0| < x_c$, Newton's method will converge to $x_* = 0$, but if $|x_0| > x_c$, Newton's method will diverge; i.e., the error $|x_k - x_*|$ will increase at each iteration. Thus Newton's method is useful to us for its fast local convergence, but we need to incorporate it into a more robust method that will be successful from farther starting points.

2.5 GLOBALLY CONVERGENT METHODS* FOR SOLVING ONE EQUATION IN ONE UNKNOWN

We will use a simple philosophy to incorporate Newton's method into a globally convergent algorithm: use Newton's method whenever it seems to be working well, otherwise fall back on a slower but sure global method. This strategy produces globally convergent algorithms with the fast local conver-

---

* For our definition of "global method," see the last paragraph of Section 1.1.

gence of Newton's method. In this section we discuss two global methods and then show how to combine a global method with Newton's method into a hybrid algorithm. We also discuss the stopping tests and other computer-dependent criteria necessary to successful computational algorithms.

The simplest global method is the method of bisection. It makes the somewhat reasonable assumption that one starts with an interval $[x_0, z_0]$ that contains a root. It sets x_1 to the midpoint of this interval, chooses the new interval to be the one of $[x_0, x_1]$ or $[x_1, z_0]$ that contains a root, and continues to halve the interval until a root is found (see Figure 2.5.1). This is expressed algebraically as:

given x_0, z_0 such that $f(x_0) \cdot f(z_0) < 0$,

for $k = 0, 1, 2, \ldots,$ do

$$x_{k+1} = \frac{x_k + z_k}{2}$$

$$z_{k+1} = \begin{cases} x_k, & \text{if } f(x_k) \cdot f(x_{k+1}) < 0, \\ z_k, & \text{otherwise.} \end{cases}$$

The method of bisection always works in theory, but it is guaranteed only to reduce the error bound by $\frac{1}{2}$ for each iteration. This makes the method very marginal for practical use. Programs that use bisection generally do so only until an x_k is obtained from which some variant of Newton's method will converge. The method of bisection also does not extend naturally to multiple dimensions.

A method more indicative of how we will proceed in n-space is the following. Think of Newton's method as having suggested not only the step $x_N = x_c - f(x_c)/f'(x_c)$, but also the direction in which that step points. [Assume $f'(x_c) \neq 0$.] Although the Newton step may actually cause an increase in the absolute value of the function, its direction always will be one in which the absolute function value decreases initially (see Figure 2.5.2). This should be obvious geometrically; for the simple proof, see Exercise 16. Thus, if the Newton point x_N doesn't produce a decrease in $|f(x)|$, a reasonable strategy is to backtrack from x_N toward x_c until one finds a point x_+ for

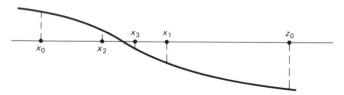

Figure 2.5.1   The method of bisection

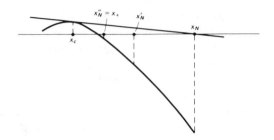

Figure 2.5.2 Backtracking from the Newton step

which $|f(x_+)| < |f(x_c)|$. A possible iteration is

$$x_+ = x_c - \frac{f(x_c)}{f'(x_c)}$$

while $|f(x_+)| \geq |f(x_c)|$ do

$$x_+ \leftarrow \frac{x_+ + x_c}{2}. \tag{2.5.1}$$

Note that this strategy does not require an initial interval bracketing a root.

Iteration (2.5.1) is an example of a hybrid algorithm, one that attempts to combine global convergence and fast local convergence by first trying the Newton step at each iteration, but always insisting that the iteration decreases some measure of the closeness to a solution. Constructing such hybrid algorithms is the key to practical success in solving multivariable nonlinear problems. Below is the general form of a class of hybrid algorithms for finding a root of one nonlinear equation; it is meant to introduce and emphasize those basic techniques for constructing globally and fast locally convergent algorithms that will be the foundation of all the algorithms in this book.

ALGORITHM 2.5.1 General hybrid quasi-Newton algorithm for solving one nonlinear equation in one unknown:

given $f : \mathbb{R} \to \mathbb{R}$, x_0,

for $k = 0, 1, 2, \ldots$, do

    1. *decide* whether to stop; if not:

    2. *make a local model* of f around x_k, and find the point x_N that solves (or comes closest to solving) the model problem.

    3. (a) *decide* whether to take $x_{k+1} = x_N$, if not,

        (b) choose x_{k+1} using a *global strategy* (make more conservative use of the solution to the model problem).

Step 1 is discussed below; it requires our first use of computer-dependent and problem-dependent tolerances. Step 2 usually involves calculating the Newton step, or a variant without derivatives (see Section 2.6). Equation (2.5.1) is an example of Step 3(a)–(b). We will see in Chapter 6 that the criterion in Step 3(a) has to be chosen with only a little bit of care to assure the global convergence in most cases of the hybrid algorithm to a solution.

Deciding when to stop is a somewhat ad hoc process that can't be perfect for every problem, yet it calls for considerable care. Since there may be no computer-representable x_* such that $f(x_*) = 0$, one must decide when one is "close enough." This decision usually takes two parts: "Have you approximately solved the problem?" or "Have the last two (or few) iterates stayed in virtually the same place?" The first question is represented by a test such as, "Is $|f(x_+)| < \tau_1$?" where the tolerance τ_1 is chosen to reflect the user's idea of being close enough to zero for this problem. For example, τ_1 might be set to $(\text{macheps})^{1/2}$. Naturally this test is very sensitive to the scale of $f(x)$, and so it is important that a routine instruct the user to choose τ_1, or scale f, so that an x_+ that satisfies $|f(x_+)| < \tau_1$ will be a satisfactory solution to the problem. Partly to guard against this condition's being too restrictive, the second question is included and it is tested by a relation such as, "Is $(|x_+ - x_c|/|x_+|) < \tau_2$?" A reasonable tolerance is $\tau_2 = (\text{macheps})^{1/2}$, which corresponds to stopping whenever the left half of the digits of x_c and x_+ agree, though any τ_2 greater than macheps can be selected. Since x_+ might be close to zero, the second test is usually modified to something like, "Is $(|x_+ - x_c|/\max\{|x_+|, |x_c|\} < \tau_2$?" A better test uses a user-supplied variable typ$x$ containing the typical size of x in the place of the $|x_c|$ in the denominator (see Exercise 17), so that the stopping condition on a CDC machine might be,

$$\text{if} \quad |f(x_+)| \le 10^{-5} \quad \text{or} \quad \frac{|x_+ - x_c|}{\max\{\text{typ}x, |x_+|\}} \le 10^{-7}, \quad \text{stop.}$$

In practice, $f(x_+)$ usually gets small before the step does in any problem for which local convergence is fast, but for a problem on which convergence is only linear, the step may become small first. The reader can already see that the choice of stopping rules is quite a can of worms, especially for poorly scaled problems. We will treat it more completely in Chapter 7.

2.6 METHODS WHEN DERIVATIVES ARE UNAVAILABLE

In many practical applications, $f(x)$ is not given by a formula; rather it is the output from some computational or experimental procedure. Since $f'(x)$ usually is not available then, our methods that use values of $f'(x)$ to solve $f(x) = 0$ must be modified to require only values of $f(x)$.

We have been using $f'(x)$ in modeling f near the current solution estimate x_c by the line tangent to f at x_c. When $f'(x)$ is unavailable, we replace this model by the secant line that goes through f at x_c and at some nearby point $x_c + h_c$ (see Figure 2.6.1). It is easy to calculate that the slope of this line is

$$a_c = \frac{f(x_c + h_c) - f(x_c)}{h_c} \tag{2.6.1}$$

and so the model we obtain is the line

$$\hat{M}_c(x) = f(x_c) + a_c(x - x_c).$$

Therefore, what we have done is equivalent to replacing the derivative $f'(x_c)$ in our original model $M_c(x) = f(x_c) + f'(x_c)(x - x_c)$ by the approximation a_c. The quasi-Newton step to the zero of $\hat{M}_c(x)$ then becomes

$$\hat{x}_N = x_c - \frac{f(x_c)}{a_c}.$$

Two questions immediately arise: "Will this method work?" and, "How should h_c be chosen?" We know from calculus that as $h_c \to 0$, a_c will converge to $f'(x_c)$. If h_c is chosen to be a small number, a_c is called a *finite-difference approximation* to $f'(x_c)$. It seems reasonable that the finite-difference Newton method, the quasi-Newton method based on this approximation, should work almost as well as Newton's method, and we will see later that it does. However, this technique requires an additional evaluation of f at each iteration, and if $f(x)$ is expensive, the extra function evaluation is undesirable. In this case, h_c is set to $(x_- - x_c)$, where x_- is the previous iterate, so that $a_c = (f(x_-) - f(x_c))/(x_- - x_c)$ and no additional function evaluations are used. The resultant quasi-Newton algorithm is called a *secant method*. While it may seem totally ad hoc, it also turns out to work well; the convergence of the secant method is slightly slower than a properly chosen finite-difference method, but it is usually more efficient in terms of total function evaluations required to achieve a specified accuracy.

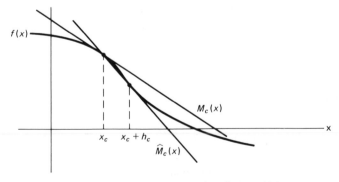

Figure 2.6.1 A secant approximation to $f(x)$

Example 2.6.1 contains samples of the convergence of the secant method and the finite-difference Newton's method (with $h_c = 10^{-7} x_c$) on the problem $f(x) = x^2 - 1$, $x_0 = 2$, that was solved by Newton's method in Example 2.4.4. Notice how similar Newton's method and the finite-difference Newton's method are; the secant method is a bit slower.

EXAMPLE 2.6.1   Finite-difference Newton's method and the secant method applied to $f(x) = x^2 - 1$ (CDC, single precision).

Finite-Difference N.M. $(h_k = 10^{-7} \cdot x_k)$		Secant Method $(x_1$ chosen by f.d.N.M.$)$
2	x_0	2
1.2500000266453	x_1	1.2500000266453
1.0250000179057	x_2	1.0769230844910
1.0003048001120	x_3	1.0082644643823
1.0000000464701	x_4	1.0003048781354
1.0	x_5	1.0000012544523
	x_6	1.0000000001912
	x_7	1.0

There is a great deal of insight to be gained from an analysis of the effect of the difference step h_c on the convergence rate of the resultant finite-difference Newton method. Let us take

$$x_+ = \hat{x}_N = x_c - \frac{f(x_c)}{a_c}.$$

Then

$$x_+ - x_* = x_c - x_* - \frac{f(x_c)}{a_c}$$
$$= a_c^{-1}[f(x_*) - f(x_c) - a_c(x_* - x_c)]$$
$$= a_c^{-1}[f(x_*) - \hat{M}_c(x_*)]$$
$$= a_c^{-1}\{f(x_*) - f(x_c) - f'(x_c)(x_* - x_c) + [f'(x_c) - a_c](x_* - x_c)\}$$
$$= a_c^{-1}\left\{\int_{x_c}^{x_*} [f'(z) - f'(x_c)]\,dz + [f'(x_c) - a_c](x_* - x_c)\right\}.$$

If we define $e_c = |x_c - x_*|$ and $e_+ = |x_+ - x_*|$, then we have

$$e_+ \le |a_c^{-1}|\left(\frac{\gamma}{2} e_c^2 + |f'(x_c) - a_c|e_c\right), \tag{2.6.2}$$

under the same Lipschitz continuity assumption $f' \in \mathrm{Lip}_\gamma(D)$ as in Lemma 2.4.2.

The reader can see that (2.6.2) is similar to our error bound for Newton's method in Section 2.4, except that the right-hand side of (2.6.2) has $|a_c^{-1}|$ in the place of $|f'(x_c)^{-1}|$, and an additional term involving the difference between $f'(x_c)$ and its approximation a_c. Notice also that the above analysis so far is independent of the value of $a_c \neq 0$. Now let us define a_c by (2.6.1) and again bring in the assumption of Lipschitz continuity of f' from Section 2.4. Then we have an easy corollary which tells us how close, as a function of h_c, the finite-difference approximation (2.6.1) is to $f'(x_c)$.

COROLLARY 2.6.2 Let $f : D \to \mathbb{R}$ for an open interval D and let $f' \in \text{Lip}_\gamma(D)$. Let x_c, $x_c + h_c \in D$, and define a_c by (2.6.1). Then

$$|a_c - f'(x_c)| \le \frac{\gamma |h_c|}{2}. \qquad (2.6.3)$$

Proof. From Lemma 2.4.2,

$$|f(x_c + h_c) - f(x_c) - h_c f'(x_c)| \le \frac{\gamma |h_c|^2}{2}$$

Dividing both sides by $|h_c|$ gives the desired result. $\qquad \square$

Substituting (2.6.3) into (2.6.2) gives

$$e_+ \le \frac{\gamma}{2|a_c|} (e_c + |h_c|)e_c.$$

Now, if we bring in from Theorem 2.4.3 the assumption that $|f'(x)| \ge \rho > 0$ in a neighborhood of x, it is easy to show from (2.6.3), for $|h_c|$ sufficiently small, and $x_c \in D$, that $|a_c^{-1}| \le 2\rho^{-1}$. This gives

$$e_+ \le \frac{\gamma}{\rho} (e_c + |h_c|)e_c. \qquad (2.6.4)$$

At this point it is really rather easy to finish a proof of the following theorem.

THEOREM 2.6.3 Let $f : D \to \mathbb{R}$ for an open interval D and let $f' \in \text{Lip}_\gamma(D)$. Assume that $|f'(x)| \ge \rho$ for some $\rho > 0$ and for every $x \in D$. If $f(x) = 0$ has a solution $x_* \in D$, then there exist positive constants η, η' such that if $\{h_k\}$ is a real sequence with $0 < |h_k| \le \eta'$, and if $|x_0 - x_*| < \eta$, then the sequence $\{x_k\}$ defined by

$$x_{k+1} = x_k - \frac{f(x_k)}{a_k}, \qquad a_k = \frac{f(x_k + h_k) - f(x_k)}{h_k}, \qquad k = 0, 1, \dots,$$

is defined and converges q-linearly to x_*. If $\lim_{k \to \infty} h_k = 0$, then the convergence is q-superlinear. If there exists some constant c_1 such that

$$|h_k| \le c_1 |x_k - x_*|,$$

or equivalently, a constant c_2 such that

$$|h_k| \le c_2 |f(x_k)|, \tag{2.6.5}$$

then the convergence is q-quadratic. If there exists some constant c_3 such that

$$|h_k| \le c_3 |x_k - x_{k-1}|, \tag{2.6.6}$$

then the convergence is at least two-step q-quadratic; this includes the secant method $h_k = x_{k-1} - x_k$ as a special case.

We began this discussion with a claim that insight is contained in the existence and simplicity of the analysis. In particular, the finite-difference idea seemed somewhat ad hoc when we introduced it. However, not only can we observe its excellent computational behavior in practice, but with analysis hardly more complicated than for Newton's method we can characterize exactly how it will converge as a function of the stepsizes $\{h_k\}$. This is just one of many instances in numerical analysis where theoretical analysis is an ally to algorithmic development.

If order of convergence were the only consideration in choosing $\{h_k\}$ in the finite-difference Newton's method, then we would just set $h_c = c_2 |f(x_c)|$ for some constant c_2, and our analysis would be complete, since it is easy to show for x_c sufficiently close to x_* that $|f(x_c)| \le \hat{c} |x_c - x_*|$. (See Exercise 18.) However, finite-precision arithmetic has a very important effect that we want to look at now. Obviously in practice h_c cannot be too small in relation to x_c. For example, if $\mathrm{fl}(x_c) \ne 0$ and $|h_c| < |x_c| \cdot$ macheps, then $\mathrm{fl}(x_c + h_c) = \mathrm{fl}(x_c)$ and so the numerator of (2.6.1) will be computed to be zero, since the subroutine that evaluates f will have been called with the same floating-point argument both times.

Even if $|h_c|$ is large enough that $\mathrm{fl}(x_c + h_c) \ne \mathrm{fl}(x_c)$, there remains the loss of accuracy in the floating-point subtraction necessary to form the numerator of (2.6.1). After all, since f is continuously differentiable, if f is evaluated at two nearby points, then the resulting values of $f(x_c + h_c)$ and $f(x_c)$ should also be nearly equal. We might even have $\mathrm{fl}(f(x_c + h_c)) = \mathrm{fl}(f(x_c))$. If $|h_c|$ is small, we certainly should expect to have several of the leftmost digits of the mantissa the same. For example, suppose on a base-10 machine with 5-digit mantissa we have $f(x_c) = 1.0001$ and $f(x_c + h_c) = 1.0010$ returned by some hypothetical function subroutine with $h_c = 10^{-4}$. In this case, $f(x_c + h_c) - f(x_c)$ would be computed as 9×10^{-4} and a_c as 9. Thus we have lost most of the significant digits in taking the difference of the function values.

In practice, furthermore, one often has progressively less confidence in the digits of the function values as one scans to the right. This is due not only to finite-precision arithmetic, but also to the fact that the function values are sometimes themselves just the approximate results returned by a numerical routine. Thus in the above example the difference $f(x_c + h_c) - f(x_c) = 9 \times 10^{-4}$ may reflect only random fluctuation in the rightmost digits of f and have no significance at all. The result is that the slope of the model of f at x_c may not even have the same sign as $f'(x_c)$.

The obvious way to compute a_c more accurately is to take $|h_c|$ large enough so that not all the most trustworthy leading digits are the same for $f(x_c + h_c)$ and $f(x_c)$. There is a limit to how large $|h_c|$ can be, since our whole object in computing a_c is to use it as an approximation to $f'(x_c)$, and (2.6.3) indicates that this approximation deteriorates as $|h_c|$ increases. A good compromise is to try to balance the nonlinearity error caused by taking $|h_c|$ too large with the finite-precision and function evaluation errors from allowing $|h_c|$ too small. We develop this as an exercise (Exercise 19) and discuss it further in Section 5.4.

A simple rule for the case when f is assumed to be computed accurately to machine precision is to perturb roughly half the digits of x_c:

$$|h_c| = \sqrt{\text{macheps}} \cdot \max \{\text{typ}x, |x_c|\},$$

where typ$x$ is the typical size of x used in Section 2.5. This often-used rule is generally satisfactory in practice. If the accuracy of the f-subroutine is suspect, then $|h_c|$ should be large enough so that only half the reliable digits of $f(x_c + h_c)$ and $f(x_c)$ will be the same. If this $|h_c|$ is large enough to cast doubt on the utility of (2.6.1) as an approximation to $f'(x_c)$, then one remedy is to use the central difference,

$$a_c = \frac{f(x_c + h_c) - f(x_c - h_c)}{2h_c}. \tag{2.6.7}$$

This gives a more accurate derivative approximation for a given h_c (see Exercise 20) than the forward difference approximation (2.6.1), but it has the disadvantage of doubling the expense in function evaluations.

2.7 MINIMIZATION OF A FUNCTION OF ONE VARIABLE

We conclude our study of one-dimensional problems by discussing minimization of a function of one variable. It turns out that this problem is so closely related to solving one nonlinear equation in one unknown that we virtually know already how to compute solutions.

First of all, one must again admit that one cannot practically answer questions of existence and uniqueness. Consider, for example, the fourth-degree polynomial of Figure 2.1.1. Its *global minimizer*, where the function takes its absolute lowest value, is at $x \cong 0.943$, but it also has a *local minimizer*, a minimizing point in an open region, at $x \cong 4.60$. If we divide the function by x, it becomes a cubic with a local minimum at $x \cong 4.58$, but no finite global minimizer since

$$\lim_{x \to -\infty} f(x) = -\infty.$$

In general, it is practically impossible to know if you are at the global minimum of a function. So, just as our algorithms for nonlinear equations can find only one root, our optimization algorithms at best can locate one local minimum; usually this is good enough in practice. Once again, a closed-form solution is out of the question: if we could find a closed form for a global minimizer of $f(x)$, we could solve $\hat{f}(x) = 0$ by setting $f(x) = \hat{f}(x)^2$.

The reason we said earlier that the reader already knows how to solve the minimization problem is that a local minimum of a continuously differentiable function must come at a point where $f'(x) = 0$. Graphically, this just says that the function can't initially decrease in either direction from such a point. A proof of this fact suggests an algorithm, so we give it below. It will be useful to denote by $C^1(D)$ and $C^2(D)$, respectively, the sets of once and twice continuously differentiable functions from D into $\mathbb{R}$.

THEOREM 2.7.1   Let $f \in C^1(D)$ for an open interval D, and let $z \in D$. If $f'(z) \neq 0$, then for any s with $f'(z) \cdot s < 0$, there is a constant $t > 0$ for which $f(z + \lambda s) < f(z)$, for every $\lambda \in (0, t)$.

Proof. We need only choose t, using the continuity of f', so that $f'(z + \lambda s)s < 0$ and $z + \lambda s \in D$ for every $\lambda \in (0, t)$. The rest is immediate from calculus, since, for any such λ,

$$f(z + \lambda s) - f(z) = \int_0^\lambda f'(z + \alpha s)s \, d\alpha < 0. \qquad \square$$

Theorem 2.7.1 suggests that we find a minimizing point x_* of a function f by solving $f'(x) = 0$. On the other hand, by applying the theorem to $g(x) = -f(x)$, we see that this condition also identifies possible maximizing points. That is, solving $f'(x) = 0$ is necessary for finding a minimizing point for f, but not sufficient. Theorem 2.7.2 shows that a sufficient additional condition is $f''(x_*) > 0$.

THEOREM 2.7.2   Let $f \in C^2(D)$ for an open interval D and let $x_* \in D$ for which $f'(x_*) = 0$ and $f''(x_*) > 0$. Then there is some open subinterval $D' \subset D$ for which $x_* \in D'$ and $f(x) > f(x_*)$ for any other $x \in D'$.

Proof. Let D' be chosen about x_* so that $f'' > 0$ on D'. The use of a Taylor series with remainder does not lead to later problems in the multidimensional instance of the minimization problem, so we note that for any $x \in D'$ there is an $\bar{x} \in (x_*, x) \subset D'$ for which

$$f(x) - f(x_*) = f'(x_*)(x - x_*) + \tfrac{1}{2} f''(\bar{x})(x - x_*)^2.$$

Thus we are finished, since $f'(x_*) = 0$ and $\bar{x} \in D'$ imply

$$f(x) - f(x_*) = \tfrac{1}{2}(x - x_*)^2 f''(\bar{x}) > 0. \qquad \square$$

Now that we can recognize a solution, let us decide how to compute one. The easiest way to the class of algorithms we will use is to think of solving $f'(x) = 0$ by applying the hybrid Newton's method strategy we studied in Section 2.5, only making sure that we find a minimizer and not a maximizer by incorporating into the globalizing strategy the requirement that $f(x_k)$ decreases as k increases.

An iteration of the hybrid method starts by applying Newton's method, or a modification discussed in Section 2.6, to $f'(x) = 0$ from the current point x_c. The Newton step is

$$x_+ = x_c - \frac{f'(x_c)}{f''(x_c)}. \tag{2.7.1}$$

It is important to note the meaning of this step in terms of model problems. Since (2.7.1) is derived by making an affine model of $f'(x)$ around x_c, it is equivalent to having made a quadratic model of $f(x)$ around x_c,

$$m_c(x) = f(x_c) + f'(x_c)(x - x_c) + \tfrac{1}{2} f''(x_c)(x - x_c)^2,$$

and setting x_+ to the critical point of this model. A quadratic model is more appropriate than an affine model of $f(x)$ for either maximization or minimization because it has at most one extreme point. Thus the step (2.7.1) will find the extreme point of a quadratic function in one iteration; also, by exactly the same proof as in Theorem 2.4.3, it will converge locally and q-quadratically to an extreme point x_* of $f(x)$ if $f''(x_*) \neq 0$ and f'' is Lipschitz continuous near x_*.

Our global strategy for minimization will differ from that in Section 2.5 in that, rather than deciding to use the Newton point x_N by the condition $|f'(x_N)| < |f'(x_c)|$, which measures progress toward a zero of $f'(x)$, we will want $f(x_N) < f(x_c)$, which indicates progress toward a minimum. If $f(x_N) \geq f(x_c)$ but $f'(x_c)(x_N - x_c) < 0$, then Theorem 2.7.1 shows that $f(x)$ must initially decrease in the direction from x_c toward x_N, and so we can find an acceptable next point x_+ by backtracking from x_N toward x_c. From (2.7.1), $f'(x_c)(x_N - x_c)$ will be negative if and only if $f''(x_c)$ (or its approximation) is positive. That is, if the local model used to derive the Newton step has a minimum and not a

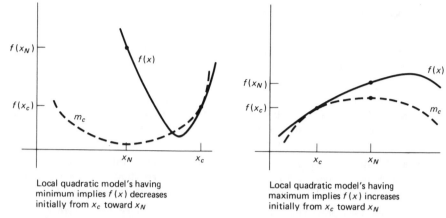

Local quadratic model's having
minimum implies $f(x)$ decreases
initially from x_c toward x_N

Local quadratic model's having
maximum implies $f(x)$ increases
initially from x_c toward x_N

Figure 2.7.1   Correspondence of the local quadratic model to the initial slope from x_c toward x_N

maximum, then it is guaranteed to provide an appropriate step *direction* (see Figure 2.7.1). On the other hand, if $f''(x_c) < 0$ and $f'(x_c)(x_N - x_c) > 0$, then $f(x)$ initially increases going from x_c toward x_N, and we should take a step in the opposite direction. One strategy is to try a step of length $|x_N - x_c|$ and then backtrack, if necessary, until $f(x_+) < f(x_c)$. More advanced global strategies for minimization are discussed in Chapter 6.

The stopping criteria for optimization are a bit different than those for solving nonlinear equations. Again one asks, "Have we approximately solved the problem?" or, "Have the iterates stayed in virtually the same place?" The second question is tested by the same relative/absolute step-size condition as in Section 2.5, "Is $|x_+ - x_c|/\max \{typx, |x_+|\} < \tau_1$?" The first can simply be "Is $|f'(x_+)| < \tau_2$?" but once again this is dependent upon the scale of f, and now also on the scale of x. This time, however, something can be done about it: one can ask for the relative rate of change in f, $f'(x)/f(x)$, to have absolute value less than τ_2, or even for the relative change in f divided by the relative change in x,

$$\frac{f'(x_+)}{f(x_+)} x_+,$$

to be less than τ_2 in absolute value. The last test is appealing because it tries to account for scaling in both f and x.

Finally, there is the question of what to do if the derivatives $f'(x)$ are not available. The solution is so similar to the discussion in Section 2.6 that we defer consideration of this problem to the multidimensional case (see also Exercise 21).

2.8 EXERCISES

1. Carry out one iteration of Newton's method from $x = 2$ for the function $f_1(x)$ given in Section 2.1. Which root are you converging to? What happens if you start from $x = 1$?

2. Approximately how many iterations will the q-linearly convergent sequence $x_k = 1 + (0.9)^k$, $k = 0, 1, \ldots$, take to converge to 1 on a CDC machine? (Do this on paper, not on the computer!) Is q-linear convergence with constant 0.9 satisfactory for general computational algorithms?

3. Consider the sequence $x_k = 1 + 1/k!$, $k = 0, 1, \ldots$. Does this sequence converge q-linearly to 1? Does it converge q-linearly with any constant $c > 0$? What type of convergence is this? Does $\{x_k\}$ converge to 1 with q-order l for any constant $l > 1$?

4. Prove that the r-order of a sequence is always at least as large as its q-order. Give a counterexample to the converse.

5. Prove that $\{x_k\}$ has q-order at least p if and only if it has a sequence of error bounds $\{b_k\}$ with $b_k \geq e_k = |x_k - x_*|$ for which $\{b_k\}$ converges to zero with q-order at least p and

$$\overline{\lim} \frac{b_k}{e_k} < \infty.$$

6. (Harder) There is an interesting relationship between l-step q-order and "1-step" r-order. See if you can discover and prove this result. To find the result and an application see Gay (1979).

7. On the IBM 360-370 series, extended-precision multiplication is hardwired, but extended-precision division is done by doing double-precision division, and then using a Newton's method iteration that requires only multiplication and addition. To understand this, (a) derive a suitable Newton's method iteration for calculating $x_* = 1/a$ for a given real number a, and (b) carry out three iterations for $a = 9$ given $x_0 = 0.1$. What sort of convergence are you getting?

8. Analyze the convergence rate of the bisection method.

9. Write, debug, and test a program for solving one nonlinear equation in one unknown. It should follow the hybrid strategy explained in Section 2.5. For the local step, use either Newton's method, finite-difference Newton's method, or the secant method. For the global strategy, use bisection, backtracking, a strategy that interpolates between x_N and x_k, or another strategy of your own choosing. Run your program on:
   (a) $f(x) = \sin x - \cos 2x$, $x_0 = 1$
   (b) $f(x) = x^3 - 7x^2 + 11x - 5$, $x_0 = 2, 7$
   (c) $f(x) = \sin x - \cos x$, $x_0 = 1$
   (d) $f_2(x)$ given in Section 2.1, $x_0 = 0, 2$.
   (If you use bisection, pick an appropriate initial interval for each problem, and constrain all your iterates to remain in that interval.) What rate of convergence do you observe in each case? It won't always be quadratic. Why?

10. Use your program from Exercise 9 to find the point $c > 0$ from which Newton's method for arc tan $x = 0$ will cycle (see Figure 2.4.1).

11. Modify your program from Exercise 9 to find the minimum of a function of one variable. (The modifications are explained in Section 2.7.)

12. What rate of convergence (to $x_* = 0$) does Newton's method have when solving $x^2 = 0$? $x^3 = 0$? How about $x + x^3 = 0$? $x + x^4 = 0$? Which assumptions of Theorem 2.4.3, if any, are violated in each case?

13. Prove that if $f(x_*) = f'(x_*) = 0, f''(x_*) \neq 0$, Newton's method converges q-linearly to x_* with $\lim_{k \to \infty} e_{k+1}/e_k = \frac{1}{2}$. [If you wish, generalize this result to $f(x_*) = f'(x_*) = \cdots = f^n(x_*) = 0, f^{n+1}(x_*) \neq 0$.]

14. Prove that if $f(x_*) = 0, f'(x_*) \neq 0, f''(x_*) = 0, f'''(x_*) \neq 0$, Newton's method converges q-cubicly to x_* [If you wish, generalize this result to $f'(x_*) \neq 0, f(x_*) = f^2(x_*) = \cdots = f^n(x_*) = 0, f^{n+1}(x_*) \neq 0$.]

15. Derive a cubicly convergent method for solving $f(x) = 0$ by modeling f by a quadratic Taylor series approximation at each step. What are the major problems in using this method? Do you think a quadratic model is appropriate for this problem? (See also "Müller's method" in most introductory numerical analysis texts.)

16. Given $f \in C^1(D)$, $x_c \in D$ with $f'(x_c) \neq 0$, and $d = -f(x_c)/f'(x_c)$, show that there exists some $t > 0$ such that $|f(x_c + \lambda d)| < |f(x_c)|$ for all $\lambda \in (0, t)$. [Hint: Choose t so that $f'(x_c + \lambda d) \cdot f'(x_c) > 0$ for all $\lambda \in (0, t)$, and then use the techniques of proof of Theorem 2.7.1.] (Assume $f(x_c)$ is not 0.)

17. Consider solving $x^2 = 0$ by Newton's method starting from $x_0 = 1$. Will any two successive iterates satisfy $|x_{k+1} - x_k|/\max\{|x_k|, |x_{k+1}|\} < 10^{-7}$? What about the test involving "typx" suggested at the end of Section 2.6?

18. Given $f \in C^1(D)$ and $x_* \in D$ with $f(x_*) = 0$, show that there exists a constant $\alpha > 0$ and an interval $\hat{D} \subset D$ containing x_* such that for all $x \in \hat{D}, |f(x)| \leq \alpha |x - x_*|$. [Hint: Choose $\hat{D}$ so that $|f'(x)| \leq \alpha$ for all $x \in \hat{D}$.]

19. Suppose $f: D \to \mathbb{R}$ for an open interval D and assume $f' \in \text{Lip}_\gamma(D)$. Suppose that the computer routine for evaluating $f(x)$ has a total error bounded by $\eta |f(x)|$, where η is a nonnegative constant. Develop an upper bound on the total error in $[f(x_c + h_c) - f(x_c)]/h_c$ as an approximation to $f'(x_c)$, as a function of h_c [and γ, η, $f(x_c)$, and $f(x_c + h_c)$]. Find the value of h_c that minimizes this bound for fixed γ, η, and $f(x_c)$. [Assume here that $f(x_c + h_c) \cong f(x_c)$.] Also comment on the computational utility of finite differences if $f'(x)$ is extremely small in comparison to $f(x)$.

20. Suppose D is an open interval containing $x_c - h_c$, $x_c + h_c \in \mathbb{R}$, and let $f'': D \to \mathbb{R}$ satisfy $f'' \in \text{Lip}_\gamma(D)$. Prove that the error in the central difference formula (2.6.7) as an approximation to $f'(x_c)$ is bounded by $\gamma |h_c|^2/6$. [Hint: Expand the techniques of Lemma 2.4.2 to show that

$$\left| f(z) - \left[f(x) + f'(x)(z - x) + \frac{f''(x)(z - x)^2}{2} \right] \right| \leq \frac{\gamma |z - x|^3}{6},$$

or derive a similar bound from the Taylor series with remainder.]

21. Suppose you are solving a minimization problem in one variable in which $f(x)$ and $f'(x)$ are analytically available but expensive to evaluate, and $f''(x)$ is not analytically available. Suggest a local method for solving this problem.

A second program note

It was evident in Chapter 2 that we like to derive iterative algorithms for nonlinear problems from considering the solutions to properly chosen models of the problems. The model used must be from a class for which the solution of the model problem is possible by an effective numerical procedure, but also it must be chosen to adequately model the nonlinear problems so that the iteration sequence will converge to a solution.

In Chapter 2, elementary calculus gave all the necessary techniques for building the relevant linear and quadratic models and analyzing their approximation errors with respect to a realistic class of nonlinear problems. The solution of the model problem required only a couple of arithmetic operations. This simplicity was our reason for writing Chapter 2. There seems no pressing need for another exposition of scalar iterations, but we were able to present, without the complications of multivariable calculus and linear algebra, ideas that will extend to the multivariable case.

In multiple dimensions, our models will involve derivatives of multivariable functions, and the solution of a system of linear equations will be needed to solve the model problem. Chapter 3 presents material from computational linear algebra relevant to the solution of the model problems or the extraction of useful information from them. Chapter 4 is a review of some multivariable calculus theorems that we will find useful in setting up multivariable models and analyzing their approximation properties.

There is a subtle and interesting relationship between the models we build and the means we have to extract information from them. If it is cheap to form and to solve the model problem, the model need not be as good as if the model solution were very expensive. This is just a way of stating the obvious fact that we can afford to do more iterations of a cheap method than of an expensive one. Part of our interest lies in using problem structure, which is not too expensive to incorporate into the model, in order to improve the model or to facilitate the extraction of x_+ from it. Another interest is in considering how much problem structure we can safely ignore in the model for the facilitation of its solution and still expect meaningful convergence of the iteration. We will give examples of both approaches at appropriate times.

3

Numerical Linear
Algebra Background

This chapter discusses the topics in numerical linear algebra needed to implement and analyze algorithms for multivariable nonlinear problems. In Section 3.1 we introduce vector and matrix norms to measure the sizes of these objects in our algorithms, and norm properties useful in analyzing our methods. We then describe in Section 3.2 the various matrix factorizations used by our algorithms for solving systems of n linear equations in n unknowns. In Section 3.3 we briefly discuss problem sensitivity to the errors involved in solving such systems. Our point of view is that the user may obtain the appropriate algorithms from some subroutine library, but the serious user needs to understand the principles behind them well enough to know the tools of numerical linear algebra that can help, which routines to use for specific problems, and the costs and errors involved. For reference, Section 3.5 summarizes properties of eigenvalues that we will use in this book, especially their relation to positive definiteness. In Sections 3.4 and 3.6 we discuss the updating of matrix factorizations and the solution of overdetermined systems of linear equations, respectively. Since these topics will not be needed until Chapters 8 and 11, respectively, their consideration can be postponed until then.

Excellent references for this material are the books by J. Wilkinson

(1965), G. W. Stewart (1973), G. Strang (1976), G. H. Golub, and C. Van Loan (1983).

3.1 VECTOR AND MATRIX NORMS AND ORTHOGONALITY

In Chapter 2 we found it useful to make certain assumptions on f, such as the absolute value of a derivative being bounded below, or the Lipschitz condition (2.4.1), that gave a bound on the magnitude of the derivative change divided by the magnitude of the corresponding argument change. In the sequel, we will need similar conditions, but in a context where the absolute value no longer can measure magnitude because the arguments and derivatives are no longer scalar. Similarly, conditions like the stopping criterion $|f(x_k)| \leq \tau_1$ will have to be replaced by criteria suitable for the case where $F(x_k)$ is a vector.

In Chapter 4 we will see that function values will be vectors or scalars, and derivatives will be matrices or vectors. This section will introduce norms of vectors and matrices as the appropriate generalizations of the absolute value of a real number.

There are many different norms, appropriate to different purposes, and we will limit our discussion to just a few. We will especially be interested in the so-called Euclidean or l_2 norm, which coincides with the usual notion of vector magnitude. Also in this section is a discussion of orthogonal matrices. These matrices have a special relationship to the Euclidean norm that makes them very important for numerical linear algebra.

Recall that $\mathbb{R}^n$ denotes the vector space of real $n \times 1$ vectors. We will generally use Greek lower-case letters for real numbers, elements of $\mathbb{R}$, and lower-case roman letters for vectors. Matrices will be denoted by capital roman letters, and the set of $n \times m$ real matrices will be denoted by $\mathbb{R}^{n \times m}$. The superscript T will denote the matrix transpose. We will use the same symbol 0 for the zero vector as well as the zero scalar. Thus the reader will easily see from (3.1.1b) below that $\|0\| = 0$.

Definition 3.1.1.   A norm on $\mathbb{R}^n$ is a real-valued function $\|\cdot\|$ on $\mathbb{R}^n$ that obeys:

$$\|v\| \geq 0 \text{ for every } v \in \mathbb{R}^n, \text{ and } \|v\| = 0$$

$$\text{only if } v = 0 \in \mathbb{R}^n. \tag{3.1.1a}$$

$$\|\alpha v\| = |\alpha| \cdot \|v\| \text{ for every } v \in \mathbb{R}^n \text{ and } \alpha \in \mathbb{R}. \tag{3.1.1b}$$

$$\|v + w\| \leq \|v\| + \|w\| \text{ for every } v, w \in \mathbb{R}^n. \tag{3.1.1c}$$

The three vector norms most suitable for our purposes are: for $v = (v_1, v_2, \ldots, v_n)^T \in \mathbb{R}^n$,

$$\| v \|_\infty = \max_{1 \le i \le n} | v_i | \qquad (l_\infty \text{ or sup norm}) \qquad (3.1.2a)$$

$$\| v \|_1 = \sum_{i=1}^n | v_i | \qquad (l_1 \text{ or LAR (Least Absolute Residual) norm}) \qquad (3.1.2b)$$

$$\| v \|_2 = \left(\sum_{i=1}^n (v_i)^2 \right)^{1/2} \qquad (l_2, \text{ Euclidean, or least-squares norm}). \qquad (3.1.2c)$$

These are all instances of the general class of l_p vector norms for $1 \le p < \infty$,

$$\| v \|_p = \left(\sum_{i=1}^n | v_i |^p \right)^{1/p}. \qquad (3.1.3)$$

The reader can show that (3.1.2a) is consistent with (3.1.3) in the limit as p approaches ∞ (Exercise 1).

In our discussions of convergence, we will be speaking of a vector sequence $\{x_k\}$ converging to a vector x_*. We will be thinking of a certain norm when we make such a statement, and we will mean

$$\lim_{k \to \infty} \| x_k - x_* \| = 0.$$

We will also speak of the rate of convergence of $\{x_k\}$ to x_* and we will mean the rate of convergence of $\{\| x_k - x_* \|\}$ to 0. It would be unfortunate if it were necessary to specify the norm, as we would have to if it were possible to converge in one norm but not in another. It would mean that whether a particular algorithm converged in practice might depend on what norm its stopping criteria used. Fortunately in $\mathbb{R}^n$ we have no such problem, owing to the following result that is not true in infinite-dimensional space. We should point out that q-linear convergence is a norm-dependent property.

THEOREM 3.1.2   Let $\| \cdot \|$ and $\| | \cdot | \|$ be any two norms on $\mathbb{R}^n$. There exist positive constants α and β such that

$$\alpha \| v \| \le \| | v | \| \le \beta \| v \| \qquad (3.1.4)$$

for any $v \in \mathbb{R}^n$. Furthermore, if $\{v_k\}$ is any sequence in $\mathbb{R}^n$, then for $v_* \in \mathbb{R}^n$,

$$\lim_{k \to \infty} \| v_k - v_* \| = 0$$

if and only if for each i, $1 \le i \le n$, the sequence $(v_k)_i$ of ith components of v_k converges to the ith component $(v_*)_i$ of v_*.

We leave the proof as an exercise, since it is not central to our topic. An easier exercise is the particular set of relations

$$\| v \|_1 \geq \| v \|_2 \geq \| v \|_\infty, \qquad \| v \|_1 \leq \sqrt{n} \, \| v \|_2, \qquad \| v \|_2 \leq \sqrt{n} \, \| v \|_\infty, \qquad (3.1.5)$$

from which (3.1.4) can be derived for the l_1, l_2, and l_∞ norms. The relation $\| v \|_1 \leq \sqrt{n} \, \| v \|_2$ can be proven easily, using the Cauchy-Schwarz inequality:

$$v^T w = \sum_{i=1}^{n} v_i w_i \leq \| v \|_2 \| w \|_2 \qquad \text{for } v, w \in \mathbb{R}^n.$$

The important conclusion from relations (3.1.4) and (3.1.5) is that the performance of an algorithm is unlikely to be affected seriously by a choice of norm, and so this choice is usually based on convenience, or appropriateness to a particular problem.

We mentioned earlier that we will need to measure matrices as well as vectors. This certainly can be done by thinking of a matrix $A = (a_{ij})$ as corresponding to the vector $(a_{11}, a_{21}, \ldots, a_{n1}, a_{12}, \ldots, a_{nn})^T$ and applying any vector norm to this "long" vector. In fact, the useful Frobenius norm is just the l_2 norm of A written as a long vector:

$$\| A \|_F = \left(\sum_{j=1}^{n} \sum_{i=1}^{n} |a_{ij}|^2 \right)^{1/2}. \qquad (3.1.6)$$

It is usual to call these "matrix norms," and they are relevant when we view the matrix as a set of pigeonholes filled with quantitative information. This will relate to their function in the local models on which the iterations of our nonlinear methods will be based.

We also will want to measure matrices in the context of their roles as operators. If v has a certain magnitude $\| v \|$, then we will want to be able to bound $\| Av \|$, the magnitude of the image of v under the operator A. This will be useful in analyzing convergence; the reader can look back to Section 2.4 and see that this was exactly the use of the bound, involving the symbol ρ, for the model derivative. Operator norms must depend on the particular vector norms we use to measure v and Av.

A natural definition of the norm of A induced by a given vector norm is

$$\| A \| = \max_{v \in \mathbb{R}^n, \, v \neq 0} \left\{ \frac{\| Av \|}{\| v \|} \right\} \qquad (3.1.7)$$

—that is, the maximum amount A can stretch any vector in the given vector norm $\| \cdot \|$. It is not necessary at all to use the same norm on v and Av, but we have no need of such generality.

Norm (3.1.7) is called the operator norm induced by vector norm $\| \cdot \|$, and for any l_p vector norm it is denoted by $\| A \|_p$. It is easily shown to obey the three defining properties of a matrix norm, which are simply (3.1.1) with A,

$B \in \mathbb{R}^{n \times n}$ replacing v, $w \in \mathbb{R}^n$. Although it appears that definition (3.1.7) may be of little use computationally, since it involves the solution of an optimization problem, it is an exercise to show that the l_1, l_2, and l_∞ induced matrix norms are given by

$$\| A \|_1 = \max_{1 \le j \le n} \{ \| a_{.j} \|_1 \} \qquad (3.1.8a)$$

$$\| A \|_2 = (\text{maximum eigenvalue of } A^T A)^{1/2} \qquad (3.1.8b)$$

$$\| A \|_\infty = \max_{1 \le i \le n} \{ \| a_{i.} \|_1 \}. \qquad (3.1.8c)$$

Here $a_{.j}$ denotes the jth column of A and $a_{i.}$ its ith row. Thus $\| A \|_1$ and $\| A \|_\infty$ are easy to compute, while $\| A \|_2$ is useful in analysis.

Sometimes a linear transformation of the problem will cause us to use a weighted Frobenius norm:

$$\| W_1 A W_2 \|_F, \qquad W_1 \text{ and } W_2 \; n \times n \text{ nonsingular matrices} \qquad (3.1.9)$$

("nonsingular" means "invertible"). Although the Frobenius and the l_p induced norms obey the consistency condition

$$\| AB \| \le \| A \| \cdot \| B \|, \qquad (3.1.10)$$

the weighted Frobenius norm does not, except in special cases.

Several properties of matrix norms and matrix-vector products are of particular importance in analyzing our algorithms, and these properties are listed below. Proofs of some are given as exercises with appropriate hints as necessary. The n-dimensional identity matrix is denoted by I, with n understood from the context. For $M \in \mathbb{R}^{n \times n}$, the trace of M is defined as

$$\text{tr}(M) = \sum_{i=1}^{n} M_{ii}.$$

THEOREM 3.1.3   Let $\| \cdot \|$ and $\|\| \cdot \|\|$ be any norms on $\mathbb{R}^{n \times n}$. There exist positive constants α, β such that

$$\alpha \| A \| \le \|\| A \|\| \le \beta \| A \| \qquad (3.1.11)$$

for every $A \in \mathbb{R}^{n \times n}$. In particular,

$$n^{-1/2} \| A \|_F \le \| A \|_2 \le \| A \|_F \qquad (3.1.12)$$

and, for $p = 1$ or $p = \infty$,

$$n^{-1/2} \| A \|_p \le \| A \|_2 \le n^{1/2} \| A \|_p. \qquad (3.1.13)$$

The Frobenius norm of A satisfies

$$\| A \|_F = [\text{tr}(A^T A)]^{1/2} \qquad (3.1.14)$$

and for any $B \in \mathbb{R}^{n \times n}$,

$$\| AB \|_F \leq \min \{ \| A \|_2 \| B \|_F , \ \| A \|_F \| B \|_2 \}. \tag{3.1.15}$$

Furthermore, for any $v, w \in \mathbb{R}^n$,

$$\| Av \|_2 \leq \| A \|_F \cdot \| v \|_2 \tag{3.1.16}$$

and

$$\| vw^T \|_F = \| vw^T \|_2 = \| v \|_2 \cdot \| w \|_2 . \tag{3.1.17}$$

If A is nonsingular, then the operator norm induced by $\| \cdot \|$ satisfies

$$\| A^{-1} \| = \frac{1}{\displaystyle\min_{v \in \mathbb{R}^n, \ v \neq 0} \frac{\| Av \|}{\| v \|}}. \tag{3.1.18}$$

The next theorem says that matrix inversion is continuous in norm. Furthermore it gives a relation between the norms of the inverses of two nearby matrices that will be useful later in analyzing algorithms.

THEOREM 3.1.4   Let $\| \cdot \|$ be any norm on $\mathbb{R}^{n \times n}$ that obeys (3.1.10) and $\| I \| = 1$ and let $E \in \mathbb{R}^{n \times n}$. If $\| E \| < 1$, then $(I - E)^{-1}$ exists and

$$\| (I - E)^{-1} \| \leq \frac{1}{1 - \| E \|}. \tag{3.1.19}$$

If A is nonsingular and $\| A^{-1}(B - A) \| < 1$, then B is nonsingular and

$$\| B^{-1} \| \leq \frac{\| A^{-1} \|}{1 - \| A^{-1}(B - A) \|}. \tag{3.1.20}$$

Proof.   We will give the idea rather than the detail. Inequality (3.1.20) follows from the direct application of (3.1.19) with one use of (3.1.10), and so we leave it as an exercise. The proof of (3.1.19) is very much like the proof that

$$\sum_{k=0}^{\infty} (\| E \|)^k = \frac{1}{1 - \| E \|}.$$

The outline is to show that $(I + E + E^2 + \cdots + E^k) = S_k$ defines a Cauchy sequence in $\mathbb{R}^{n \times n}$ and hence converges. Then it is easy to show that $(\lim_{k \to \infty} S_k)$ is $(I - E)^{-1}$, and (3.1.19) follows from

$$\| (I - E)^{-1} \| = \lim_{k \to \infty} \| S_k \| \leq \sum_{i=0}^{\infty} \| E \|^i. \qquad \square$$

Finally we introduce the concept of orthogonal vectors and matrices. Some simple properties of orthogonal matrices in the l_2 norm indicate that they will be useful computationally. We have been denoting the inner product of two vectors v, $w \in \mathbb{R}^n$ as $v^T w$, but we will occasionally find some special notation useful. In the following two definitions, we juxtapose two meanings for v_i to warn the reader again to interpret it in context.

Definition 3.1.5.   Let v, $w \in \mathbb{R}^n$; the inner product of v and w is defined to be

$$\langle v, w \rangle \equiv v^T w = \sum_{i=1}^{n} v_i w_i = \|v\|_2 \cdot \|w\|_2 \cdot \cos \theta, \qquad (3.1.21)$$

where θ is the angle between v and w, if $v \neq 0 \neq w$. If $\langle v, w \rangle = 0$, then v and w are said to be *orthogonal* or *perpendicular*.

Definition 3.1.6.   Vectors $v_1, \ldots, v_k \in \mathbb{R}^n$ are said to be *orthogonal* or mutually orthogonal if $\langle v_i, v_j \rangle = 0$ for $i \neq j$. If $\langle v_i, v_i \rangle = 1$ in addition, then the set $\{v_1, \ldots, v_k\}$ is said to be *orthonormal*, and if $k = n$, it is an orthonormal basis for $\mathbb{R}^n$.

If $Q \in \mathbb{R}^{n \times p}$, then Q is said to be an *orthogonal matrix* if

$$Q^T Q = I \quad \text{or} \quad QQ^T = I. \qquad (3.1.22)$$

An equivalent definition to (3.1.22) is that the columns or rows of an orthogonal matrix form an orthonormal set of vectors. The following theorem contains some properties of orthogonal matrices that make them important in practice. Proofs are left as exercises.

> THEOREM 3.1.7   If Q, $\hat{Q} \in \mathbb{R}^{n \times n}$ are orthogonal matrices, then $Q\hat{Q}$ is orthogonal. Also, for any $v \in \mathbb{R}^n$ and $A \in \mathbb{R}^{n \times n}$,
>
> $$\|Qv\|_2 = \|v\|_2, \qquad (3.1.23)$$
>
> $$\|QA\|_2 = \|AQ\|_2 = \|A\|_2, \qquad (3.1.24)$$
>
> and
>
> $$\|Q\|_2 = 1. \qquad (3.1.25)$$

Equation (3.1.23) shows that an orthogonal matrix corresponds to our geometric concept of a rotation-reflection operator, since it can change only the direction, but not the size, of a vector v. Equation (3.1.24) says that an orthogonal matrix also does not change the size of a matrix. These are two reasons why orthogonal matrices are useful in matrix factorizations.

3.2 SOLVING SYSTEMS OF LINEAR EQUATIONS—MATRIX FACTORIZATIONS

Multidimensional nonlinear algorithms almost always require the simultaneous solution of at least one system of n linear equations in n unknowns,

$$Ax = b, \quad A \in \mathbb{R}^{n \times n}, \; b, \; x \in \mathbb{R}^n, \tag{3.2.1}$$

at each iteration, usually to find the Newton point by solving the model problem or some modification of it. Luckily, excellent subroutine libraries for this problem are available: in the United States primarily in the LINPACK and IMSL libraries, and in Britain in the NAG and Harwell libraries. These libraries include algorithms to solve (3.2.1) for general matrices A and special algorithms more efficient for cases when A has particular kinds of structure such as symmetry and positive definiteness—common attributes in systems arising from minimization applications. Because of the availability of these routines and the careful effort and testing invested in them, our approach is that a person who solves nonlinear problems need not, and probably should not, write a personal linear equations solver; rather, that person should know which canned routine to use for the various types of problems (3.2.1) he or she may encounter, and the costs and possible problems involved in using these routines. It is also important to have a basic understanding of the structure of these routines in order to interface with them easily and to be aware of techniques useful in other parts of nonlinear algorithms.

It is common practice to write the form $A^{-1}b$ for economy of notation in mathematical formulas—for example, $x_{k+1} = x_k - A_k^{-1} F(x_k)$ in the Newton iteration for $F(x) = 0$. The trouble is that a reader unfamiliar with numerical computation might assume that we actually compute A^{-1} and take its product with the vector b. On most computers it is always more effective to calculate $A^{-1}b$ by solving the linear system $Ax = b$ using matrix factorization methods; this may not be true on some "vector machines."

Matrix factorization techniques are based on decomposing A into

$$A = A_1 \cdot A_2 \cdot \ldots \cdot A_m$$

where each A_l is of a form for which (3.2.1) is easy. For us, $m \leq 5$ with $m = 2$ or 3 most often. Once we have A factored, or while we are factoring it, we try to determine questions of singularity or invertibility of A. If A is nearly singular, then problem (3.2.1) is not numerically well posed, and perhaps it is more important to know this than to compute an unreliable solution. More will be said about this in the next section.

If we decide to proceed with the solution of $Ax = b$, then we peel off the factors to solve in order:

$$A_1 b_1 = b,$$

$$A_2 b_2 = b_1,$$

$$\vdots$$

$$A_m b_m = b_{m-1},$$

and $x = b_m$ is the desired solution. To verify this, note that each b_l equals $A_l^{-1} A_{l-1}^{-1} \cdots A_1^{-1} b$, so that $x = A_m^{-1} A_{m-1}^{-1} \cdots A_1^{-1} b = A^{-1} b$.

Below we list the six most important choices of A_l and briefly mention special features of solving (3.2.1) with each choice.

1. A_l is a permutation matrix P. A permutation matrix has the same rows (and columns) as the identity matrix, although not in the same order. For any matrix M, PM is then just the same permutation of the rows of M, and MP permutes the columns. Also $P^{-1} = P^T$ is a permutation matrix. P is usually stored as a vector $p = (p_1, \ldots, p_n)^T$, a permutation of $(1, 2, \ldots, n)^T$, where the convention is that the ith row of P is the p_ith row of I. It is easy to see that $Px = b$ is solved by $x_{p_i} = b_i$, $i = 1, \ldots, n$.

2. A_l is an orthogonal matrix denoted by Q or U. Since $Q^{-1} = Q^T$, $Qx = b$ is solved by forming $Q^T b$, and, since we often generate Q as a product of elementary orthogonal transformations, often this can be done in an efficient way that doesn't require an explicit computation of Q.

3. A_l is a nonsingular diagonal matrix D, $d_{ii} = d_i \neq 0$, $i = 1, \ldots, n$, $d_{ij} = 0$, $i \neq j$. $Dx = b$ is solved by $x_i = b_i/d_i$, $i = 1, \ldots, n$. D is stored as a vector.

4. A_l is a nonsingular block diagonal matrix D_B, which consists of 1×1 and 2×2 invertible blocks on the main diagonal and zeros elsewhere; for example,

$$D_B = \begin{bmatrix} x & x & & & \\ x & x & & & 0 \\ & & x & x & \\ & & x & x & \\ 0 & & & & x \end{bmatrix}.$$

$D_B x = b$ is solved by solving the corresponding 1×1 and 2×2 linear systems.

5. A_l is a nonsingular lower triangular matrix L, $l_{ii} \neq 0$, $i = 1, \ldots, n$, $l_{ij} = 0$,

$1 \le i < j \le n$; schematically,

$$
L = \begin{bmatrix}
x & & & 0 \\
x & x & & \\
x & x & x & \\
x & x & x & x
\end{bmatrix}.
$$

If $l_{ii} = 1$, $i = 1, \ldots, n$, L is called unit lower triangular. $Lx = b$ is solved by forward substitution; use the first equation to solve for x_1, then substitute into the second equation to solve for x_2, then the third equation for x_3, and so on.

6. A_l is a nonsingular upper triangular matrix denoted by U or R, the transpose of a lower triangular matrix. $Ux = b$ is solved by back substitution, using equations n through 1 to solve for x_n through x_1, in that order.

The arithmetic costs of solving linear systems with the above types of matrices are given in Table 3.2.1. All are small in comparison to the cost of matrix factorizations. Note that solving $Px = b$ just involves n assignments.

Table 3.2.1   Arithmetic cost in solving linear systems, leading terms

Matrix	Multiplications/Divisions	Additions/Subtractions
Q	n^2	n^2
D	n	—
D_B	$\le 5n/2$	$\le n$
L or U	$n^2/2$	$n^2/2$

We will be interested in factorizations for three types of matrices: no special structure, symmetric, and symmetric and positive definite. A $\in \mathbb{R}^{n \times n}$ is symmetric if $A = A^T$. A is positive definite if $v^T A v > 0$ for all nonzero $v \in \mathbb{R}^n$, or equivalently if A is symmetric, if all its eigenvalues are positive.

For a general square matrix A, two important factorizations are the *PLU* and the *QR*. The *PLU* decomposition yields $A = P \cdot L \cdot U$ or $P^T A = LU$ for a permutation matrix P, a unit lower triangular L, and an upper triangular U. The decomposition is found by Gaussian elimination with partial pivoting, or Doolittle reduction, using standard row operations (equivalent to premultiplying A by $L^{-1}P^{-1}$) to transform A into U, and simultaneously yields the decomposition. The *QR* or *QRP* decomposition, $A = QRP$, Q orthogonal, R upper triangular, and P a permutation matrix, is obtained by transforming A to R by premultiplying A by a series of $n - 1$ orthogonal matrices Q_i. Each Q_i zeros out the elements of the ith column of $Q_{i-1} \cdot \ldots \cdot Q_1 A$ below the main diagonal, while leaving the first $i - 1$ columns unchanged. Q_i is called a

Householder transformation and is of the form

$$Q_i = I - u_i u_i^T,$$

where $(u_i)_j = 0$, $j = 1, \ldots, i - 1$, and the remaining elements of u_i are selected so that Q_i is orthogonal and induces the desired zeros in column i. The permutation matrix, which is not always used, is formed from the column permutations necessary to move the column with the largest sum of squares below row $i - 1$ into column i at the ith iteration.

The advantage of the QR decomposition is that, since $\| R \|_2 = \| QR \|_2 = \| AP^T \|_2 = \| A \|_2$, it doesn't magnify size differences of elements in A when forming R. This makes it very stable numerically and is one reason Householder and other orthogonal transformations are important to numerical linear algebra. On the other hand, the PLU decomposition is generally quite accurate and is half as expensive as the QR, so both factorizations are used in practice. In our secant algorithms, we will use the QR without column pivots on the derivative matrices because it is cheaper to update, after low-rank changes to the matrix, than the PLU. In fact, we will recommend using the QR algorithm in any algorithm for solving dense systems of nonlinear equations for an implementation reason that is discussed in Section 6.5. A QR decomposition algorithm is given in Algorithm A3.2.1 in the appendix.

When A is symmetric and positive definite, a more efficient algorithm is the Cholesky decomposition, $A = LL^T$, L lower triangular. L is found simply by writing the $(n^2 + n)/2$ equations expressing each element in the lower triangle of A in terms of the elements in L:

$$a_{11} = (l_{11})^2,$$

$$a_{21} = l_{11} \cdot l_{21},$$

$$\vdots$$

$$a_{n1} = l_{11} l_{n1},$$

$$a_{22} = (l_{21})^2 + (l_{22})^2,$$

$$a_{32} = l_{21} l_{31} + l_{22} l_{32},$$

$$\vdots$$

$$a_{n2} = l_{21} l_{n1} + l_{22} l_{n2},$$

$$a_{33} = (l_{31})^2 + (l_{32})^2 + (l_{33})^2,$$

$$\vdots$$

and solving in this order for $l_{11}, l_{21}, \ldots, l_{n1}, l_{22}, l_{32}, \ldots, l_{n2}, l_{33}, \ldots, l_{n3}, \ldots$, using the equation for a_{ij} to solve for l_{ij}. The algorithm is very stable numerically; it involves taking n square roots to find the diagonal elements l_{jj}, $i = 1, \ldots, n$, which will all be of positive numbers if and only if A is positive definite. Sometimes the decomposition is expressed as $A = LDL^T$, L unit lower triangular, D diagonal with positive diagonal elements. This form of the decomposition does not require any square roots. A Cholesky decomposition algorithm is a special case of Algorithm A5.5.2 in the appendix.

If A is symmetric but indefinite—i.e., A has positive and negative eigenvalues—then A can be efficiently and stably factored into $PLD_B L^T P^T$, P a permutation matrix, L unit lower triangular, D_B a 1×1 and 2×2 block diagonal matrix. For further information on this factorization, see Bunch and Parlett (1971). Aasen (1973) gives a version in which D_B is replaced by T, a tridiagonal matrix. See Golub and Van Loan (1983) for more details. For a detailed discussion of the PLU, QR, or LL^T decompositions, see e.g. Stewart (1973) or Dongarra et al (1979).

The arithmetic costs of all the factorizations are small multiples of n^3 and are given in Table 3.2.2.

Table 3.2.2   Arithmetic cost of matrix factorization, leading terms

Factorization	Multiplications/Divisions	Additions/Subtractions
$A = PLU$	$n^3/3$	$n^3/3$
$A = QR$	$2n^3/3$	$2n^3/3$
$A = LL^T$, LDL^T	$n^3/6$	$n^3/6$
$A = PLD_B L^T P^T$	$n^3/6$	$n^3/6$
$A = PLTL^T P^T$	$n^3/6$	$n^3/6$

Nonlinear problems that give rise to sparse linear systems, where most of the matrix elements are zero, are discussed briefly in Chapter 11. In solving them, one should use the special subroutines available for sparse matrix problems; the Harwell and Yale packages are widely available.

3.3 ERRORS IN SOLVING LINEAR SYSTEMS

An iteration of a nonlinear algorithm will use the solution s of a linear system $A_c s = -F(x_c)$ to determine the step or direction to the next approximate solution x_+. Therefore, it is important to know how much the computed step may be affected by the use of finite-precision arithmetic. Also, since A_c and $F(x_c)$ are sometimes approximations to the quantities one really wants to use, one is interested in how sensitive the computed step is to changes in the data A_c and $F(x_c)$. These topics are discussed in this section.

Consider two linear systems

$$A_1 x = b_1 : \begin{bmatrix} 8 & -5 \\ 4 & 10 \end{bmatrix} \begin{bmatrix} x_1 \\ x_2 \end{bmatrix} = \begin{bmatrix} 3 \\ 14 \end{bmatrix},$$

$$A_2 \bar{x} = b_2 : \begin{bmatrix} 0.66 & 3.34 \\ 1.99 & 10.01 \end{bmatrix} \begin{bmatrix} \bar{x}_1 \\ \bar{x}_2 \end{bmatrix} = \begin{bmatrix} 4 \\ 12 \end{bmatrix}.$$

Both have the solution $(1, 1)^T$. If we change b_1 to $b_1 - (0.04, 0.06)^T$, a relative change of 0.43% in the l_2 norm of b_1, the new solution to the first system is $(0.993, 0.9968)^T$, a 0.51% relative change in the l_2 norm of x. However, if we change b_2 by the same $(-0.04, -0.06)^T$, a relative change of 0.55%, the new solution to the second system is $(6, 0)^T$, a whopping relative change of 324% in $\bar{x}$. Similarly, changing the first column of A_2 to $(\frac{2}{3}, 2.00)^T$ also causes $\bar{x}$ to become $(6, 0)^T$. Clearly, the second system is very sensitive to changes in its data. Why? Graphing its two equations shows that they represent two nearly parallel lines, so that moving one a little alters their point of intersection drastically. On the other hand, the first system corresponds to lines meeting at about an 80° angle, so that a shift in either line causes a similar shift in the point at which they cross. (See Figure 3.3.1.)

Linear systems whose solutions are very sensitive to changes in their data are called *ill-conditioned*, and we will want a way to recognize such systems. It is easily shown that ill-conditioning can be detected in terms of the matrix in the system. Consider any system $Ax = b$ with A nonsingular, and let us determine the relative change in the solution x_*, given some relative change in b or in A. If we change b by Δb, then the new solution can be written as $x_* + \Delta x$, where

$$A(x_* + \Delta x) = b + \Delta b = Ax_* + \Delta b$$

so that

$$A \cdot \Delta x = \Delta b.$$

Thus in any vector norm $\| \cdot \|$ and the corresponding induced matrix operator

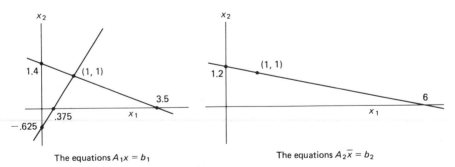

The equations $A_1 x = b_1$ The equations $A_2 \bar{x} = b_2$

Figure 3.3.1

norm

$$\| \Delta x \| \leq \| A^{-1} \| \cdot \| \Delta b \|.$$

Also from $Ax_* = b$,

$$\frac{1}{\| x_* \|} \leq \frac{\| A \|}{\| b \|}$$

so that

$$\frac{\| \Delta x \|}{\| x_* \|} \leq \| A \| \cdot \| A^{-1} \| \frac{\| \Delta b \|}{\| b \|}.$$

Similarly if we change A by ΔA, set $(A + \Delta A)\hat{x} = b$, and let $\hat{x} = x_* + \Delta \hat{x}$, then

$$(A + \Delta A)(x_* + \Delta \hat{x}) = b,$$

which yields

$$A \, \Delta \hat{x} = -\Delta A \cdot \hat{x},$$

$$\| \Delta \hat{x} \| \leq \| A^{-1} \| \cdot \| \Delta A \| \cdot \| \hat{x} \|,$$

$$\frac{\| \Delta \hat{x} \|}{\| \hat{x} \|} \leq \| A \| \cdot \| A^{-1} \| \frac{\| \Delta A \|}{\| A \|}.$$

In both cases, the relative change in x is bounded by the relative change in the data multiplied by $\| A \| \cdot \| A^{-1} \|$. This term is known as the *condition number* of A and is denoted by $\kappa_p(A)$, when using the corresponding l_p induced matrix norm. In any induced matrix norm, the condition number is the ratio of the maximum to the minimum stretch induced by A [see equations (3.1.7) and (3.1.18)] and so it is greater than or equal to 1. In our examples, $\kappa_1(A_1) = (15)(0.14) = 2.1$ and $\kappa_1(A_2) = (13.35)(300) = 4005$, indicating that the second system is far more sensitive to changes in its data than the first. Since the minimum stretch induced by a singular matrix is zero, the condition number of a singular matrix can be considered infinite. Therefore, the condition number of a nonsingular matrix is a measure of the matrix's nearness to singularity. It has the nice property of being a scale-free measure with respect to scalar multiples, since $\kappa(\alpha A) = \kappa(A)$ for any nonzero $\alpha \in \mathbb{R}$ and any matrix norm.

The condition number is also a measure of sensitivity of the solution of $Ax = b$ to finite-precision arithmetic. When a linear system is solved on a computer, it can be shown that the solution obtained is the exact solution to a perturbed system $(A + \Delta A)x = b$. This method of analysis is often called the Wilkinson backward error analysis, and the reference is Wilkinson (1965). It can be shown that some of the methods discussed in Section 3.2 limit $\| \Delta A \|/\| A \|$ to a constant times machine epsilon, and in practice this is almost always true for all of them. Therefore, the relative error in the finite-precision solution is bounded by a constant times $(\kappa(A) \cdot \text{macheps})$.

So, the condition number of A is a useful quantity in bounding how sensitive the solution of $Ax = b$ will be to either data errors or the effects of finite-precision arithmetic. If A is almost singular in this sense, one can expect trouble in solving $Ax = b$. Thus, we will look out for ill-conditioned linear systems in extracting information from our local models, because however simple and accurate the model, the solution of a model problem that is sensitive to small changes is certainly of limited use as an approximation to the solution of the nonlinear problem. If ill-conditioning occurs far from the solution to the nonlinear problem, where the model is not reckoned to be very accurate anyway, we usually just perturb the linear system into a better-conditioned one, and proceed. If ill-conditioned systems occur near the solution, where the model seems to be good, then this indicates that the solution of the nonlinear problem is itself very sensitive to small changes in its data (see Figure 3.3.2). This may indicate that the underlying problem is not well posed.

Finally, we need to discuss how we will determine in practice whether a linear system (3.2.1) is sufficiently ill-conditioned that we should avoid solving it. From the analysis of Wilkinson (1965), we see that if $\kappa(A) \geq (\text{macheps})^{-1}$, then the computed solution to (3.2.1) is likely to be entirely unreliable. In fact it is generally felt that if $\kappa(A) > (\text{macheps})^{-1/2}$, then the computed solution to (3.2.1) may not be trustworthy. Our algorithms will check for such a condition and then perturb poorly conditioned models into ones that are better behaved.

A problem is that calculating $\kappa(A)$ involves finding $\| A^{-1} \|$, and not only may this be unreliable, but its expense is rarely justified. Therefore what is done is to estimate $\kappa(A)$ by

$$condest = \| M \| \cdot invest,$$

where M is either A or the factor of A most likely to have the same conditioning as A, and $invest$ is an estimate of $\| M^{-1} \|$. In our application, we will want

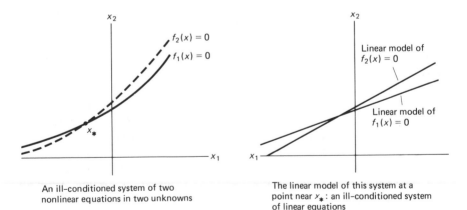

An ill-conditioned system of two nonlinear equations in two unknowns

The linear model of this system at a point near x_*: an ill-conditioned system of linear equations

Figure 3.3.2   The correspondence between ill conditioning in linear and nonlinear systems of equations

to estimate the condition number of a matrix A that has been factored into $Q \cdot R$. It is an easy exercise to show that in this case,

$$\frac{1}{n} \kappa_1(A) \le \kappa_1(R) \le n\kappa_1(A).$$

Therefore, we estimate $\kappa_1(R)$, since the l_1 norm of a matrix is easy to compute. The algorithm we use is an instance of a class of condition number estimates given by Cline, Moler, Stewart, and Wilkinson (1979) and is given by Algorithm A3.3.1 in Appendix A.

The algorithm first computes $\| R \|_1$. The technique then used to estimate $\| R^{-1} \|_1$ is based on the inequality

$$\| R^{-1} \|_1 \ge \frac{\| R^{-1}z \|_1}{\| z \|_1},$$

for any nonzero z. The algorithm chooses z by solving $R^T z = e$, where e is a vector of ± 1's with the signs chosen to maximize the magnitude of $z_1, \ldots, z_n$ in succession. It then solves $Ry = z$ and returns $condest = \| R \|_1 (\| y \|_1/\| z \|_1)$. Although $\| y \|_1/\| z \|_1$ is only guaranteed to be a lower bound on $\| R^{-1} \|_1$, in practice it is quite close. One explanation is that the process for obtaining y is related to the inverse power method for finding the largest eigenvalue of $(R^T R)^{-1}$. Another is that it turns out to have a knack for extracting any large elements of R^{-1}. The reader who wants to establish more intuition about this is urged to work through the example given in Exercise 19.

3.4   UPDATING MATRIX FACTORIZATIONS

Later in the book we will consider multivariable generalizations of the secant method for which the successive model derivatives A_c, $A_+ \in \mathbb{R}^{n \times n}$, are related in an especially simple way. In this section we see how to use the two most important relationships between A_c and A_+ to lessen the work in obtaining factorizations of A_+ from those for A_c.

Specifically, we want to look first in some detail at the problem of obtaining the factorization of $Q_+ R_+$ of

$$A_+ = A_c + uv^T, \tag{3.4.1}$$

where $u, v \in \mathbb{R}^n$, and A_c is some nonsingular matrix with no special structure, for which we already have the decomposition $Q_c R_c$. The idea of the algorithm is not hard at all. For $w = Q_c^T u$, we write

$$A_+ = Q_c R_c + uv^T = Q_c(R_c + wv^T),$$

and then we form the QR decomposition

$$R_c + wv^T = \tilde{Q}\tilde{R}. \tag{3.4.2}$$

Then $R_+ = \tilde{R}$ and $Q_+ = Q_c \tilde{Q}$. The advantage is that the QR decomposition of $R_c + wv^T$ is much cheaper than the $5n^3/3$ operations we would have to spend for $Q_+ R_+$ if we made no use of the previous decomposition.

The tool we need for the factorization (3.4.2) is an orthogonal transformation called a *Jacobi rotation*, a two-dimensional rotation imbedded in an n-dimensional matrix. A Jacobi rotation is used to zero out one element of a matrix while only changing two rows of the matrix. The two-dimensional rotation matrix and th n-dimensional Jacobi rotation matrix are defined in Definition 3.4.1, and the important properties are stated in Lemma 3.4.2.

Definition 3.4.1.   A two-dimensional rotation matrix is an $R(\phi) \in \mathbb{R}^{2 \times 2}$,

$$R(\phi) = \begin{bmatrix} \cos \phi & -\sin \phi \\ \sin \phi & \cos \phi \end{bmatrix}, \quad \phi \in \mathbb{R},$$

or equivalently,

$$\hat{R}(\alpha, \beta) = \frac{1}{\sqrt{\alpha^2 + \beta^2}} \begin{bmatrix} \alpha & -\beta \\ \beta & \alpha \end{bmatrix}, \quad \alpha, \beta \in \mathbb{R}, |\alpha| + |\beta| \neq 0.$$

A Jacobi rotation is a matrix $J(i, j, \alpha, \beta) \in \mathbb{R}^{n \times n}$ such that

$$1 \leq i < j \leq n, \quad \alpha, \beta \in \mathbb{R}, \quad |\alpha| + |\beta| \neq 0,$$

$$[J(i, j, \alpha, \beta)]_{ii} = [J(i, j, \alpha, \beta)]_{jj} = \frac{\alpha}{\sqrt{\alpha^2 + \beta^2}} = \bar{\alpha},$$

$$-[J(i, j, \alpha, \beta)]_{ij} = [J(i, j, \alpha, \beta)]_{ji} = \frac{\beta}{\sqrt{\alpha^2 + \beta^2}} = \bar{\beta},$$

$$[J(i, j, \alpha, \beta)]_{kk} = 1, \quad 1 \leq k \leq n, \quad i \neq k \neq j,$$

$$J(i, j, \alpha, \beta) = 0, \quad \text{otherwise.}$$

Schematically,

$$J(i, j, \alpha, \beta) = \begin{bmatrix} 1 & & & & & & & & & \\ & 1 & & & & & & & & \\ & & \ddots & & & & & & & \\ & & & \bar{\alpha} & & & -\bar{\beta} & & & \\ & & & & 1 & 0 & & & & \\ & & & & & 1 & & & & \\ & & & 0 & & & 1 & & & \\ & & & \bar{\beta} & & & & \bar{\alpha} & & \\ & & & & & & & & 1 & \\ & & & & & & & & & \ddots \\ 0 & & & & & & & & & 1 \end{bmatrix}$$

with the ith and jth columns marked (↓) and the ith row and jth row marked (←).

LEMMA 3.4.2   Let $n \geq 2$, and $R(\phi)$, $\hat{R}(\alpha, \beta)$, $J(i, j, \alpha, \beta)$ be defined above. Then:

(a) For all $\phi \in R$ and nonzero $v \in \mathbb{R}^2$, $R(\phi)$ is orthogonal, and $R(\phi) \cdot v$ rotates v by ϕ counterclockwise; $\hat{R}(v_2, v_1) \cdot v = (0, \|v\|_2)^T$, and $\hat{R}(v_1, -v_2) \cdot v = (\|v\|_2, 0)^T$.

(b) For all integers $i, j \in [1, n]$, $i < j$, and $\alpha, \beta \in \mathbb{R}$, $|\alpha| + |\beta| \neq 0$, $J(i, j, \alpha, \beta)$ is orthogonal. If $M_+ = J(i, j, \alpha, \beta) \cdot M$, $M \in \mathbb{R}^{n \times n}$, then $(M_+)_{k \cdot} = (M)_{k \cdot}$ for $1 \leq k \leq n$, $k \neq i$ or j; rows i and j of M_+ are each linear combinations of rows i and j of M.

(c) If $M_{jl} \cdot M_{il} \neq 0$, then $[J(i, j, M_{jl}, M_{il}) \cdot M]_{il} = 0$
and $[J(i, j, M_{il}, -M_{jl}) \cdot M]_{jl} = 0$.

To update the QR decomposition of A_c into a QR decomposition of $A_+ = Q_c R_c + uv^T$, Jacobi rotations are used to decompose $R_c + Q_c^T uv^T = R_c + wv^T$ into $\tilde{Q}\tilde{R}$ as follows. First, $n - 1$ Jacobi rotations are applied to zero out in succession rows $n, n - 1, \ldots, 2$ of wv^T, by combining rows i and $i - 1$ to zero out row i. The effect of each rotation on R_c is to alter some existing elements and to introduce one new element directly below the diagonal in the $(i, i - 1)$ position. Thus the $n - 1$ rotations transform $R_c + wv^T$ into an upper triangular matrix augmented by a diagonal below the main diagonal (an "upper Hessenberg" matrix). Now $n - 1$ additional Jacobi rotations are applied, successively zeroing out the $(i, i - 1)$ element, $i = 2, 3, \ldots, n$, by combining rows $i - 1$ and i. The result is an upper triangular matrix $\tilde{R}$, so that $R_c + wv^T = \tilde{Q}\tilde{R}$ and $(\tilde{Q})^T$ is the product of the $2n - 2$ rotations.

The reader can verify that the entire QR update process requires only $O(n^2)$ operations. The most expensive part is getting $Q_+ = Q_c \tilde{Q}$ collected to save for the next step. For the other important special relationship between A_c and A_+, Q_+ is unnecessary, as we will now see.

When we are using local quadratic modeling to solve the unconstrained minimization problem, we will prefer the Hessians, second-derivative matrices in $\mathbb{R}^{n \times n}$, of the quadratic models to be symmetric and positive definite. We will find extremely reasonable circumstances for the next model Hessian A_+ to be chosen to have these properties if the current model Hessian A_c does. Furthermore, the derivation of A_+ from A_c will be very suggestive. It will say that if $L_c L_c^T$ is the Cholesky factorization of A_c, then $A_+ = J_+ J_+^T$, where

$$J_+ = L_c + vu^T.$$

But this tells us exactly how to get the Cholesky decomposition

$$A_+ = L_+ L_+^T.$$

We simply apply the above technique to obtain the decomposition $Q_+ R_+$ of $J_+^T = L_c^T + uv^T$ from the QR decomposition $Q_c R_c = I \cdot L_c^T$ of L_c^T. We then have immediately that

$$A_+ = J_+ J_+^T = R_+^T Q_+^T Q_+ R_+ = R_+^T R_+ = L_+ L_+^T.$$

This update process also requires $O(n^2)$ operations. In this special case, however, we don't have the n^2 operations necessary to form $w = Q_c^T u$, since $Q_c = I$, nor do we have to accumulate Q_+, since only $R_+ = L_+^T$ is needed. The complete algorithm is given in Algorithm A3.4.1 in the appendix.

As an alternative to the QR factorization and the updates discussed in this secton, Gill, Golub, Murray, and Saunders (1974) suggest using the decomposition $A = LDV$, where L is unit lower triangular, D is diagonal, and the rows of V are an orthogonal, but not necessarily orthonormal, basis for $\mathbb{R}^n$. In this case, $VV^T = \tilde{D}$, a diagonal matrix. The computation of $L_+ D_+ V_+ = L_c D_c V_c + uv^T$, for u, $v \in \mathbb{R}^n$, is a bit cheaper than the QR update described above. In the symmetric positive definite case, this corresponds to carrying along the LDL^T decomposition of $\{A_k\}$. These algorithms are collected in Goldfarb (1976).

The sequencing of symmetric indefinite factorizations has been studied extensively by Sorensen (1977), and his somewhat complex algorithm seems completely satisfactory. The main source of the complication is in the permutation matrices, and this is also why no algorithm for updating the PLU decomposition of a matrix A_c into the decomposition $P_+ L_+ U_+$ of $A_+ = A_c + uv^T$ is known that is satisfactory in the current context.

3.5 EIGENVALUES AND POSITIVE DEFINITENESS

In this section we state the properties of eigenvalues and eigenvectors that we have used or will use in the text. We also summarize the definitions of positive definite, negative definite, and indefinite symmetric matrices and their characterizations in terms of eigenvalues. This characterization provides insight into the shapes of the multivariable quadratic models introduced in Chapter 4 and used thereafter.

Most of the theorems are stated without proof. Proofs can be found in any of the references listed at the beginning of this chapter.

Definition 3.5.1. Let $A \in \mathbb{R}^{n \times n}$. The *eigenvalues* and *eigenvectors* of A are the real or complex scalars λ and n-dimensional vectors v such that $Av = \lambda v, v \neq 0$.

Definition 3.5.2. Let $A \in \mathbb{R}^{n \times n}$ be symmetric. A is said to be *positive definite* if $v^T Av > 0$ for every nonzero $v \in \mathbb{R}^n$. A is said to be *positive semidefinite* if $v^T Av \geq 0$ for all $v \in \mathbb{R}^n$. A is said to be *negative definite* or *negative* *semidefinite* if $-A$ is positive definite or positive semidefinite, respectively. A is said to be *indefinite* if it is neither positive semidefinite nor negative semidefinite.

THEOREM 3.5.3   Let $A \in \mathbb{R}^{n \times n}$ be symmetric. Then A has n real eigenvalues $\lambda_1, \ldots, \lambda_n$, and a corresponding set of eigenvectors $v_1, \ldots, v_n$ that form an orthonormal basis for $\mathbb{R}^n$.

THEOREM 3.5.4   Let $A \in \mathbb{R}^{n \times n}$ be symmetric. Then A is positive definite if, and only if, all its eigenvalues are positive.

Proof.   Let A have eigenvalues $\lambda_1, \ldots, \lambda_n$, and corresponding orthonormal eigenvectors $v_1, \ldots, v_n$. Suppose $\lambda_j \leq 0$ for some j. Then $v_j^T A v_j = v_j^T (\lambda_j v_j) = \lambda_j v_j^T v_j = \lambda_j \leq 0$, which shows that A is not positive definite. Now suppose every λ_i is positive. Since $\{v_i\}$ is a basis for $\mathbb{R}^n$, any nonzero $v \in \mathbb{R}^n$ can be written as

$$v = \sum_{i=1}^{n} \alpha_i v_i,$$

where at least one α_i is nonzero. Then

$$v^T A v = \sum_{i=1}^{n} \sum_{j=1}^{n} \lambda_i \alpha_i \alpha_j v_i^T v_j = \sum_{i=1}^{n} \lambda_i \alpha_i^2 > 0,$$

owing to the orthonormality of the v_i's. Thus A is positive definite.   □

THEOREM 3.5.5   Let $A \in \mathbb{R}^{n \times n}$ be symmetric. Then A is positive semidefinite if, and only if, all its eigenvalues are nonnegative. A is negative definite or negative semidefinite if, and only if, all its eigenvalues are negative or nonpositive, respectively. A is indefinite if and only if it has both positive and negative eigenvalues.

The following definitions and theorems will be needed in Section 5.5.

THEOREM 3.5.6   Let $A \in \mathbb{R}^{n \times n}$ be symmetric with eigenvalues $\lambda_1, \ldots, \lambda_n$. Then

$$\| A \|_2 = \max_{1 \leq i \leq n} | \lambda_i |.$$

If A is nonsingular, then

$$\kappa_2(A) = \frac{\max_{1 \leq i \leq n} |\lambda_i|}{\min_{1 \leq j \leq n} |\lambda_j|}.$$

THEOREM 3.5.7   Let $A \in \mathbb{R}^{n \times n}$ have eigenvalue λ_i. Then $\lambda_i + \alpha$ is an eigenvalue of $A + \alpha I$, for any real α.

Definition 3.5.8.   Let $A \in \mathbb{R}^{n \times n}$ be symmetric. A is said to be *strictly diagonally dominant* if, for $i = 1, \ldots, n$,

$$a_{ii} - \sum_{\substack{j=1 \\ j \neq i}}^{n} |a_{ij}| > 0$$

THEOREM 3.5.9   GERSCHGORIN

Let $A \in \mathbb{R}^{n \times n}$ be symmetric with eigenvalues $\lambda_1, \ldots, \lambda_n$. Then

$$\min_{1 \leq i \leq n} \lambda_i \geq \min_{1 \leq i \leq n} \left\{ a_{ii} - \sum_{\substack{j=1 \\ j \neq i}}^{n} |a_{ij}| \right\},$$

$$\max_{1 \leq k \leq n} \lambda_k \leq \max_{1 \leq k \leq n} \left\{ a_{kk} + \sum_{\substack{j=1 \\ j \neq k}}^{n} |a_{kj}| \right\}.$$

COROLLARY 3.5.10   If A is strictly diagonally dominant, then A is positive definite.

3.6 LINEAR LEAST SQUARES

The final linear algebra topic we discuss is the linear least-squares problem:

$$\text{given} \quad A \in \mathbb{R}^{m \times n}, m \geq n, b \in \mathbb{R}^m$$

$$\min_{x \in \mathbb{R}^n} \quad \| Ax - b \|_2, \tag{3.6.1}$$

which is the special case of the nonlinear least-squares problem:

$$\text{given} \quad r_i(x): \mathbb{R}^n \longrightarrow \mathbb{R}, i = 1, \ldots, m,$$

$$\min_{x \in \mathbb{R}^n} \quad \sum_{i=1}^{m} r_i(x)^2,$$

where $r_i(x) = (a_i.)x - b_i$ for each $i = 1, \dots, m$. The reasons for studying problem (3.6.1) are that one sometimes needs to solve it as a subproblem of the nonlinear least-squares problem and that an understanding of the linear problem aids the understanding of its nonlinear counterpart. In connection with the linear least-squares problem, we introduce the singular value decomposition, a powerful numerical linear algebra tool that is helpful in many situations.

A common occurrence of the linear least-squares problem is in trying to fit m data points (t_i, y_i) with a function $f(x, t)$ that is linear in its n free parameters $x_1, \dots, x_n$. For example, suppose one would like to fit the three pairs (1, 2), (2, 3), (3, 5) with the function $f(x, t) = x_1 t + x_2 e^t$ (see Figure 3.6.1). To get an exact fit $f(x, t_i) = y_i$, $i = 1, 2, 3$, would require $Ax = b$, where

$$A = \begin{bmatrix} 1 & e \\ 2 & e^2 \\ 3 & e^3 \end{bmatrix}, \quad x = \begin{bmatrix} x_1 \\ x_2 \end{bmatrix}, \quad \text{and} \quad b = \begin{bmatrix} 2 \\ 3 \\ 5 \end{bmatrix}. \tag{3.6.2}$$

Since $Ax = b$ is an overdetermined system of equations, we don't expect to solve it in the usual sense. Instead we choose x to minimize some measure of the vector of residuals $Ax - b$. The l_2 norm is chosen because the resultant problem (3.6.1) is mathematically well behaved and also has a statistical justification.

Note that finding the solution to this instance of (3.6.1) is equivalent to asking "Which linear combination of $(1, 2, 3)^T$ and $(e, e^2, e^3)^T$ comes closest to $(2, 3, 5)^T$ in the l_2 norm?" In general, problem (3.6.1) can be interpreted as "Find the linear combination of the columns $a_{.1}, \dots, a_{.n}$ of A that is closest to b in the l_2 norm." Geometrically, this means finding the point in the n-dimensional subspace $C(A)$ spanned by the columns of A that is closest to the

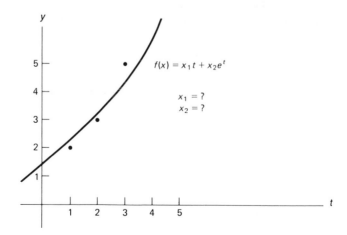

Figure 3.6.1   A sample linear least squares problem

vector b in the Euclidean norm (see Figure 3.6.2). This interpretation leads to an easy solution of the linear least-squares problem.

We know that the closest point to b in $C(A)$ will be $Ax_* \in C(A)$, such that $Ax_* - b$ is perpendicular to the entire subspace $C(A)$. Thus x_* must satisfy $(a_{\cdot i})^T(Ax_* - b) = 0$, $i = 1, \ldots, n$, or equivalently, $A^T(Ax_* - b) = 0$. Clearly, Ax_* is unique, and if the columns of A are linearly independent, the value of x_* that yields the point Ax_* is also unique and is the solution to the system of equations

$$(A^TA)x_* = A^Tb \qquad (3.6.3)$$

which is nonsingular.

This information is summarized in Theorem 3.6.1, and its algebraic proof is left as an easy exercise.

THEOREM 3.6.1   Let $m \geq n > 0$, $A \in \mathbb{R}^{m \times n}$, $b \in \mathbb{R}^m$. Then the solution to

$$\min_{x \in \mathbb{R}^n} \| Ax - b \|_2$$

is the set of points $\{x_* : A^T(Ax_* - b) = 0\}$. If the columns of A are linearly independent, then x_* is unique, A^TA is nonsingular, and $x_* = (A^TA)^{-1}A^Tb$.

Equations (3.6.3) are known as the *normal equations*. Although they define x_* uniquely when A has full column rank, they are not always the way x_* should be calculated. This is because forming the matrix A^TA can cause underflows and overflows and can square the conditioning of the problem in

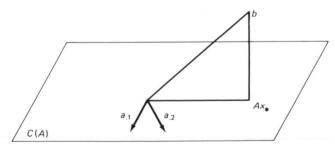

Figure 3.6.2   The solution to the linear least squares problem

comparison to a method that uses A directly. When A has full column rank, we can use the QR decomposition of A,

$$ (3.6.4) $$

$Q \in \mathbb{R}^{m \times m}$ orthogonal, $R \in \mathbb{R}^{m \times n}$ upper triangular, obtained using the same techniques as the QR decomposition of a square matrix. This decomposition also yields a numerically stable orthonormalization of the columns of A. The following theorem shows how to use (3.6.4) to solve (3.6.1).

THEOREM 3.6.2   Let $m \geq n > 0$, $b \in \mathbb{R}^m$, $A \in \mathbb{R}^{m \times n}$ with full column rank. Then there exists a decomposition $A = QR$ of the form (3.6.4), where Q is an $m \times m$ orthogonal matrix, $R \in \mathbb{R}^{m \times n}$ is upper triangular, and R_u, the top n rows of R, is a nonsingular upper triangular matrix. The unique solution to (3.6.1) is

$$ x_* = R_u^{-1}(Q^T b)_u, $$

where $(Q^T b)_u^T = ((Q^T b)_1, \ldots, (Q^T b)_n)$;

the minimum of $\| Ax - b \|_2^2$ is $\displaystyle\sum_{i=n+1}^{m} [(Q^T b)_i]^2$.

Proof. The existence of the QR decomposition of A follows from its derivation by Householder transformations, and the nonsingularity of R_u follows from the full column rank of A. Using equation (3.1.23), $\| Ax - b \|_2 = \| QRx - b \|_2 = \| Rx - Q^T b \|_2$, so (3.6.1) can be rewritten

$$ \min_{x \in \mathbb{R}^n} \| Rx - Q^T b \|_2. $$

Then if $(Q^T b)_l^T = ((Q^T b)_{n+1}, \ldots, (Q^T b)_m)$, we have

$$ \| Rx - Q^T b \|_2^2 = \| R_u x - (Q^T b)_u \|_2^2 + \| (Q^T b)_l \|_2^2 $$

which clearly is minimized when $x = R_u^{-1}(Q^T b)_u$.   □

If A doesn't have full rank, there are multiple answers x to problem (3.6.1), and so the problem can be further restricted to find the smallest of these answers in the l_2 norm:

given $A \in \mathbb{R}^{m \times n}, b \in \mathbb{R}^m,$

$$\min_{x \in \mathbb{R}^n} \quad \{ \| x \|_2 : \| Ax - b \|_2 \leq \| A\hat{x} - b \|_2 \ \forall \ \hat{x} \in \mathbb{R}^n \}. \tag{3.6.5}$$

Problem (3.6.5) is of interest whether m is greater than, equal to, or less than n, and it can be solved in all cases by using the singular value decomposition (SVD) of A. The SVD is a matrix factorization that is more costly to obtain than the factorizations we have discussed so far, but also more powerful. It is defined in Definition 3.6.3, and its factors are characterized in Theorem 3.6.4. Theorem 3.6.5 shows how to use the SVD to solve problem (3.6.5).

Definition 3.6.3.   Let $A \in \mathbb{R}^{m \times n}$ and let $k = \min \{m, n\}$. The singular value decomposition of A is $A = UDV^T$, where $U \in \mathbb{R}^{m \times m}$ and $V \in \mathbb{R}^{n \times n}$ are orthogonal matrices, $D \in \mathbb{R}^{m \times n}$ is defined by $d_{ii} = \sigma_i \geq 0$, $i = 1, \ldots, k$, $d_{ij} = 0$, $i \neq j$. The quantities $\sigma_1, \ldots, \sigma_k$ are called the *singular values* of A.

Notice that we are using U to denote an orthogonal matrix, even though we used the same symbol in Section 3.3 for upper triangular matrices. There should be no confusion for the alert reader in this well-established notation. Furthermore, the use of the symbols U and V is traditional for matrices whose columns are to be thought of as eigenvectors, and the following lemma will show that this is the case.

THEOREM 3.6.4   Let $A \in \mathbb{R}^{m \times n}$. Then the SVD of A exists, the diagonal elements σ_i of D are the nonnegative square roots of the eigenvalues of AA^T if $m \leq n$, or of $A^T A$ if $m \geq n$, and the columns of U and V are the eigenvectors of AA^T and $A^T A$, respectively. The number of nonzero singular values equals the rank of A.

Proof.   An existence proof for the SVD can be found in the references given at the beginning of the chapter. Since $AA^T = UDD^T U^T$ and $A^T A = VD^T D V^T$, we have $(AA^T) \cdot U = U \cdot (DD^T)$, $(A^T A)V = V(D^T D)$. Thus, if $u_{.j}$ and $v_{.j}$ are the respective jth columns of U and V, then:

$$(AA^T)u_{.j} = \lambda_j u_{.j}, \quad j = 1, \ldots, m; \quad (A^T A)v_{.j} = \lambda_j v_{.j}, \quad j = 1, \ldots, n,$$

where the λ_j's are the diagonal elements of DD^T and $D^T D$, $\lambda_j = (\sigma_j)^2$, $j = 1, \ldots, \min \{m, n\}$; and $\lambda_j = 0, j = \min \{m, n\} + 1, \ldots, \max \{m, n\}$. Since multiplying a matrix by a nonsingular matrix doesn't change its rank, rank (A) = rank (D), the number of nonzero σ_j's. $\square$

THEOREM 3.6.5.   Let $A \in \mathbb{R}^{m \times n}$ have the SVD $A = UDV^T$, with U, D, V as defined in Definition 3.6.3. Let the pseudoinverse of A be defined as

$$A^+ = VD^+U^T, \qquad D^+ = \begin{cases} d_{ii}^+ = \begin{cases} 1/\sigma_i, & \sigma_i > 0 \\ 0, & \sigma_i = 0 \end{cases} \\ d_{ij}^+ = 0, & i \neq j, \end{cases} \qquad (3.6.6)$$

$D^+ \in \mathbb{R}^{n \times m}$. Then the unique solution to problem (3.6.5) is $x_* = A^+b$.

Proof.   From $A = UDV^T$ and equation (3.1.23), (3.6.5) is equivalent to

$$\min_{x \in \mathbb{R}^n} \{ \| V^T x \|_2 \,|\, \| DV^T x - U^T b \|_2 \leq \| DV^T \hat{x} - U^T b \|_2 \,\forall\, \hat{x} \in \mathbb{R}^n \},$$

or equivalently, for $z = V^T x$,

$$\min_{z \in \mathbb{R}^n} \{ \| z \|_2 \,|\, \| Dz - U^T b \|_2 \leq \| D\hat{z} - U^T b \|_2 \,\forall\, \hat{z} \in \mathbb{R}^n \}. \qquad (3.6.7)$$

Let k equal the number of nonzero σ_i's in D. Then

$$\| Dz - U^T b \|_2^2 = \sum_{i=1}^{k} (\sigma_i z_i - (U^T b)_i)^2 + \sum_{i=k+1}^{m} ((U^T b)_i)^2$$

which is minimized by any z such that $z_i = (U^T b)_i / \sigma_i$, $i = 1, \ldots, k$. Among all such z, $\| z \|_2$ is minimized by the one for which $z_i = 0$, $i = k+1, \ldots, m$. Thus the solution to (3.6.7) is $z = D^+ U^T b$, so that $x_* = VD^+ U^T b = A^+ b$.   □

The singular value decomposition is found by an iterative process closely related to the algorithm for finding the eigenvalues and eigenvectors of a symmetric matrix, so it is more costly to obtain than our other matrix decompositions. It is the recommended method for solving (3.6.5) when A is not known to have full row or column rank. If $m < n$ and A has full row rank, (3.6.5) is solved more efficiently by the LQ decomposition (see Exercise 27). The SVD is also the most reliable technique for computing the rank of a matrix, or equivalently, the number of linearly independent vectors in a set of vectors. The SVD can also be used to determine the sensitivity of $x = A^{-1}b$ for a square matrix A with respect to data changes Δb. Also, if A is nonsingular, then $\kappa_2(A) = \sigma_1 / \sigma_n$, and the SVD is the most dependable way of computing $\kappa_2(A)$.

One final note on determining the rank of a matrix from its SVD is that D really gives us more information about linear independence of the columns of A than we can pack into the single integer whose value is the rank of A. For example, if $n = 3$ and $\sigma_1 = 1$, $\sigma_2 = .1$, $\sigma_3 = 0$, then we would agree that A has rank 2. If σ_3 were changed to macheps, we would probably still feel that all the information was conveyed by saying rank $(A) = 2$. What about $\sigma_1 = 1$, $\sigma_3 =$

macheps, and $\sigma_2 = (\text{macheps})^{1/2}$ or 100 macheps, and so on? As n get larger, there are more awkward distributions of the singular values to worry about. No one really knows what to do, but it is certainly reasonable to look for "breaks" in the singular values and count all singular values of about the same magnitude as either all zero, or all nonzero.

It is important that the user of canned programs understand these issues. The reason is that, for example, in ROSEPACK and EISPACK, when (3.6.1) is solved by use of the SVD, once rank decisions have been made, A is modified to have the "small" singular values set to zero and the modified problem is solved. However, nothing is done behind the user's back, and it is only necessary to read the program documentation to understand what is done. It is safe to say that for internal efficiency reasons, any subroutine to compute an SVD will return zero values for any singular values that are negligible in the context of the computation. Dongarra et al. (1979) gives a particularly succinct explanation of a typical test for negligibility.

3.7 EXERCISES

1. Prove for any $x \in \mathbb{R}^n$ that

$$\lim_{p \to \infty} \| x \|_p = \| x \|_\infty .$$

   Notice that we really mean that for every real sequence $\{p_k\}$ that converges to ∞ we have

$$\lim_{k \to \infty} \| x \|_{p_k} = \| x \|_\infty .$$

2. (Hard) Prove Theorem 3.1.2.

3. Prove relations (3.1.5).

4. Prove that for any $p \geq 1$, the l_p induced matrix operator norm (3.1.7) obeys the three defining properties of a norm [(3.1.1) using matrices] plus consistency [(3.1.10)].

5. Prove that (3.1.8a) correctly evaluates the l_1 induced matrix norm. [*Hint:* Show that for all $v \in \mathbb{R}^n$,

$$\frac{\| Av \|_1}{\| v \|_1} \leq \max_{1 \leq j \leq n} \| a_{.j} \|_1,$$

   with equality for at least one v.] Do the same for the l_∞ induced matrix norm.

6. Prove (3.1.8b). (Use the techniques of proof of Theorem 3.5.4.)

7. Show that $\text{tr}(A^T B)$ is the standard inner product of A and B treated as "long vectors." Relate this to the Frobenius norm; i.e., prove (3.1.14).

8. Prove that $\| A \| = \| A^T \|$ for both the l_2 and Frobenius norms. Notice that $(AB)_{.j} = AB_{.j}$ and see if these facts let you prove (3.1.15, 3.1.16).

9. Prove (3.1.17).

10. Complete the proof of Theorem 3.1.4.

11. Prove for any $A \in \mathbb{R}^{n \times n}$ and any orthonormal set of vectors $v_i \in \mathbb{R}^n$, $i = 1, \ldots, n$, that

$$\| A \|_F^2 = \sum_{i=1}^{n} \| A v_i \|_2^2.$$

12. Prove Theorem 3.1.7.

13. Solve the block diagonal system: $Ax = b$,

$$A = \begin{bmatrix} 3 & 1 & 0 & 0 & 0 \\ 1 & 2 & 0 & 0 & 0 \\ 0 & 0 & 2 & 0 & 0 \\ 0 & 0 & 0 & 2 & -1 \\ 0 & 0 & 0 & -1 & 4 \end{bmatrix} \qquad b = \begin{bmatrix} 5 \\ 5 \\ 6 \\ 3 \\ 16 \end{bmatrix}.$$

14. Derive the Householder transformation: given $v \in \mathbb{R}^n$, $v \neq 0$, find $u \in \mathbb{R}^n$ such that $(I - uu^T)$ is orthogonal and $(I - uu^T)v = (\alpha, 0, \ldots, 0)^T$, $\alpha \neq 0$. [*Hint:* u must have the form $\beta(\gamma, v_2, \ldots, v_n)^T$, $\beta, \gamma \in \mathbb{R}$.]

15. Apply Householder transformations to factor A into QR,

$$A = \begin{bmatrix} 1 & -1 \\ 1 & 0 \\ 1 & 1 \\ 1 & 2 \end{bmatrix}.$$

16. Is $\left(\begin{smallmatrix} 1 & 2 \\ 2 & 3 \end{smallmatrix}\right)$ positive definite? Use the Cholesky factorization to find out.

17. Find the Cholesky factorization of

$$\begin{bmatrix} 4 & 6 & -2 \\ 6 & 10 & 1 \\ -2 & 1 & 22 \end{bmatrix}.$$

What if the 10 is changed to a 7?

18. Prove that if $A = Q \cdot R$, then

$$\frac{1}{n} \kappa_1(A) \leq \kappa_1(R) \leq n\kappa_1(A).$$

Also prove that $\kappa_2(A) = \kappa_2(R)$.

19. Calculate the l_1 condition number of

$$R = \begin{bmatrix} 1 & 1 \\ 0 & 10^{-7} \end{bmatrix}.$$

Then estimate its condition number by Algorithm A.3.3.1.

20. Program the upper triangular condition estimator as a module for later use.

21. The simple form $\hat{R}(\alpha, \beta)$ we gave for Jacobi rotations is subject to unnecessary

finite-precision inaccuracies from α or β being small, or large, or of differing magnitudes. Write an accurate procedure to evaluate $\hat{R}(\alpha, \beta)$.

22. Write a Jacobi rotation that would introduce a zero into the (3, 1) position of

$$\begin{bmatrix} 1 & 3 & 5 \\ 2 & 4 & 6 \\ 3 & 5 & 9 \end{bmatrix}.$$

23. Prove Theorem 3.6.1: Show that if $A^T(Ax_* - b) = 0$, then for all $x \in \mathbb{R}^n$, $\|Ax - b\|_2 \geq \|Ax_* - b\|_2$, with strict inequality if A has full column rank and $x \neq x_*$. Show also that if A has full column rank, then $A^T A$ is nonsingular.

24. Given the QR decomposition of a rectangular matrix A with full column rank (as defined in Theorem 3.6.2), show that the first n columns of Q form an orthonormal basis for the space spanned by the columns of A.

25. Fit a line $x_1 + x_2 t$ to the data $(t_i, y_i) = (-1, 3), (0, 2), (1, 0), (2, 4)$. Do the problem using QR and the normal equations. [*Hint*: Did you do Exercise 15?]

26. Given $n > m > 0$, $b \in \mathbb{R}^m$, $A \in \mathbb{R}^{m \times n}$ with full row rank, show how to solve the problem

$$\min_{x \in \mathbb{R}^m} \{\|x\|_2 : Ax = b\}$$

using the decomposition $A = LQ$, $L \in \mathbb{R}^{m \times n}$ lower triangular, $Q \in \mathbb{R}^{n \times n}$ orthogonal. [*Hint*: Use techniques similar to those used in the proofs of Theorems 3.6.2 and 3.6.5.]

27. Show how to carry out the LQ decomposition in the last exercise using Householder transformations.

28. Given the singular value decomposition $A = UDV^T$ of a nonsingular $A \in \mathbb{R}^{n \times n}$, prove that $\kappa_2(A) = \sigma_1/\sigma_n$. If u_j and v_j are the columns of U and V, respectively, show also that a change $\Delta b = \alpha u_j$ to the right-hand side of a linear system $Ax = b$ will produce a change $\Delta x = (\alpha/\sigma_j)v_j$ to the answer.

4

Multivariable Calculus Background

This chapter presents the material from multivariable calculus prerequisite to our study of the numerical solution of nonlinear problems. In Section 4.1 we discuss derivatives, line integrals, Taylor series, and related properties we will use to derive and analyze our methods. Section 4.2 introduces the formulas for constructing finite-difference approximations to derivatives of multivariable functions, and Section 4.3 presents the necessary and sufficient conditions characterizing an unconstrained minimizer. Ortega and Rheinboldt (1970) is a detailed reference for this material and for the theoretical treatment of nonlinear algebraic problems in general.

4.1 DERIVATIVES AND MULTIVARIABLE MODELS

In our study of function minimization in one variable, we made use of first- and second-order derivatives in order to build a local quadratic model of the function to be minimized. In this way we managed to have our model problem match the real problem in value, slope, and curvature at the point currently held to be the best estimate of the minimizer. Taylor series with remainder was a useful analytic device for bounding the approximation error in the quadratic model, and it will be equally useful here.

We begin this section with a no-frills exposition of Taylor series up to order two for a real-valued function of a real vector variable. Then, in order to set the stage for a consideration of nonlinear simultaneous equations, we consider the problem of providing a local affine model for a function $F: \mathbb{R}^n \longrightarrow \mathbb{R}^m$. In this case, the Taylor series approach is quite unsatisfactory because the mean value theorem fails. It was in anticipation of this fact that we used Newton's theorem in Chapter 2 to derive our affine models. The second part of this section will be devoted to building the machinery to make the same analysis in multiple dimensions.

Consider first a continuous function $f: \mathbb{R}^n \longrightarrow \mathbb{R}$, $n > 1$. If the first partial derivatives of f with respect to the n variables exist and are continuous, the column vector of these n partials is called the *gradient* of f at x, $\nabla f(x)$, and it functions like the first derivative of a one-variable function. In particular, there is a corresponding mean value theorem that says that the difference of the function values at two points is equal to the inner product of the difference between the points and the gradient of some point on the line connecting them. Similarly, if the n^2 second partial derivatives of f also exist and are continuous, the matrix containing them is called the *Hessian* of f at x, $\nabla^2 f(x)$, and it too satisfies a mean value theorem in the form of a Taylor series with remainder. The exact definitions and lemmas are given below. The reason that these important results from single-variable calculus hold for functions of multiple variables is because they involve the value of a real-valued function at two points in $\mathbb{R}^n$, x and $x + p$, and the line connecting these points can be parameterized using one variable. Thus the results follow directly from the corresponding theorems in one-variable calculus.

We will denote the open and closed line segment connecting x, $\bar{x} \in \mathbb{R}^n$, by $(x, \bar{x})$ and $[x, \bar{x}]$, respectively, and we remind the reader that $D \subset \mathbb{R}^n$ is called a *convex set*, if for every x, $\bar{x} \in D$, $[x, \bar{x}] \subset D$.

Definition 4.1.1. A continuous function $f: \mathbb{R}^n \longrightarrow \mathbb{R}$ is said to be continuously differentiable at $x \in \mathbb{R}^n$, if $(\partial f / \partial x_i)(x)$ exists and is continuous, $i = 1, \ldots, n$; the *gradient* of f at x is then defined as

$$\nabla f(x) = \left[\frac{\partial f}{\partial x_1}(x), \ldots, \frac{\partial f}{\partial x_n}(x) \right]^T \tag{4.1.1}$$

The function f is said to be continuously differentiable in an open region $D \subset \mathbb{R}^n$, denoted $f \in C^1(D)$, if it is continuously differentiable at every point in D.

LEMMA 4.1.2   Let $f: \mathbb{R}^n \longrightarrow R$ be continuously differentiable in an open convex set $D \subset \mathbb{R}^n$. Then, for $x \in D$ and any nonzero perturbation $p \in \mathbb{R}^n$, the *directional derivative* of f at x in the direction of p, defined by

$$\frac{\partial f}{\partial p}(x) \equiv \lim_{\varepsilon \to 0} \frac{f(x + \varepsilon p) - f(x)}{\varepsilon},$$

exists and equals $\nabla f(x)^T p$. For any $x, x + p \in D$,

$$f(x + p) = f(x) + \int_0^1 \nabla f(x + tp)^T p \, dt \equiv f(x) + \int_x^{x+p} \nabla f(z) \, dz, \qquad (4.1.2)$$

and there exists $z \in (x, x + p)$ such that

$$f(x + p) = f(x) + \nabla f(z)^T p. \qquad (4.1.3)$$

Proof. We simply parametrize f along the line through x and $(x + p)$ as the function of one variable

$$g : \mathbb{R} \longrightarrow \mathbb{R}, \; g(t) = f(x + tp)$$

and appeal to the calculus of one variable. Define $x(t) = x + tp$. Then by the chain rule, for $0 \le \alpha \le 1$,

$$\frac{dg}{dt}(\alpha) = \sum_{i=1}^n \frac{\partial f(x(t))}{\partial x(t)_i}(x(\alpha)) \frac{dx(t)_i}{dt}(\alpha)$$

$$= \sum_{i=1}^n \frac{\partial f}{\partial x_i}(x(\alpha)) \cdot p_i$$

$$= \nabla f(x + \alpha p)^T p. \qquad (4.1.4)$$

Substituting $\alpha = 0$ reduces (4.1.4) to

$$\frac{\partial f(x)}{\partial p} = \nabla f(x)^T p.$$

By the fundamental theorem of calculus or Newton's theorem,

$$g(1) = g(0) + \int_0^1 g'(t) \, dt$$

which, by the definition of g and (4.1.4), is equivalent to

$$f(x + p) = f(x) + \int_0^1 \nabla f(x + tp)^T p \, dt$$

and proves (4.1.2). Finally, by the mean value theorem for functions of one variable,

$$g(1) = g(0) + g'(\xi), \qquad \xi \in (0, 1)$$

which by the definition of g and (4.1.4), is equivalent to

$$f(x + p) = f(x) + \nabla f(x + \xi p)^T p, \qquad \xi \in (0, 1),$$

and proves (4.1.3). $\square$

EXAMPLE 4.1.3 Let $f: \mathbb{R}^2 \longrightarrow \mathbb{R}$, $f(x) = x_1^2 - 2x_1 + 3x_1x_2^2 + 4x_2^3$, $x_c = (1, 1)^T$, $p = (-2, 1)^T$. Then

$$\nabla f(x) = \begin{bmatrix} 2x_1 - 2 + 3x_2^2 \\ 6x_1x_2 + 12x_2^2 \end{bmatrix},$$

$f(x_c) = 6$, $f(x_c + p) = 23$, $\nabla f(x_c) = (3, 18)^T$. In the notation of Lemma 4.1.2, $g(t) = f(x_c + tp) = f(1 - 2t, 1 + t) = 6 + 12t + 7t^2 - 2t^3$, and the reader can verify that (4.1.3) is true for $z = x_c + tp$ with $t = (7 - \sqrt{19})/6 \cong .44$.

Definition 4.1.4 A continuously differentiable function $f: \mathbb{R}^n \longrightarrow \mathbb{R}$ is said to be *twice continuously differentiable* at $x \in \mathbb{R}^n$, if $(\partial^2 f/\partial x_i \partial x_j)(x)$ exists and is continuous, $1 \le i, j \le n$; the *Hessian* of f at x is then defined as the $n \times n$ matrix whose i, j element is

$$\nabla^2 f(x)_{ij} = \frac{\partial^2 f(x)}{\partial x_i \partial x_j}, \qquad 1 \le i, \ j \le n. \tag{4.1.5}$$

The function f is said to be twice continuously differentiable in an open region $D \subset \mathbb{R}^n$ denoted $f \in C^2(D)$, if it is twice continuously differentiable at every point in D.

LEMMA 4.1.5 Let $f: \mathbb{R}^n \longrightarrow \mathbb{R}$ be twice continuously differentiable in an open convex set $D \subset \mathbb{R}^n$. Then for any $x \in D$ and any nonzero perturbation $p \in \mathbb{R}^n$, the second directional derivative of f at x in the direction of p, defined by

$$\frac{\partial^2 f}{\partial p^2}(x) = \lim_{\varepsilon \to 0} \frac{\dfrac{\partial f}{\partial p}(x + \varepsilon p) - \dfrac{\partial f}{\partial p}(x)}{\varepsilon},$$

exists and equals $p^T \nabla^2 f(x) p$. For any x, $x + p \in D$, there exists $z \in (x, x + p)$ such that

$$f(x + p) = f(x) + \nabla f(x)^T p + \tfrac{1}{2} p^T \nabla^2 f(z) p. \tag{4.1.6}$$

Proof. The proof is analogous to the proof of Lemma 4.1.2. □

The Hessian is always symmetric as long as f is twice continuously differentiable. This is the reason we were interested in symmetric matrices in our chapter on linear algebra.

EXAMPLE 4.1.6 Let f, x_c, and p be given by Example 4.1.3. Then

$$\nabla^2 f(x) = \begin{bmatrix} 2 & 6x_2 \\ 6x_2 & 6x_1 + 24x_2 \end{bmatrix}, \qquad \nabla^2 f(x_c) = \begin{bmatrix} 2 & 6 \\ 6 & 30 \end{bmatrix}.$$

The reader can verify that (4.1.6) is true for $z = x_c + tp$, $t = \frac{1}{3}$.

Lemma 4.1.5 suggests that we might model the function f around a point x_c by the quadratic model

$$m_c(x_c + p) = f(x_c) + \nabla f(x_c)^T p + \tfrac{1}{2} p^T \, \nabla^2 f(x_c) p, \tag{4.1.7}$$

and this is precisely what we will do. In fact, it shows that the error in this model is given by

$$f(x_c + p) - m_c(x_c + p) = \tfrac{1}{2} p^T (\nabla^2 f(z) - \nabla^2 f(x_c)) p$$

for some $z \in (x_c, x_c + p)$. In Corollary 4.1.14 we will give an alternative error bound.

Now let us proceed to the less simple case of $F: \mathbb{R}^n \longrightarrow \mathbb{R}^m$, where $m = n$ in the nonlinear simultaneous equations problem and $m > n$ in the nonlinear least-squares problem. It will be convenient to have the special notation e_i^T for the ith row of the identity matrix. There should be no confusion with the natural log base. Since the value of the ith component function of F can be written $f_i(x) = e_i^T F(x)$, consistency with the product rule makes it necessary that $f_i'(x) = e_i^T F'(x)$, the ith row of $F'(x)$. Thus $F'(x)$ must be an $m \times n$ matrix whose ith row is $\nabla f_i(x)^T$. The following definition makes this official.

Definition 4.1.7. A continuous function $F: \mathbb{R}^n \longrightarrow \mathbb{R}^m$ is continuously differentiable at $x \in \mathbb{R}^n$ if each component function f_i, $i = 1, \ldots, m$ is continuously differentiable at x. The derivative of F at x is sometimes called the *Jacobian* (matrix) of F at x, and its transpose is sometimes called the gradient of F at x. The common notations are:

$$F'(x) \in \mathbb{R}^{m \times n}, \qquad F'(x)_{ij} = \frac{\partial f_i}{\partial x_j}(x), \qquad F'(x) = J(x) = \nabla F(x)^T.$$

F is said to be continuously differentiable in an open set $D \subset \mathbb{R}^n$, denoted $F \in C^1(D)$, if F is continuously differentiable at every point in D.

EXAMPLE 4.1.8   Let $F: \mathbb{R}^2 \longrightarrow \mathbb{R}^2$, $f_1(x) = e^{x_1} - x_2$, $f_2(x) = x_1^2 - 2x_2$. Then

$$J(x) = \begin{pmatrix} e^{x_1} & -1 \\ 2x_1 & -2 \end{pmatrix}.$$

For the remainder of this book we will denote the Jacobian matrix of F at x by $J(x)$. Also, we will often speak of the Jacobian of F rather than the Jacobian matrix of F at x. The only possible confusion might be that the latter term is sometimes used for the determinant of $J(x)$. However, no confusion should arise here, since we will find little use for determinants.

Now comes the big difference from real-valued functions: there is no mean value theorem for continuously differentiable vector-valued functions. That is, in general there may not exist $z \in \mathbb{R}^n$ such that $F(x + p) = F(x) + J(z)p$. Intuitively the reason is that, although each function f_i satisfies $f_i(x + p) = f_i(x) + \nabla f_i(z_i)^T p$, the points z_i may differ. For example, consider the function of Example 4.1.8. There is no $z \in \mathbb{R}^n$ for which $F(1, 1) = F(0, 0) + J(z)(1, 1)^T$, as this would require

$$\begin{bmatrix} e - 1 \\ -1 \end{bmatrix} = \begin{bmatrix} 1 \\ 0 \end{bmatrix} + \begin{bmatrix} e^{z_1} & -1 \\ 2z_1 & -2 \end{bmatrix} \begin{bmatrix} 1 \\ 1 \end{bmatrix},$$

or $e^{z_1} = e - 1$ and $2z_1 = 1$, which is clearly impossible.

Although the standard mean value theorem is impossible, we will be able to replace it in our analysis by Newton's theorem and the triangle inequality for line integrals. Those results are given below. The integral of a vector-valued function of a real variable can be interpreted as the vector of Riemann integrals of each component function.

LEMMA 4.1.9 Let $F: \mathbb{R}^n \longrightarrow \mathbb{R}^m$ be continuously differentiable in an open convex set $D \subset \mathbb{R}^n$. For any $x, x + p \in D$,

$$F(x + p) - F(x) = \int_0^1 J(x + tp)p \, dt \equiv \int_x^{x+p} F'(z) \, dz. \qquad (4.1.8)$$

Proof. The proof comes right from Definition 4.1.7 and a component-by-component application of Newton's formula (4.1.2). □

Notice that (4.1.8) looks like a mean value theorem if we think of it as

$$F(x + p) - F(x) = \left[\int_0^1 J(x + tp) \, dt \right] p,$$

where the integral of the matrix-valued function is now interpreted componentwise. We will occasionally find this form useful. To use (4.1.8) in our analysis, we need to be able to bound the integral in terms of the integrand. The next lemma is just what we need.

LEMMA 4.1.10 Let $G : D \subset \mathbb{R}^n \to \mathbb{R}^{m \times n}$, where D is an open convex set with $x, x + p \in D$. For any norm on $\mathbb{R}^{m \times n}$, if G is integrable on $[x, x + p]$,

$$\left\| \int_0^1 G(x + tp)p \, dt \right\| \leq \int_0^1 \| G(x + tp)p \| \, dt. \qquad (4.1.9)$$

Proof. The proof is an exercise but it comes from writing the integral as a vector Riemann sum and then applying the triangle inequality (3.1.1c). Then a limiting argument can be applied.   □

Lemma 4.1.9 suggests that we model $F(x_c + p)$ by the affine model

$$M_c(x_c + p) = F(x_c) + J(x_c)p, \qquad (4.1.10)$$

and this is what we will do. To produce a bound on the difference between $F(x_c + p)$ and $M_c(x_c + p)$, we need to make an assumption about the continuity of $J(x)$ just as we did in Section 2.4. Then we derive a bound analogous to the one given in Lemma 2.4.2 for functions of one variable.

Definition 4.1.11.   Let m, $n > 0$, $G: \mathbb{R}^n \longrightarrow \mathbb{R}^{m \times n}$, $x \in \mathbb{R}^n$, let $\|\cdot\|$ be a norm on $\mathbb{R}^n$, and $\|\|\cdot\|\|$ a norm on $\mathbb{R}^{m \times n}$. G is said to be *Lipschitz continuous* at x if there exists an open set $D \subset \mathbb{R}^n$, $x \in D$, and a constant γ such that for all $v \in D$,

$$\|\| G(v) - G(x) \|\| \le \gamma \| v - x \|. \qquad (4.1.11)$$

The constant γ is called a Lipschitz constant for G at x. For any specific D containing x for which (4.1.11) holds, G is said to be Lipschitz continuous at x in the neighborhood D. If (4.1.11) holds for every $x \in D$, then $G \in \text{Lip}_\gamma(D)$.

Note that the value of γ depends on the norms in (4.1.11), but the existence of γ does not.

LEMMA 4.1.12   Let $F: \mathbb{R}^n \longrightarrow \mathbb{R}^m$ be continuously differentiable in the open convex set $D \subset \mathbb{R}^n$, $x \in D$, and let J be Lipschitz continuous at x in the neighborhood D, using a vector norm and the induced matrix operator norm and the constant γ. Then, for any $x + p \in D$,

$$\| F(x + p) - F(x) - J(x)p \| \le \frac{\gamma}{2} \| p \|^2. \qquad (4.1.12)$$

Proof.   By Lemma 4.1.9,

$$F(x + p) - F(x) - J(x)p = \left[\int_0^1 J(x + tp)p \, dt \right] - J(x)p$$

$$= \int_0^1 [J(x + tp) - J(x)]p \, dt.$$

Using (4.1.9), the definition of a matrix operator norm, and the Lipschitz continuity of J at x in neighborhood D, we obtain

$$\| F(x + p) - F(x) - J(x)p \|$$

$$\leq \int_0^1 \| J(x + tp) - J(x) \| \, \| p \| \, dt$$

$$\leq \int_0^1 \gamma \| tp \| \, \| p \| \, dt$$

$$= \gamma \| p \|^2 \int_0^1 t \, dt$$

$$= \frac{\gamma}{2} \| p \|^2. \qquad \square$$

EXAMPLE 4.1.13   Let $F : \mathbb{R}^2 \longrightarrow \mathbb{R}^2$, $f_1(x) = x_1^2 - x_2^2 - \alpha$, $f_2(x) = 2x_1x_2 - \beta$, $\alpha, \beta \in \mathbb{R}$. [The system of equations $F(x) = 0$ expresses the solution to $x_1 + ix_2 = \sqrt{\alpha + i\beta}$.] Then

$$J(x) = \begin{bmatrix} 2x_1 & -2x_2 \\ 2x_2 & 2x_1 \end{bmatrix},$$

and the reader can easily verify that, for any $x, v \in \mathbb{R}^2$ and using the l_1 norm, $J(x)$ obeys (4.1.11) with $\gamma = 2$.

Using Lipschitz continuity, we can also obtain a useful bound on the error in the quadratic model (4.1.7) as an approximation to $f(x_c + p)$. This is given below. The proof is similar to that of Lemma 4.1.12 and is left as an exercise.

LEMMA 4.1.14.   Let $f: \mathbb{R}^n \longrightarrow \mathbb{R}$ be twice continuously differentiable in an open convex set $D \subset \mathbb{R}^n$, and let $\nabla^2 f(x)$ be Lipschitz continuous at x in the neighborhood D, using a vector norm and the induced matrix operator norm and the constant γ. Then, for any $x + p \in D$,

$$\left| f(x + p) - \left(f(x) + \nabla f(x)^T p + \frac{1}{2} p^T \nabla^2 f(x) p \right) \right| \leq \frac{\gamma}{6} \| p \|^3. \quad (4.1.13)$$

We conclude this section with two inequalities, related to (4.1.12), that we will use to bound $\| F(v) - F(u) \|$ in our convergence proofs. The proof of Lemma 4.1.15, which generalizes Lemma 4.1.12, is left as an exercise.

LEMMA 4.1.15   Let F, J satisfy the conditions of Lemma 4.1.12. Then, for any $v, u \in D$,

$\| F(v) - F(u) - J(x)(v - u) \|$

$$\leq \gamma \frac{\| v - x \| + \| u - x \|}{2} \| v - u \|. \qquad (4.1.14)$$

LEMMA 4.1.16   Let F, J satisfy the conditions of Lemma 4.1.12, and assume that $J(x)^{-1}$ exists. Then there exists $\varepsilon > 0$, $0 < \alpha < \beta$, such that

$$\alpha \| v - u \| \leq \| F(v) - F(u) \| \leq \beta \| v - u \|, \qquad (4.1.15)$$

for all $v, u \in D$ for which $\max \{ \| v - x \|, \| u - x \| \} \leq \varepsilon$.

Proof.   Using the triangle inequality and (4.1.14),

$$\| F(v) - F(u) \| \leq \| J(x)(v - u) \| + \| F(v) - F(u) - J(x)(v - u) \|$$

$$\leq \left[\| J(x) \| + \frac{\gamma}{2} (\| v - x \| + \| u - x \|) \right] \| v - u \|$$

$$\leq \left[\| J(x) \| + \gamma \varepsilon \right] \| v - u \|,$$

which proves the second part of (4.1.15) with $\beta = \| J(x) \| + \gamma \varepsilon$. Similarly,

$$\| F(v) - F(u) \| \geq \| J(x)(v - u) \| - \| F(v) - F(u) - J(x)(v - u) \|$$

$$\geq \left[\left(\frac{1}{\| J(x)^{-1} \|} \right) - \frac{\gamma}{2} (\| v - x \| + \| u - x \|) \right] \| v - u \|$$

$$\geq \left[\frac{1}{\| J(x)^{-1} \|} - \gamma \varepsilon \right] \| v - u \|.$$

Thus if $\varepsilon < (1 / \| J(x)^{-1} \| \gamma)$, the first part of (4.1.15) holds with

$$\alpha = \left(\frac{1}{\| J(x)^{-1} \|} \right) - \gamma \varepsilon > 0. \qquad \square$$

4.2 MULTIVARIABLE FINITE-DIFFERENCE DERIVATIVES

In the preceding section we saw that the Jacobian, gradient, and Hessian will be useful quantities in forming models of multivariable nonlinear functions. In many applications, however, these derivatives are not analytically available. In this section we introduce the formulas used to approximate these derivatives by finite differences, and the error bounds associated with these formulas. The choice of finite-difference stepsize in the presence of finite-precision arithmetic and the use of finite-difference derivatives in our algorithms are discussed in Sections 5.4 and 5.6.

Recall that, for a function f of a single variable, the finite-difference approximation to $f'(x)$ was given by

$$a = \frac{f(x + h) - f(x)}{h},$$

where h is a small quantity. From the error bound on the affine model of $f(x + h)$ we had $|a - f'(x)| = O(h)$ for h sufficiently small. In the case when $F: \mathbb{R}^n \longrightarrow \mathbb{R}^m$, it is reasonable to use the same idea to approximate the (i, j)th component of $J(x)$ by the *forward difference* approximation

$$a_{ij} = \frac{f_i(x + he_j) - f_i(x)}{h},$$

where e_j denotes the jth unit vector. This is equivalent to approximating the jth column of $J(x)$ by

$$A_{.j} = \frac{F(x + he_j) - F(x)}{h}. \tag{4.2.1}$$

Again, one would expect $\| A_{.j} - (J(x))_{.j} \| = O(h)$, for h sufficiently small, and this is verified in Lemma 4.2.1.

LEMMA 4.2.1   Let $F: \mathbb{R}^n \longrightarrow \mathbb{R}^m$ satisfy the conditions of Lemma 4.1.12, using a vector norm $\| \cdot \|$ for which $\| e_j \| = 1$, $j = 1, \ldots, n$. Let $A \in \mathbb{R}^{n \times m}$, with $A_{.j}$ defined by (4.2.1), $j = 1, \ldots, n$. Then

$$\| A_{.j} - J(x)_{.j} \| \leq \frac{\gamma}{2} |h|. \tag{4.2.2}$$

If the norm being used is the l_1 norm, then

$$\| A - J(x) \|_1 \leq \frac{\gamma}{2} |h|. \tag{4.2.3}$$

Proof.   Substituting $p = he_j$ into (4.1.12) gives

$$\| F(x + he_j) - F(x) - J(x)he_j \| \leq \frac{\gamma}{2} \| he_j \|^2 = \frac{\gamma}{2} |h|^2.$$

Dividing both sides by h gives (4.2.2). Since the l_1 norm of a matrix is the maximum of the l_1 norms of its columns, (4.2.3) follows immediately. $\square$

It should be mentioned that, although (4.2.3) assumes that one stepsize h is used in constructing all of A, finite-precision arithmetic will cause us to use a different stepsize for each column of A. Notice also that finite-difference approximation of the Jacobian requires n additional evaluations of the function F, so it can be relatively expensive if evaluation of F is expensive or n is large.

When the nonlinear problem is minimization of a function $f : \mathbb{R}^n \longrightarrow \mathbb{R}$, we may need to approximate the gradient $\nabla f(x)$ and/or the Hessian $\nabla^2 f(x)$. Approximation of the gradient is just a special case of the approximation of $J(x)$ discussed above, with $m = 1$. In some cases, finite-precision arithmetic causes us to seek a more accurate finite-difference approximation using the central difference approximation given in Lemma 4.2.2. Notice that this approximation requires twice as many evaluations of f as forward differences.

LEMMA 4.2.2   Let $f : \mathbb{R}^n \longrightarrow \mathbb{R}$ satisfy the conditions of Lemma 4.1.14 using a vector norm with the property that $\| e_i \| = 1$, $i = 1, \ldots, n$. Assume $x + h e_i, x - h e_i \in D$, $i = 1, \ldots, n$. Let $a \in \mathbb{R}^n$, a_i given by

$$a_i = \frac{f(x + h e_i) - f(x - h e_i)}{2h}. \tag{4.2.4}$$

Then

$$|a_i - \nabla f(x)_i| \le \frac{\gamma h^2}{6}. \tag{4.2.5}$$

If the norm being used is the l_∞ norm, then

$$\| a - \nabla f(x) \|_\infty \le \frac{\gamma h^2}{6}. \tag{4.2.6}$$

Proof.   Define the real quantities α and β by

$$\alpha = f(x + h e_i) - f(x) - h \nabla f(x)_i - h^2 \nabla^2 f(x)_{ii}/2,$$

$$\beta = f(x - h e_i) - f(x) + h \nabla f(x)_i - h^2 \nabla^2 f(x)_{ii}/2,$$

Applications of (4.1.13) with $p = \pm h e_i$ yield

$$|\alpha| \le \frac{\gamma h^3}{6} \quad \text{and} \quad |\beta| \le \frac{\gamma h^3}{6},$$

and from the triangle inequality,

$$|\alpha - \beta| \le \frac{\gamma h^3}{3},$$

so (4.2.5) follows from $\alpha - \beta = 2h(a_i - \nabla f(x)_i)$. The definition of the l_∞ norm and (4.2.5) imply (4.2.6). $\square$

On some occasions ∇f is analytically available but $\nabla^2 f$ is not. In this case, $\nabla^2 f$ can be approximated by applying formula (4.2.1) to the function $F \triangleq \nabla f$, followed by $\hat{A} := \frac{1}{2}(A + A^T)$, since the approximation to $\nabla^2 f$ should be symmetric. We know from Lemma 4.2.1 that $\| A - \nabla^2 f(x) \| = O(h)$ for h sufficiently small and it is easy to show that $\| \hat{A} - \nabla^2 f(x) \|_F \leq \| A - \nabla^2 f(x) \|_F$ (see Exercise 12). Thus $\| \hat{A} - \nabla^2 f(x) \| = O(h)$.

If ∇f is not available, it is possible to approximate $\nabla^2 f$ using only values of $f(x)$. This is described in Lemma 4.2.3. Once again the error is $O(h)$ for h sufficiently small. Notice, however, that $(n^2 + 3n)/2$ additional evaluations of f are required.

LEMMA 4.2.3   Let f satisfy the conditions of Lemma 4.2.2. Assume x, $x + he_i$, $x + he_j$, $x + he_i + he_j \in D$, $1 \leq i, j \leq n$. Let $A \in \mathbb{R}^{n \times n}$, A_{ij} given by

$$A_{ij} = \frac{f(x + he_i + he_j) - f(x + he_i) - f(x + he_j) + f(x)}{h^2}. \quad (4.2.7)$$

Then

$$| A_{ij} - \nabla^2 f(x)_{ij} | \leq \tfrac{5}{3}\gamma h. \quad (4.2.8)$$

If the matrix norm being used is the l_1, l_∞, or Frobenius norm, then

$$\| A - \nabla^2 f(x) \| \leq \tfrac{5}{3}\gamma nh. \quad (4.2.9)$$

Proof. The proof is very similar to that of Lemma 4.2.2. To prove (4.2.8), first substitute $p = he_i + he_j$, $p = he_i$, and $p = he_j$ into (4.1.13). Call the expressions inside the absolute-value signs, on the three resulting left-hand sides, α, β, and η, respectively. Then $\alpha - \beta - \eta = (A_{ij} - \nabla^2 f(x)_{ij})h^2$. Using $| \alpha - \beta - \eta | \leq | \alpha | + | \beta | + | \eta |$ and dividing by h^2 yields (4.2.8). Then (4.2.9) follows immediately from (3.1.8a), (3.1.8c), and the definition of the Frobenius norm. $\square$

4.3 NECESSARY AND SUFFICIENT CONDITIONS FOR UNCONSTRAINED MINIMIZATION

In this section we derive the first- and second-order necessary and sufficient conditions for a point x_* to be a local minimizer of a continuously differentiable function $f : \mathbb{R}^n \longrightarrow \mathbb{R}$, $n \geq 1$. Naturally, these conditions will be a key to our algorithms for the unconstrained minimization problem. They are closely related to the following conditions given in Chapter 2 for the one-variable problem. In one dimension, it is necessary that $f'(x_*) = 0$ for x_* to be a local minimizer of f; it is sufficient that, in addition, $f''(x_*) > 0$, and it is necessary that $f''(x_*) \geq 0$. In multiple dimensions, $\nabla f(x_*) = 0$ is required for x_* to be a local minimizer; a sufficient condition is that, in addition, $\nabla^2 f(x_*)$ is positive definite, and it is necessary that $\nabla^2 f(x_*)$ at least be positive semidefinite.

The proof and interpretation of these conditions is virtually the same as for one-variable functions. The gradient being zero says simply that there is no direction from x_* in which $f(x)$ necessarily decreases, or increases, initially. Thus a point with zero gradient can be a minimizer or maximizer of f, or a saddle point, which is a maximizer in some cross section of the variable space and a minimizer in another (and corresponds to the seat of a saddle in two variables). Positive definitiveness of $\nabla^2 f(x_*)$ corresponds to the geometric concept of strict local convexity, and just means that f curves up from x_* in all directions.

LEMMA 4.3.1 Let $f: \mathbb{R}^n \longrightarrow \mathbb{R}$ be continuously differentiable in an open convex set $D \subset \mathbb{R}^n$. Then $x \in D$ can be a local minimizer of f only if $\nabla f(x) = 0$.

Proof. As in the one-variable case, a proof by contradiction is better than a direct proof, because the lemma and proof taken together tell how to recognize a possible minimizer and how to improve the current candidate if it can't be a minimizer.

If $\nabla f(x) \neq 0$, then choose any $p \in \mathbb{R}^n$ for which $p^T \nabla f(x) < 0$. There certainly is such a p, since $p = -\nabla f(x)$ would do. To keep the notation simple, scale p by a positive constant so that $x + p \in D$. Since D is open and convex, this can be done and $x + tp \in D$ for every $0 \leq t \leq 1$.

Now by the continuity of ∇f on D, there is some $\bar{t} \in [0, 1]$ such that $\nabla f(x + tp)^T p < 0$ for every $t \in [0, \bar{t}]$. Now we apply Newton's theorem (Lemma 4.1.2) to give

$$f(x + \bar{t}p) = f(x) + \int_{t=0}^{\bar{t}} \nabla f(x + tp)^T p \, dt$$

and so $f(x + \bar{t}p) < f(x)$. □

A class of algorithms called *descent methods* are characterized by their being based on the previous proof. If $\nabla f(x_c) \neq 0$, they choose a descent direction p as above, and then a point x_+ by descending in the direction p as above.

THEOREM 4.3.2 Let $f: \mathbb{R}^n \longrightarrow \mathbb{R}$ be twice continuously differentiable in the open convex set $D \subset \mathbb{R}^n$, and assume there exists $x \in D$ such that

$\nabla f(x) = 0$. If $\nabla^2 f(x)$ is positive definite, then x is a local minimizer of f. If $\nabla^2 f(x)$ is Lipschitz continuous at x, then x can be a local minimizer of f only if $\nabla^2 f(x)$ is positive semidefinite.

Proof.   If $\nabla^2 f(x)$ is positive definite, then by the continuity of $\nabla^2 f$ there exists an open convex subset of D containing x, such that $\nabla^2 f(z)$ is positive definite for all z in that subset. To keep it simple, let us rename that set D. Then for any $p \in \mathbb{R}^n$ for which $x + p \in D$, by Lemma 4.1.5, for some $z \in [x, x + p]$,

$$f(x + p) = f(x) + \nabla f(x)^T p + \tfrac{1}{2} p^T \nabla^2 f(z) p$$
$$= f(x) + \tfrac{1}{2} p^T \nabla^2 f(z) p > f(x)$$

which proves that x is a local minimizer of f. The necessity for $\nabla^2 f(x)$ to be positive semidefinite is left to an exercise. $\square$

The proof of the following corollary is an easy exercise. We include it because it is a clean statement of that part of Theorem 4.3.2 most relevant to the conditions for Newton's method to find a local minimizer by finding a zero of ∇f.

COROLLARY 4.3.3   Let $f : \mathbb{R}^n \longrightarrow \mathbb{R}$ be twice continuously differentiable in the open convex set $D \subset \mathbb{R}^n$. If $x_* \in D$ and $\nabla f(x_*) = 0$ and if $\nabla^2 f$ is Lipschitz continuous at x_* with $\nabla^2 f(x_*)$ nonsingular, then x_* is a local minimizer of f if, and only if, $\nabla^2 f(x_*)$ is positive definite.

The necessary and sufficient conditions for x_* to be a local maximizer of f are simply the conditions for x_* to be a local minimizer of $-f$: $\nabla f(x_*) = 0$ is necessary, and in addition, $\nabla^2 f(x_*)$ negative semidefinite is necessary; $\nabla^2 f(x_*)$ negative definite is sufficient. If $\nabla^2 f(x_*)$ is indefinite, x_* cannot be a local minimizer or maximizer.

The sufficient conditions for minimization partially explain our interest in symmetric and positive definite matrices in Chapter 3. They become more important because our minimization algorithms commonly model f at perturbations p of x_c by a quadratic function

$$m_c(x_c + p) = f(x_c) + \nabla f(x_c)^T p + \tfrac{1}{2} p^T H p ,$$

and a slight modification of the proof of Theorem 4.3.2 shows that m_c has a unique minimizer if, and only if, H is positive definite. Thus we will often use symmetric and positive definite matrices H in our models. It is important to understand the shapes of multivariable quadratic functions: they are strictly convex or convex, respectively bowl- or trough-shaped in two dimensions, if H

is positive definite or positive semidefinite; they are strictly concave or concave (turn the bowl or trough upside down) if H is negative definite or negative semidefinite; and they are saddle-shaped (in n dimensions) if H is indefinite.

4.4 EXERCISES

1. Calculate the Jacobian of the following system of three equations in three unknowns:

$$F(x) = \begin{bmatrix} x_1^2 + x_2^3 - x_3^4 \\ x_1 x_2 x_3 \\ 2x_1 x_2 - 3x_2 x_3 + x_1 x_3 \end{bmatrix}.$$

2. Let $f : \mathbb{R}^2 \longrightarrow \mathbb{R}$, $f(x) = x_1^3 x_2^4 + x_2/x_1$. Find $\nabla f(x)$ and $\nabla^2 f(x)$.

3. Prove Lemma 4.1.5.

4. An alternate definition of the Jacobian of $F : \mathbb{R}^n \longrightarrow \mathbb{R}^n$ at $x \in \mathbb{R}^n$ is the linear operator $\hat{J}(x)$, if it exists, for which

$$\lim_{t \to 0} \left[\frac{\| F(x + tp) - F(x) - \hat{J}(x)tp \|}{t} \right] = 0 \qquad \text{for all } p \in \mathbb{R}^n. \qquad (4.5.1)$$

Prove that if each component of F is continuously differentiable in an open set D containing x, and J is defined in Definition 4.1.5, then $J(x) = \hat{J}(x)$. *Note:* The existence of $\hat{J}(x)$ in (4.5.1) is the definition of Gateaux differentiability of F at x; a stronger condition on F is Frechet differentiability (at x), which requires $\bar{J}(x) \in \mathbb{R}^{m \times n}$ to satisfy

$$\lim_{p \to 0} \frac{\| F(x + p) - F(x) - \bar{J}(x)p \|}{\| p \|} = 0.$$

[For further information, see Ortega and Rheinboldt (1970).]

5. Let

$$f(x) = \begin{cases} x \sin \dfrac{1}{x}, & x \neq 0, \\ 0, & x = 0. \end{cases}$$

Prove that $f(x)$ is Lipschitz continuous but not differentiable at $x = 0$.

6. Prove Lemma 4.1.14. [*Hint:* You will need a double-integral extension of Newton's theorem.]

7. Prove Lemma 4.1.15 (the techniques are very similar to those used to prove Lemma 4.1.12).

8. If equation (4.1.11) in the definition of Lipschitz continuity is replaced by

$$\| G(v) - G(x) \| \leq \gamma \| v - x \|^\alpha,$$

for some $\alpha \in (0, 1]$, then G is said to be Hölder continuous at x. Show that if the assumption of Lipschitz continuity in Lemma 4.1.12 is replaced with Hölder continuity, then the lemma remains true if (4.1.12) is changed to

$$\| F(x + p) - F(x) - J(x)p \| \leq \frac{\gamma}{1 + \alpha} \| p \|^{1 + \alpha}.$$

9. Given $F: \mathbb{R}^2 \longrightarrow \mathbb{R}^2$,

$$F(x) = \begin{pmatrix} 3x_1^2 - 2x_2 \\ x_2^3 - \dfrac{1}{x_1} \end{pmatrix},$$

calculate $J(x)$ at $x = (1, 1)^T$. Then use formula (4.2.1) to calculate (by hand or hand calculator) a finite-difference approximation to $J(x)$ at $x = (1, 1)^T$, with $h = 0.01$. Comment on the accuracy of the approximation.

10. Extend the central difference formula for $\nabla f(x)$ to find a central difference approximation to $\nabla^2 f(x)_{11}$ that uses only function values and has asymptotic accuracy $O(h^2)$. (This approximation is used in discretizing some partial differential equations.) You may assume that the third derivative of $f(x)$ is Lipschitz continuous.

11. Extend your techniques from Exercise 10 to find an approximation to $\nabla^2 f(x)_{ij}, i \neq j$, that uses only function values and has asymptotic accuracy $O(h^2)$. How does the cost of this approximation in function evaluations compare to (4.2.7)?

12. Let $A, M \in \mathbb{R}^{n \times n}$, M symmetric, and let $\hat{A} = (A + A^T)/2$. Prove that $\| M - \hat{A} \|_F \leq \| M - A \|_F$.

13. Complete the proof of Theorem 4.3.2.

14. Prove Corollary 4.3.3.

15. Prove that a quadratic function of n variables, $f(x) = \frac{1}{2}x^T Ax + b^T x + c$, $A \in \mathbb{R}^{n \times n}$ symmetric, $b \in \mathbb{R}^n$, $c \in \mathbb{R}$, has a unique minimizer if, and only if, A is positive definite. What is the minimizer? How would you find it? Recall that $\nabla f(x) = Ax + b$, $\nabla^2 f(x) = A$.

16. Use the necessary and sufficient conditions for unconstrained minimization to prove that the solution(s) to the linear least-squares problem

$$\min_{x \in \mathbb{R}^n} \| Ax - b \|_2, \quad A \in \mathbb{R}^{m \times n}, \quad b \in \mathbb{R}^m,$$

is (are) $x_* \in \mathbb{R}^n$ satisfying $(A^T A)x_* = A^T b$. [*Hint*: For the case when A doesn't have full column rank, generalize Exercise 15 to consider positive semidefinite matrices.]

Another program note

Chapters 5 through 9 contain the heart of this book, the development of algorithms for the multivariable nonlinear equations and unconstrained minimization problems. In Chapter 5 we derive and discuss multidimensional Newton's method, the quickly convergent local method that will be the basic model for our algorithms. Chapter 6 presents the leading techniques that are used along with Newton's method to assure the global convergence of nonlinear algorithms. Chapter 7 discusses the practical considerations of stopping, scaling, and testing necessary to complete the development of good computer software. Chapters 8 and 9 discuss the multidimensional secant methods that will be used in place of Newton's method when derivatives are unavailable and function evaluation is expensive.

It will turn out that the structure of the hybrid algorithm for multivariable problems is basically the same as was given for one-variable problems in Chapter 2. This similarity continues at a more detailed level in the definition and analysis of the local step, except that finite-difference derivatives become more expensive and secant approximations far less obvious. However, the most significant difference is in the diversity and increased complexity of the global strategies for multivariable problems.

5

Newton's Method for Nonlinear Equations and Unconstrained Minimization

This chapter begins our consideration of multivariable problems by discussing local algorithms for systems of nonlinear equations and unconstrained minimization. We start by deriving Newton's method for systems of nonlinear equations and discussing its computer implementation and its good and bad features.

In Sections 5.2 and 5.3 we use two different and important approaches to show that Newton's method is locally q-quadratically convergent for most problems, although it will not necessarily achieve global convergence. Section 5.4 discusses the finite-difference version of Newton's method for nonlinear equations, including the selection of finite-difference stepsizes in practice. We conclude the chapter by discussing the version of Newton's method for multidimensional unconstrained minimization problems, (Section 5.5) and the use of finite-difference derivatives for these problems (Section 5.6).

5.1 NEWTON'S METHOD FOR SYSTEMS OF NONLINEAR EQUATIONS

The most basic problem studied here is the solution of a system of nonlinear equations:

$$\text{given } F: \mathbb{R}^n \longrightarrow \mathbb{R}^n, \quad \text{find } x_* \in \mathbb{R}^n \text{ such that } F(x_*) = 0 \qquad (5.1.1)$$

where F is assumed to be continuously differentiable. In this section we derive Newton's method for problem (5.1.1) and discuss its requirements, implementation, and local and global convergence characteristics. These factors motivate the work in the remainder of this book, which modifies and builds around Newton's method to create globally robust and locally fast algorithms. Once again we ignore the questions of the existence or uniqueness of solutions to (5.1.1), and assume that in practice this will not usually be a problem.

Newton's method for problem (5.1.1) again is derived by finding the root of an affine approximation to F at the current iterate x_c. This approximation is created using the same techniques as for the one-variable problem. Since we

$$F(x_c + p) = F(x_c) + \int_{x_c}^{x_c + p} J(z) \, dz. \tag{5.1.2}$$

approximate the integral in (5.1.2) by the linear term $J(x_c) \cdot p$ to get the affine approximation to F at perturbations p of x_c,

$$M_c(x_c + p) = F(x_c) + J(x_c) \cdot p. \tag{5.1.3}$$

(M_c again stands for current model.) Next we solve for the step s^N that makes $M_c(x_c + s^N) = 0$, giving the Newton iteration for (5.1.1). Solve

$$J(x_c)s^N = -F(x_c),$$
$$x_+ = x_c + s^N. \tag{5.1.4}$$

An equivalent way to view this procedure is that we are finding a simultaneous zero of the affine models of the n component functions of F given by

$$(M_c)_i(x_c + s^N) = f_i(x_c) + \nabla f_i(x_c)^T s^N, \quad i = 1, \ldots, n.$$

Since x_+ is not expected to equal x_*, but only to be a better estimate than x_c, we make the Newton iteration (5.1.4) into an algorithm by applying it iteratively from a starting guess x_0.

ALGORITHM 5.1.1   NEWTON'S METHOD FOR SYSTEMS OF NONLINEAR EQUATIONS

Given $F: \mathbb{R}^n \longrightarrow \mathbb{R}^n$ continuously differentiable and $x_0 \in \mathbb{R}^n$: at each iteration k, solve

$$J(x_k)s_k = -F(x_k),$$
$$x_{k+1} = x_k + s_k.$$

For example, let

$$F(x) = \begin{bmatrix} x_1 + x_2 - 3 \\ x_1^2 + x_2^2 - 9 \end{bmatrix}$$

which has roots at $(3, 0)^T$ and $(0, 3)^T$, and let $x_0 = (1, 5)^T$. Then the first two iterations of Newton's method are

$$J(x_0)s_0 = -F(x_0): \quad \begin{bmatrix} 1 & 1 \\ 2 & 10 \end{bmatrix} s_0 = -\begin{bmatrix} 3 \\ 17 \end{bmatrix}, \quad s_0 = \begin{bmatrix} -\frac{13}{8} \\ -\frac{11}{8} \end{bmatrix},$$

$$x_1 = x_0 + s_0 = (-.625, 3.625)^T,$$

$$J(x_1)s_1 = -F(x_1): \quad \begin{bmatrix} 1 & 1 \\ -\frac{5}{4} & \frac{29}{4} \end{bmatrix} s_1 = -\begin{bmatrix} 0 \\ \frac{145}{32} \end{bmatrix}, \quad s_1 = \begin{bmatrix} \frac{145}{272} \\ -\frac{145}{272} \end{bmatrix},$$

$$x_2 = x_1 + s_1 \cong (-.092, 3.092)^T.$$

Newton's method seems to be working well here; x_2 is already quite close to the root $(0, 3)^T$. This is the main advantage of Newton's method: if x_0 is close enough to a solution x_* and $J(x_*)$ is nonsingular, it is shown in Section 5.2 that the x_k's generated by Algorithm 5.1.1 converge q-quadratically to x_*. Newton's method will also work well on nearly linear problems, since it finds the zero of a (nonsingular) affine function in one iteration. Notice also that if any component functions of F are affine, each iterate generated by Newton's method will be a solution of these equations, since the affine models it uses will always be exact for these functions. For instance, f_1 is affine in the above example, and

$$f_1(x_1) = f_1(x_2)(= f_1(x_3) \ldots) = 0.$$

On the other hand, Newton's method naturally will be no better at converging from bad starting guesses for multivariable nonlinear problems than it was for one-variable problems. For example, if

$$F(x) = \begin{bmatrix} e^{x_1} - 1 \\ e^{x_2} - 1 \end{bmatrix}$$

for which $x_* = (0, 0)^T$, and $x_0 = (-10, -10)^T$, then

$$x_1 = (-11 + e^{10}, -11 + e^{10})^T$$

$$\cong (2.2 \times 10^4, 2.2 \times 10^4)^T,$$

which isn't a very good step! Therefore the convergence characteristics of Newton's method indicate how to use it in multidimensional algorithms: we will always want to use it in at least the final iterations of any nonlinear algorithm to take advantage of its fast local convergence, but it will have to be modified in order to converge globally.

There are also two fundamental problems in implementing the iteration in Algorithm 5.1.1. First, the Jacobian of F may not be analytically available. This occurs commonly in real applications—for example, when F itself is not given in analytic form. Thus, approximating $J(x_k)$ by finite differences or by a less expensive method is a very important problem; it is the subject of Section 5.4 and Chapter 8. Second, $J(x_k)$ may be singular or ill-conditioned, so that the linear system $J(x_k)s_k = -F(x_k)$ cannot be reliably solved for the step s_k. As we saw in Section 3.3, our matrix factorization methods can be used to detect ill-conditioning reasonably well, and so a possible modification to Algorithm 5.1.1 is to perturb $J(x_k)$ just enough to make it well-conditioned and proceed with the iteration. However, such a modification of $J(x_k)$ has no good justification in terms of the underlying problem (5.1.1), and so we don't recommend it. Instead, when $J(x_k)$ is badly conditioned, we prefer to proceed directly to a global method for nonlinear equations (Section 6.5) that corrresponds to perturbing the linear model (5.1.3) in a way meaningful to the nonlinear equations problem.

The advantages and disadvantages of Newton's method discussed above are the key to the development of our multidimensional algorithms, since they point out the properties we will wish to retain and the areas where improvement or modification may be necessary. They are summarized in Table 5.1.2.

Table 5.1.2   Advantages and disadvantages of Newton's method for systems of nonlinear equations

---

Advantages

1. Q-quadratically convergent from good starting guesses if $J(x_*)$ is nonsingular.
2. Exact solution in one iteration for an affine F (exact at each iteration for any affine component functions of F).

Disadvantages

1. Not globally convergent for many problems.
2. Requires $J(x_k)$ at each iteration.
3. Each iteration requires the solution of a system of linear equations that may be singular or ill-conditioned.

5.2 LOCAL CONVERGENCE OF NEWTON'S METHOD

In this section we prove the local q-quadratic convergence of Newton's method for systems of nonlinear equations and discuss its implications. The techniques of the proof are similar to those in the one-variable case; they also form the prototype for most of the convergence proofs in this book. Given a vector norm $\|\cdot\|$, we define $N(x, r)$ as the open neighborhood of radius r around x—i.e., $N(x, r) = \{\bar{x} \in \mathbb{R}^n : \|\bar{x} - x\| < r\}$.

THEOREM 5.2.1   Let $F: \mathbb{R}^n \longrightarrow \mathbb{R}^n$ be continuously differentiable in an open convex set $D \subset \mathbb{R}^n$. Assume that there exists $x_* \in \mathbb{R}^n$ and $r, \beta > 0$, such that $N(x_*, r) \subset D$, $F(x_*) = 0$, $J(x_*)^{-1}$ exists with $\|J(x_*)^{-1}\| \leq \beta$, and $J \in \text{Lip}_\gamma (N(x_*, r))$. Then there exists $\varepsilon > 0$ such that for all $x_0 \in N(x_*, \varepsilon)$ the sequence $x_1, x_2, \ldots$ generated by

$$x_{k+1} = x_k - J(x_k)^{-1} F(x_k), \qquad k = 0, 1, \ldots$$

is well defined, converges to x_*, and obeys

$$\|x_{k+1} - x_*\| \leq \beta \gamma \|x_k - x_*\|^2, \qquad k = 0, 1, \ldots. \qquad (5.2.1)$$

Proof.   We choose ε so that $J(x)$ is nonsingular for any $x \in N(x_*, \varepsilon)$, and then show that, since the local error in the affine model used to produce each iterate of Newton's method is at most $O(\|x_k - x_*\|^2)$, the convergence is q-quadratic. Let

$$\varepsilon = \min \left\{ r, \frac{1}{2\beta\gamma} \right\}. \qquad (5.2.2)$$

We show by induction on k that at each step (5.2.1) holds, and also that

$$\|x_{k+1} - x_*\| \leq \tfrac{1}{2} \|x_k - x_*\|$$

and so

$$x_{k+1} \in N(x_*, \varepsilon). \qquad (5.2.3)$$

We first show that $J(x_0)$ is nonsingular. From $\|x_0 - x_*\| \leq \varepsilon$, the Lipschitz continuity of J at x_*, and (5.2.2), it follows that

$$\|J(x_*)^{-1}[J(x_0) - J(x_*)]\| \leq \|J(x_*)^{-1}\| \, \|J(x_0) - J(x_*)\|$$

$$\leq \beta \cdot \gamma \|x_0 - x_*\| \leq \beta \cdot \gamma \cdot \varepsilon \leq \tfrac{1}{2}.$$

Thus by the perturbation relation (3.1.20), $J(x_0)$ is nonsingular and

$$\|J(x_0)^{-1}\| \leq \frac{\|J(x_*)^{-1}\|}{1 - \|J(x_*)^{-1}[J(x_0) - J(x_*)]\|}$$

$$\leq 2\|J(x_*)^{-1}\|$$

$$\leq 2 \cdot \beta. \qquad (5.2.4)$$

Therefore x_1 is well defined and

$$x_1 - x_* = x_0 - x_* - J(x_0)^{-1} F(x_0)$$

$$= x_0 - x_* - J(x_0)^{-1}[F(x_0) - F(x_*)]$$

$$= J(x_0)^{-1}[F(x_*) - F(x_0) - J(x_0)(x_* - x_0)].$$

Notice that the term in brackets is just the difference between $F(x_*)$ and the affine model $M_c(x)$ evaluated at x_*. Therefore, using Lemma 4.1.12 and (5.2.4),

$$\| x_1 - x_* \| \leq \| J(x_0)^{-1} \| \, \| F(x_*) - F(x_0) - J(x_0)(x_* - x_0) \|$$

$$\leq 2\beta \cdot \frac{\gamma}{2} \| x_0 - x_* \|^2$$

$$= \beta\gamma \| x_0 - x_* \|^2.$$

This proves (5.2.1). Since $\| x_0 - x_* \| \leq 1/(2\beta\gamma)$,

$$\| x_1 - x_* \| \leq \tfrac{1}{2} \| x_0 - x_* \|,$$

which shows (5.2.3) and completes the case $k = 0$. The proof of the induction step proceeds identically. $\square$

The constants γ and β can be combined into one constant $\gamma_{Rel} = \gamma \cdot \beta$, a Lipschitz constant measuring the relative nonlinearity of F at x_*, since

$$\| J(x_*)^{-1}[J(x) - J(x_*)] \| \leq \| J(x_*)^{-1} \| \, \| J(x) - J(x_*) \|$$

$$\leq \beta\gamma \| x - x_* \|$$

$$= \gamma_{Rel} \| x - x_* \|$$

for $x \in N(x_*, r)$. Therefore Theorem 5.2.1 can also be viewed as saying that the radius of convergence of Newton's method is inversely proportional to the relative nonlinearity of F at x_*. The relative nonlinearity of a function is the key factor determining its behavior in our algorithms, and all our convergence theorems could easily be restated and proven in terms of this concept. We have chosen to use absolute nonlinearity only because it is standard and therefore perhaps less confusing.

The reader might wonder whether the bounds on the region of convergence of Newton's method given by Theorem 5.2.1 are tight in practice. This question is studied in the exercises, where it is illustrated that the bound ε in Theorem 5.2.1 is a worst-case estimate in that it shows reasonably well how far the region of quadratic convergence extends in the direction from x_* in which F is most nonlinear. On the other hand, in directions from x_* in which F is less nonlinear, the region of convergence of Newton's method may be much greater. This is just another price we pay for using norms in our analysis.

Ortega and Rheinboldt (1970, p. 428) give an excellent history of the convergence analysis for Newton's method, including the Kantorovich analysis, which is introduced in the next section.

5.3 THE KANTOROVICH
AND CONTRACTIVE MAPPING
THEOREMS

In this section we state a second convergence result for Newton's method, a different and powerful result introduced by L. Kantorovich (1948). We also show that its assumptions and method of proof are related to a classical result on the convergence of iterative algorithms called the contractive mapping theorem. While we will not directly apply either of these results in the remainder of this book, they are two beautiful and powerful theorems that a person studying nonlinear problems should be aware of.

The Kantorovich theorem differs from Theorem 5.2.1 mainly in that it makes no assumption about the existence of a solution to $F(x_*) = 0$. Rather, it shows that if $J(x_0)$ is nonsingular, J is Lipschitz continuous in a region containing x_0, and the first step of Newton's method is sufficiently small relative to the nonlinearity of F, then there must be a root in this region, and furthermore it is unique. The price paid for these broader assumptions is the exhibition of only an r-quadratic rate of convergence. The reader will see as the book proceeds that a separate proof is required to get the best rate-of-convergence results for many methods.

THEOREM 5.3.1 KANTOROVICH

Let $r > 0$, $x_0 \in \mathbb{R}^n$, $F: \mathbb{R}^n \longrightarrow \mathbb{R}^n$, and assume that F is continuously differentiable in $N(x_0, r)$. Assume for a vector norm and the induced operator norm that $J \in \text{Lip}_\gamma (N(x_0, r))$ with $J(x_0)$ nonsingular, and that there exist constants $\beta, \eta \geq 0$ such that

$$\| J(x_0)^{-1} \| \leq \beta, \quad \| J(x_0)^{-1} F(x_0) \| \leq \eta.$$

Define $\gamma_{\text{Rel}} = \beta \gamma$, $\alpha = \gamma_{\text{Rel}} \eta$. If $\alpha \leq \frac{1}{2}$ and $r \geq r_0 \equiv (1 - \sqrt{1 - 2\alpha})/(\beta \gamma)$, then the sequence $\{x_k\}$ produced by

$$x_{k+1} = x_k - J(x_k)^{-1} F(x_k), \quad k = 0, 1, \ldots,$$

is well defined and converges to x_*, a unique zero of F in the closure of $N(x_0, r_0)$. If $\alpha < \frac{1}{2}$, then x_* is the unique zero of F in $N(x_0, r_1)$, where $r_1 \equiv \min [r, (1 + \sqrt{1 - 2\alpha})/(\beta \gamma)]$ and

$$\| x_k - x_* \| \leq (2\alpha)^{2^k} \frac{\eta}{\alpha}, \quad k = 0, 1, \ldots. \tag{5.3.1}$$

The proof of Theorem 5.3.1 is quite long and is sketched in Exercises 6 through 8. The basic idea behind it is simple and very clever. Plot the iterates $x_0, x_1, \ldots$ in n-space, and stretch a rope through them in order. Mark the spot on the rope where it meets each x_k, and then pull it taut with perhaps a little stretching. What one has produced is a linear segment of successive upper bounds on the distances $\| x_{k+1} - x_k \|$, $k = 0, 1, \ldots$. The proof of the Kantorovich theorem basically shows that for any sequence $\{x_k\}$ generated under the assumptions of Theorem 5.3.1, these distances are respectively less than or equal to the distances $|t_{k+1} - t_k|$ produced by applying Newton's method to the function

$$f: \mathbb{R} \longrightarrow \mathbb{R}, \qquad f(t) = \frac{\gamma}{2} t^2 - \frac{t}{\beta} + \frac{\eta}{\beta}$$

starting at $t_0 = 0$ (which is seen to exactly obey the assumptions of Theorem 5.3.1). Since the sequence $\{t_k\}$ converges monotonically to the smaller root $t_* = (1 - \sqrt{1 - 2\alpha})/(\beta\gamma)$ of $f(t)$, the total length of the x-rope past x_k must be less than or equal to $|t_k - t_*|$. Thus $\{x_k\}$ is a Cauchy sequence, and so it converges to some x_* with the errors $\| x_k - x_* \|$ bounded by the q-quadratically convergent sequence $|t_k - t_*|$. This technique of proof is called *majorization* and does not use the finite dimensionality of $\mathbb{R}^n$. The sequence $\{t_k\}$ is said to *majorize* the sequence $\{x_k\}$. For further development of the proof, see the exercises or, e.g., Dennis (1971).

The nice thing about the Kantorovich theorem is that it doesn't assume the existence of x_* or the nonsingularity of $J(x_*)$; indeed, there are functions F and points x_0 that meet the assumptions of Theorem 5.3.1 even though $J(x_*)$ is singular. The unfortunate part is that the r-order convergence says nothing about the improvement at each iteration, and the error bounds given by the theorem are often much too loose in practice. This is not important, since Dennis (1971) shows $\frac{1}{2}\|x_{k+1} - x_k\| \leq \|x_k - x_*\| \leq 2\|x_{k+1} - x_k\|$. Also, it is traditional to think of testing the assumptions of Theorem 5.3.1 to see whether Newton's method will converge from some x_0 for a particular function F, but in practice it is virtually always more work to estimate the Lipschitz constant γ accurately than to just try Newton's method and see whether it converges, so this is not a practical application except for special cases.

The form of the Kantorovich theorem is very similar to the form of the contractive mapping theorem. This other classical theorem considers any iterative method of the form $x_{k+1} = G(x_k)$ and states conditions on G under which the sequence $\{x_k\}$ will converge q-linearly to a point x_* from any point x_0 in a region D. Furthermore, x_* is shown to be the unique point in D such that $G(x_*) = x_*$. Thus the contractive mapping theorem is a broader but weaker result than the Kantorovich theorem.

THEOREM 5.3.2   CONTRACTIVE MAPPING THEOREM

Let $G: D \longrightarrow D$, D a closed subset of $\mathbb{R}^n$. If for some norm $\| \cdot \|$, there exists $\alpha \in [0, 1)$ such that

$$\| G(x) - G(y) \| \leq \alpha \| x - y \|, \qquad \forall\ x, y \in D, \tag{5.3.2}$$

> then:
>   (a) there exists a unique $x_* \in D$ such that $G(x_*) = x_*$;
>   (b) for any $x_0 \in D$, the sequence $\{x_k\}$ generated by $x_{k+1} = G(x_k)$, $k = 0, 1, \ldots$, remains in D and converges q-linearly to x_* with constant α;
>   (c) for any $\eta \geq \| G(x_0) - x_0 \|$,
>
> $$\| x_k - x_* \| \leq \frac{\eta \alpha^k}{1 - \alpha}, \qquad k = 0, 1, \ldots.$$

An iteration function G that satisfies (5.3.2) is said to be *contractive* in the region D. Contractivity implies that starting from any point $x_0 \in D$, the step lengths $\| x_{k+1} - x_k \|$ decrease by at least a factor α at each iteration, since

$$\| x_{k+1} - x_k \| = \| G(x_k) - G(x_{k-1}) \| \leq \alpha \| x_k - x_{k-1} \|. \tag{5.3.3}$$

From (5.3.3) it is straightforward to show that $\{x_k\}$ is majorized by a q-linearly convergent sequence, and the proof of Theorem 5.3.2 follows as an exercise. Thus the proofs of the Kantorovich theorem and the contractive mapping theorem are closely related.

The contractive mapping theorem can be applied to problem (5.1.1) by defining an iteration function $G(x)$ such that $G(x) = x$ if and only if $F(x) = 0$. An example is

$$G(x) = x - AF(x), \qquad \text{where } A \in \mathbb{R}^{n \times n} \text{ is nonsingular.} \tag{5.3.4}$$

The theorem can then be used to test whether there is any region D such that the points generated by (5.3.4) from an $x_0 \in D$ will converge to a root of F. However, the contractive mapping theorem can be used to show only linear convergence, and so it is generally inadequate for the methods we consider in this book.

5.4 FINITE-DIFFERENCE DERIVATIVE METHODS FOR SYSTEMS OF NONLINEAR EQUATIONS

In this section we discuss the effect on Newton's method of replacing the analytic Jacobian $J(x)$ with the finite-difference approximation developed in Section 4.2. It is easy to show that if the finite-difference step size is chosen properly, then the q-quadratic convergence of Newton's method is retained. We also discuss the choice of the stepsize in practice, taking into consideration the effects of finite-precision arithmetic. An example illustrates that for most problems, Newton's method using analytic derivatives and Newton's method using properly chosen finite differences are virtually indistinguishable.

THEOREM 5.4.1   Let F and x_* obey the assumptions of Theorem 5.2.1, where $\| \cdot \|$ denotes the l_1 vector norm and the corresponding induced matrix norm. Then there exist $\varepsilon, h > 0$ such that if $\{h_k\}$ is a real sequence with $0 < |h_k| \le h$ and $x_0 \in N(x_*, \varepsilon)$, the sequence $x_1, x_2, \ldots$ generated by

$$(A_k)_{.j} = \frac{F(x_k + h_k e_j) - F(x_k)}{h_k}, \qquad j = 1, \ldots, n, \qquad (5.4.1)$$

$$x_{k+1} := x_k - A_k^{-1} F(x_k), \qquad k = 0, 1, \ldots,$$

is well defined and converges q-linearly to x_*. If

$$\lim_{k \to 0} h_k = 0,$$

then the convergence is q-superlinear. If there exists some constant c_1 such that

$$|h_k| \le c_1 \| x_k - x_* \|, \qquad (5.4.2)$$

or equivalently a constant c_2 such that

$$|h_k| \le c_2 \| F(x_k) \|, \qquad (5.4.3)$$

then the convergence is q-quadratic.

Proof.   The proof is similar to that of Theorem 5.2.1 combined with that of Theorem 2.6.3 (finite-difference Newton's method in one dimension). We will choose ε and h so that A_k is nonsingular for any $x_k \in N(x_*, \varepsilon)$ and $|h_k| < h$. Let $\varepsilon \le r$ and

$$\varepsilon + h \le \frac{1}{2\beta\gamma}. \qquad (5.4.4)$$

(r, β, γ are defined in Theorem 5.2.1.) We first show by induction on k that at each step

$$\| x_{k+1} - x_* \| \le \tfrac{1}{2} \| x_k - x_* \|, \qquad x_{k+1} \in N(x_*, r). \qquad (5.4.5)$$

For $k = 0$, we first show that A_0 is nonsingular. From the triangle inequality, Lemma 4.2.1, the Lipschitz continuity of J at x_*, $\| x_0 - x_* \| \le \varepsilon$, and (5.4.4), we see that

$$\| J(x_*)^{-1}[A_0 - J(x_*)] \| \le \| J(x_*)^{-1} \| \, \| [A_0 - J(x_0)] + [J(x_0) - J(x_*)] \|$$

$$\le \beta\left(\frac{\gamma h}{2} + \gamma\varepsilon\right) \le \tfrac{1}{2}. \qquad (5.4.6)$$

We now proceed as in the proof of Theorem 5.2.1. By the perturbation relation (3.1.20), (5.4.6) implies that A_0 is nonsingular and

$$\| A_0^{-1} \| \le 2\beta. \tag{5.4.7}$$

Therefore x_1 is well defined and

$$x_1 - x_* = A_0^{-1}([F(x_*) - F(x_0) - J(x_0)(x_* - x_0)]$$

$$+ [(J(x_0) - A_0)(x_* - x_0)]).$$

Notice that this formula differs from the corresponding one in the analysis of Newton's method only in that A_0^{-1} has replaced $J(x_0)^{-1}$ and the term $(J(x_0) - A_0)(x_* - x_0)$ has been added. Now using Lemma 4.1.12, Lemma 4.2.1, (5.4.7), $\| x_0 - x_* \| \le \varepsilon$, and (5.4.4), we see that

$$\| x_1 - x_* \| \le \| A_0^{-1} \| [\| F(x_*) - F(x_0) - J(x_0)(x_* - x_0) \|$$

$$+ \| A_0 - J(x_0) \| \| x_* - x_0 \|]$$

$$\le 2\beta \left[\frac{\gamma}{2} \| x_* - x_0 \|^2 + \frac{\gamma h}{2} \| x_0 - x_* \| \right]$$

$$\le \beta\gamma(\varepsilon + h) \| x_0 - x_* \|$$

$$\le \tfrac{1}{2} \| x_0 - x_* \|,$$

which proves (5.4.5). The proof of the induction step is identical.

The proof of q-superlinear or q-quadratic convergence requires the improved bound on $\| A_k - J(x_k) \|$ and is left as an exercise. To show the equivalence of (5.4.2) and (5.4.3), one simply uses Lemma 4.1.16.   □

As in the one-variable case, Theorem 5.4.1 doesn't indicate exactly how finite-difference stepsizes should be chosen in practice. This is because there are again two conflicting sources of error in computing finite-difference derivatives. Consider the finite-difference equation (5.4.1) and for convenience let us omit the iteration subscripts k. From Lemma 4.2.1, the error in $A_{.j}$ as an approximation to $(J(x))_{.j}$ is $O(\gamma h)$ in exact arithmetic, which suggests choosing h as small as possible. However, another error in (5.4.1), caused by evaluating the numerator in finite-precision arithmetic, is made worse by small values of h. Since we divide the numerator by h to get (5.4.1), any error δ in the numerator results in an error in (5.4.1) of δ/h. This error in the numerator, which results from inaccuracies in the function values and cancellation errors in subtracting them, is reasonably assumed to be some small fraction of $F(x)$. In fact if h is too small, $x + he_j$ may be so close to x that $F(x + he_j) \cong F(x)$ and $A_{.j}$ may have zero, or very few, good digits. Thus h must be chosen to balance the $O(\gamma h)$ and $O(F(x)/h)$ errors in (5.4.1).

A reasonable aim is that if $F(x)$ has t reliable digits, then $F(x + he_j)$ should differ from $F(x)$ in the latter half of these digits. More precisely, if the relative error in computing $F(x)$ is η (η stands for relative noise), then we would like

$$\frac{\| F(x + he_j) - F(x) \|}{\| F(x) \|} \leq \sqrt{\eta}, \qquad j = 1, \ldots, n. \qquad (5.4.8)$$

In the absence of any better information, a reasonable way to accomplish (5.4.8) is to perturb each component x_j by its *own* stepsize

$$h_j = \sqrt{\eta} \, x_j, \qquad (5.4.9a)$$

which constitutes a relative change of $\sqrt{\eta}$ in x_j, and then compute $A_{.j}$ by

$$A_{.j} = \frac{F(x + h_j e_j) - F(x)}{h_j}. \qquad (5.4.9b)$$

For convenience, we have assumed to this point that we use one stepsize h for the entire finite-difference approximation, but there is no reason to do this in practice, and a uniform stepsize could be disastrous if the components of x differed widely in magnitude (see Exercise 12). Theorem 5.4.1 is easily revised to remain true when individual component stepsizes h_j are used.

In cases when $F(x)$ is given by a simple formula, it is reasonable that $\eta \cong$ macheps, so that the stepsize h_j is $\sqrt{\text{macheps}} \cdot x_j$. Example 5.4.2 compares the finite-difference Newton's method, using (5.4.9) with $\eta \cong$ macheps, to Newton's method, using analytic derivatives, starting from the same point. The reader will see that the results, including derivatives, are virtually indistinguishable, and this is usually the case in practice. For this reason, some software packages do not even allow for the provision of analytic derivatives; instead they always use finite differences. While we do not advocate this extreme, it should be stressed that perhaps the leading cause of error in using software for any nonlinear problem is incorrect coding of derivatives. Thus, if analytic derivatives are provided, we strongly recommend that they be checked for agreement with a finite-difference approximation at the initial point x_0.

If $F(x)$ is calculated by either a lengthy piece of code or an iterative procedure, η may be much larger than macheps. If η is so large that the finite-difference derivatives (5.4.9) are expected to be inaccurate, then a secant approximation to $J(x)$ should probably be used instead (see Chapter 8). This is also the case if the cost of n additional evaluations of $F(x)$ per iteration for computing A is prohibitive.

An algorithm for calculating finite-difference Jacobians is given in Algorithm A5.4.1 in Appendix A. Because equation (5.4.9a) may give an inappro-

EXAMPLE 5.4.2

$$F(x) \begin{pmatrix} x_1^2 + x_2^2 - 2 \\ e^{x_1-1} + x_2^3 - 2 \end{pmatrix}.$$

$$x_0 = (2, 3)^T, \qquad x_* = (1, 1)^T.$$

Computations on a computer with 48 base-2 digits ($\cong 14.4$ base-10 digits):

Newton's Method, Analytic Derivatives		Newton's Method, Finite Differences with $h_j = 10^{-7}\lvert x_j\rvert$
(2, 3)	x_0	(2, 3)
$\begin{pmatrix} 0.57465515807608 \\ 2.1168965612826 \end{pmatrix}$	x_1	$\begin{pmatrix} 0.57465515450268 \\ 2.1168966735234 \end{pmatrix}$
$\begin{pmatrix} 0.31178766389307 \\ 1.5241979559460 \end{pmatrix}$	x_2	$\begin{pmatrix} 0.31178738552306 \\ 1.5241981016335 \end{pmatrix}$
$\begin{pmatrix} 1.4841388323960 \\ 1.1464779176945 \end{pmatrix}$	x_3	$\begin{pmatrix} 1.4841386151178 \\ 1.1464781318492 \end{pmatrix}$
$\begin{pmatrix} 1.0592959013664 \\ 1.0348194625183 \end{pmatrix}$	x_4	$\begin{pmatrix} 1.0592958450507 \\ 1.0348195092235 \end{pmatrix}$
$\begin{pmatrix} 1.0008031050945 \\ 1.0014625483617 \end{pmatrix}$	x_5	$\begin{pmatrix} 1.0008031056081 \\ 1.0014625533494 \end{pmatrix}$
$\begin{pmatrix} 0.99999872187461 \\ 1.0000026672636 \end{pmatrix}$	x_6	$\begin{pmatrix} 0.99999872173640 \\ 1.0000026674316 \end{pmatrix}$
$\begin{pmatrix} 0.99999999999548 \\ 1.0000000000089 \end{pmatrix}$	x_7	$\begin{pmatrix} 0.99999999999535 \\ 1.0000000000091 \end{pmatrix}$
4	$(J(x_0))_{11}$	4.0000003309615
6	$(J(x_0))_{12}$	6.0000002122252
2.7182818284590	$(J(x_0))_{21}$	2.7182824169358
27	$(J(x_0))_{22}$	27.000002470838

priately small value of h_j if x_j wanders close to zero, it is modified slightly to

$$h_j = \sqrt{\eta} \max \{\lvert x_j\rvert, \text{typ}x_j\} \text{ sign } (x_j), \qquad (5.4.10)$$

where $\text{typ}x_j$ is a typical size of x_j provided by the user (see Section 7.1).

Finally, we will indicate briefly how stepsize (5.4.10) can be justified somewhat by analysis. It is reasonable to assume that x_j is an exact computer number. To simplify the analysis, we assume that $x_j + h_j$ is an exact computer number with $x_j + h_j \neq x_j$. This is the case if h_j is calculated by any stepsize rule that guarantees $|h_j| >$ macheps $|x_j|$, and then the calculation

$$\text{temp} = x_j + h_j,$$

$$h_j = \text{temp} - x_j,$$

is performed before h_j is used in (5.4.9b). This trick improves the accuracy of any finite-difference approximation, so we use it in practice. Now it is easy to see that the error from round-off and function noise in calculating $A_{\cdot j}$ by (5.4.9b) is bounded by

$$\frac{2(\eta + \text{macheps})\hat{F}}{|h_j|}, \qquad (5.4.11)$$

where $\hat{F}$ is an upper bound on $\|F(x)\|$ for $x \in [x, x + h_j e_j]$. [We omit the negligible error from the division in (5.4.9b).] Also from Theorem 4.2.1, the difference in exact arithmetic between $A_{\cdot j}$ and $J(x)_{\cdot j}$ is bounded by

$$\frac{\gamma_j |h_j|}{2}, \qquad (5.4.12)$$

where γ_j is a Lipschitz constant for which

$$\| [J(x + te_j) - J(x)]_{\cdot j} \| \leq \gamma_j |t|$$

for all $t \in [0, h_j]$. The h_j that minimizes the sum of (5.4.11) and (5.4.12) is

$$h_j^* = \left(\frac{4(\eta + \text{macheps})}{\gamma_j} \, \hat{F} \right)^{1/2}. \qquad (5.4.13)$$

Under the somewhat reasonable assumption (see Exercise 14) that

$$\gamma_j = 0 \left(\frac{\hat{F}}{(\max \{|x_j|, \text{typx}_j\})^2} \right),$$

(5.4.13) reduces to (5.4.10), modulo a constant. Note finally that this analysis can be modified to allow also for an absolute error in the computed value of $F(x)$.

5.5 NEWTON'S METHOD FOR UNCONSTRAINED MINIMIZATION

We now discuss Newton's method for

$$\min_{x \in \mathbb{R}^n} f: \mathbb{R}^n \longrightarrow \mathbb{R}, \qquad (5.5.1)$$

where f is assumed twice continuously differentiable. Just as in Chapter 2, we model f at the current point x_c by the quadratic

$$m_c(x_c + p) = f(x_c) + \nabla f(x_c)^T p + \tfrac{1}{2} p^T \nabla^2 f(x_c) p$$

and solve for the point $x_+ = x_c + s_N$, where $\nabla m_c(x_+) = 0$, a necessary condition for x_+ to be the minimizer of m_c. This corresponds to the following algorithm.

ALGORITHM 5.5.1   NEWTON'S METHOD FOR UNCONSTRAINED
MINIMIZATION

Given $f: \mathbb{R}^n \longrightarrow \mathbb{R}$ twice continuously differentiable, $x_0 \in \mathbb{R}^n$; at each iteration k,

$$\text{solve} \quad \nabla^2 f(x_k) s_k^N = -\nabla f(x_k),$$

$$x_{k+1} = x_k + s_k^N.$$

EXAMPLE 5.5.2   Let $f(x_1, x_2) = (x_1 - 2)^4 + (x_1 - 2)^2 x_2^2 + (x_2 + 1)^2$, which has its minimum at $x_* = (2, -1)^T$. Algorithm 5.5.1, started from $x_0 = (1, 1)^T$, produces the following sequence of points (to eight figures on a CDC machine in single precision):

			$f(x_k)$
$x_0 = (1.0$	, 1.0	$)^T$	6.0
$x_1 = (1.0$	, -0.5	$)^T$	1.5
$x_2 = (1.3913043$	, $-0.69565217)^T$		4.09×10^{-1}
$x_3 = (1.7459441$	, $-0.94879809)^T$		6.49×10^{-2}
$x_4 = (1.9862783$	, $-1.0482081)^T$		2.53×10^{-3}
$x_5 = (1.9987342$	, $-1.0001700)^T$		1.63×10^{-6}
$x_6 = (1.9999996$	, $-1.0000016)^T$		2.75×10^{-12}

Algorithm 5.5.1 is simply the application of Newton's method (Algorithm 5.1.1) to the system $\nabla f(x) = 0$ of n nonlinear equations in n unknowns, because it steps at each iteration to the zero of the affine model of ∇f defined by $M_k(x_k + p) = \nabla[m_k(x_k + p)] = \nabla f(x_k) + \nabla^2 f(x_k) p$. From this perspective, some of the algorithm's advantages and disadvantages for the minimization problem (5.5.1) are clearly those given in Table 5.1.2. On the good side, if x_0 is sufficiently close to a local minimizer x_* of f with $\nabla^2 f(x_*)$ nonsingular (and therefore positive definite by Corollary 4.3.3), then the sequence $\{x_k\}$ generated by Algorithm 5.5.1 will converge q-quadratically to x_*, as is evident in Example 5.5.2. Also, if f is a strictly convex quadratic, then ∇f is affine and so x_1 will be the unique minimizer of f.

On the bad side, Algorithm 5.5.1 shares the disadvantages of Newton's method for solving $F(x) = 0$, and it has one new problem as well. The old difficulties are that the method isn't globally convergent, requires the solutions of systems of linear equations, and requires analytic derivatives, in this case ∇f and $\nabla^2 f$. These issues are discussed subsequently: finite-difference approximation of derivatives for minimization in Section 5.6, global methods in Chapter 6, and secant approximations of the Hessian in Chapter 9.

We must also consider that even as a local method, Newton's method is not specifically geared to the minimization problem; there is nothing in Algorithm 5.5.1 that makes it any less likely to proceed toward a maximizer or saddle point of f, where ∇f is also zero. From the modeling point of view, the difficulty with Algorithm 5.5.1 is that each step simply goes to the critical point of the current local quadratic model, whether the critical point is a minimizer, maximizer, or saddle point of the model. This is only consistent with trying to minimize $f(x)$ if $\nabla^2 f(x_c)$ is positive definite so that the critical point is a minimizer. Even though Newton's method is intended primarily as a local method to be used when x_c is close enough to a minimizer that $\nabla^2 f(x_c)$ is positive definite, we would still like to make some useful adaptation when $\nabla^2 f(x_c)$ is not positive definite.

There seem to be two reasonable modifications to Newton's method when $\nabla^2 f(x_c)$ is not positive definite. The first is to try to use the shape of the model, in particular the "directions of negative curvature" p for which $p^T \nabla^2 f(x_c) p < 0$, to decrease $f(x)$ rapidly. A lot of effort has gone into such strategies [see, e.g., Moré and Sorensen (1979)], but they have not yet proven themselves. We recommend the second alternative, which is to change the model when necessary so that it has a unique minimizer and to use this minimizer to define the Newton step. Specifically, we will see in Chapter 6 that there are several good reasons for taking steps of the form $-(\nabla^2 f(x_c) + \mu_c I)^{-1} \nabla f(x_c)$, where $\mu_c \geq 0$ and $\nabla^2 f(x_c) + \mu_c I$ is positive definite. Therefore, when $\nabla^2 f(x_c)$ is not positive definite, we recommend changing the model Hessian to $\nabla^2 f(x_c) + \mu_c I$, where $\mu_c > 0$ is not much larger, ideally, than the smallest μ that will make $\nabla^2 f(x_c) + \mu I$ positive definite and reasonably well conditioned. Our algorithm for doing this is discussed at the end of this section. It is a modification of the modified Cholesky factorization given in Gill, Murray, and Wright (1981, p. 111). The result is the following modified version of Newton's method:

ALGORITHM 5.5.3   MODIFIED NEWTON'S METHOD FOR
UNCONSTRAINED MINIMIZATION

Given $f: \mathbb{R}^n \longrightarrow \mathbb{R}$ twice continuously differentiable, $x_0 \in \mathbb{R}^n$, at each iteration k:

Apply Algorithm A5.5.1 to $\nabla^2 f(x_k)$ to find the Cholesky decomposition of

$$H_k = \nabla^2 f(x_k) + \mu_k I, \quad \text{where} \quad \mu_k = 0 \text{ if } \nabla^2 f(x_k) \text{ is safely positive definite, and}$$

$\mu_k > 0$ is sufficiently large that H_k is safely positive definite otherwise:

Solve

$$H_k s_k^N = -\nabla f(x_k),$$

$$x_{k+1} = x_k + s_k^N.$$

If x_0 is sufficiently close to a minimizer x_* of f and if $\nabla^2 f(x_*)$ is positive definite and reasonably well conditioned, then Algorithm 5.5.3 will be equivalent to Algorithm 5.5.1 and hence q-quadratically convergent. A reexamination of the proof of Lemma 4.3.1 will show that Algorithm 5.5.3 has the additional advantage that the modified Newton direction $-H_c^{-1} \nabla f(x_c)$ is one in which any twice-differentiable $f(x)$ initially decreases from any x_c.

Algorithm 5.5.3 also has two important interpretations in terms of quadratic models of $f(x_c)$. The first is that it sets x_+ equal to the minimizer $x_c + p$ of the approximation

$$m_c(x_c + p) = f(x_c) + \nabla f(x_c)^T p + \tfrac{1}{2} p^T H_c p.$$

We will see in Section 6.4 that x_+ is also the minimizer of the unmodified model,

$$m_c(x_c + p) = f(x_c) + \nabla f(x_c)^T p + \tfrac{1}{2} p^T \nabla^2 f(x_c) p,$$

subject to a constraint

$$\| p \|_2 \le \| (x_+ - x_c) \|_2,$$

which is perhaps a more appealing explanation. Both interpretations can be used to explain the main problem with the modified Newton algorithm: it can produce overly large steps s_c^N in some situations. We will deal quite satisfactorily with this problem in Chapter 6. The length of the step can be adjusted by tinkering with the parameters in Algorithm A5.5.1, but any such adjustment is too dependent on the scale of the problem to be included in a general-purpose algorithm.

We conclude this section with a brief description of our algorithm (A5.5.1 in the appendix) for determining both a $\mu_c \ge 0$, such that $H_c = \nabla^2 f(x_c) + \mu_c I$ is positive definite and well conditioned, and the Cholesky factorization of H_c. Although this algorithm is central to our minimization software, it is not crucial that the reader study it in detail.

Clearly the smallest possible μ_c (when $\nabla^2 f(x_c)$ is not positive definite) is slightly larger than the magnitude of the most negative eigenvalue of $\nabla^2 f(x_c)$. Although this can be computed without too much trouble, we have decided to provide a much simpler algorithm that may result in a larger μ_c. We first apply the Gill and Murray modified Cholesky factorization algorithm to $\nabla^2 f(x_c)$, which results in $\nabla^2 f(x_c) + D = LL^T$, D a diagonal matrix with nonnegative diagonal elements that are zero if $\nabla^2 f(x_c)$ is safely positive definite. If $D = 0$, $\mu_c = 0$ and we are done. If $D \neq 0$, we calculate an upper bound b_1 on μ_c using the Gerschgorin circle theorem (Theorem 3.5.9). Since

$$b_2 = \max_{1 \leq i \leq n} \{d_{ii}\}$$

is also an upper bound on μ_c, we set $\mu_c = \min \{b_1, b_2\}$ and conclude the algorithm by calculating the Cholesky factorization of $H_c = \nabla^2 f(x_c) + \mu_c I$.

5.6 FINITE-DIFFERENCE DERIVATIVE METHODS FOR UNCONSTRAINED MINIMIZATION

In this section we consider the effect on Newton's method for unconstrained minimization of substituting finite-difference approximations for the analytic derivatives $\nabla f(x)$ and $\nabla^2 f(x)$, and also the practical choice of stepsize in computing these approximations. We will see that there are several important differences in theory and in practice from the use of finite-difference Jacobians studied in Section 5.4.

We first discuss the use of a finite-difference approximation to the Hessian. In Section 4.2 we saw two ways to approximate $\nabla^2 f(x_c)$ by finite differences, one when $\nabla f(x)$ is analytically available and the other when it isn't. Both approximations satisfy

$$\| A_c - \nabla^2 f(x_c) \| = O(h_c)$$

(h_c the finite-difference stepsize) under the standard assumptions and so, using the same techniques as in the proof of Theorem 5.4.1, it is easy to show that if $h_c = O(\| e_c \|)$, where $e_c = x_c - x_*$, then Newton's method using finite-difference Hessians is still q-quadratically convergent. The issue that remains is the choice of the finite-difference stepsize in practice.

If $\nabla f(x)$ is analytically available, A_c is calculated by

$$A_{.j} = \frac{\nabla f(x_c + h_j e_j) - \nabla f(x_c)}{h_j}, \qquad j = 1, \ldots, n, \qquad (5.6.1a)$$

$$A_c = \frac{A + A^T}{2}. \qquad (5.6.1b)$$

Since (5.6.1a) is the same formula as for finite-difference Jacobians, the same stepsize

$$h_j = \sqrt{\eta} \max \{|x_j|, \text{typx}_j\} \text{ sign } (x_j) \qquad (5.6.2)$$

is recommended, where $\eta \geq$ macheps is the estimated relative error in calculating $\nabla f(x)$. Algorithm A5.6.1 is used for this approximation. Recall that we expected to get about the first half of the digits of $\nabla^2 f(x_c)$ correct if $\nabla f(x)$ is calculated accurately; see Example 5.6.1 below.

When $\nabla f(x)$ is not available analytically, recall that A_c is calculated by

$$(A_c)_{ij} = \frac{[f(x_c + h_i e_i + h_j e_j) - f(x_c + h_i e_i)] - [f(x_c + h_j e_j) - f(x_c)]}{h_i h_j},$$

$$1 \leq i \leq j \leq n. \tag{5.6.3}$$

Example 5.6.1 shows that the stepsize (5.6.2) can be disastrous in this formula.

EXAMPLE 5.6.1   Let $f(x_1, x_2)$ be given by Example 5.5.2 and let $x_c = (1, 1)^T$. Then

$$\nabla^2 f(x_c) = \begin{bmatrix} 14 & -4 \\ -4 & 4 \end{bmatrix}.$$

If $\nabla^2 f(x_c)$ is approximated by equation (5.6.1) using stepsize $h_j = 10^{-7} x_j$ on a CDC computer with 48 base-2 digits ($\cong 14.4$ base-10 digits), the result is

$$A_c = \begin{bmatrix} 13.999999737280 & -4.0000001888529 \\ -4.000001888529 & 4.0000003309615 \end{bmatrix}.$$

However if $\nabla^2 f(x_c)$ is approximated by equation (5.6.3) using the same stepsize, the result is

$$A_c = \begin{bmatrix} 19.895196601283 & -5.6843418860808 \\ -5.6843418860808 & 0 \end{bmatrix}.$$

If the stepsize in (5.6.3) is changed to $h_j = 10^{-4} x_j$, the result is

$$A_c = \begin{bmatrix} 13.997603787175 & -4.0000003309615 \\ -4.0000003309611 & 3.9999974887906 \end{bmatrix}.$$

The incorrect results of the second portion of Example 5.6.1 are easy to explain. If, as in the example, $h_i = \sqrt{\text{macheps}}\; x_i$ and $h_j = \sqrt{\text{macheps}}\; x_j$, then the denominator of (5.6.3) equals (macheps $\cdot x_i x_j$). Since the numerator of (5.6.3) will probably be calculated with a finite-precision error of at least macheps $\cdot |f(x_c)|$, $(A_c)_{ij}$ will have a finite-precision error of at least $|f(x_c)|/|x_i \cdot x_j|$, which need not be small in comparison to $\nabla^2 f(x_c)_{ij}$. We have often seen equation (5.6.3) produce entirely incorrect results when used with stepsize (5.6.2); zero is a fairly common result in our experience. [If we use $h_j = 10^{-8} x_j$ in our example, we get

$$A_c = \begin{pmatrix} 85.2+ & 0 \\ 0 & 85.2+ \end{pmatrix}!]$$

The same reasoning and analysis as in Section 5.4 indicate that one should choose instead the finite-difference stepsize in equation (5.6.3) to perturb the right-hand two-thirds of the good digits of $f(x_c)$. This can usually be done by choosing

$$h_j = \eta^{1/3} \max \{|x_j|, \text{typ} x_j\} \text{ sign } (x_j), \tag{5.6.4}$$

or simply $h_j = (\text{macheps})^{1/3} x_j$ in many cases, where $\eta \geq$ macheps is again the relative error in calculating $f(x)$. This stepsize works quite well in practice, as shown in the last portion of Example 5.6.1, where we use $h_j = 10^{-4} x_j$ and get at least four good digits of each component of $\nabla^2 f(x_c)$. Algorithm A5.6.2 in Appendix A is our algorithm for computing finite-difference Hessians from function values, using equation (5.6.3) with stepsize (5.6.4). Notice that it can only reasonably be expected to get the first third of the digits of $\nabla^2 f(x)$ correct, with even fewer digits correct if $f(x)$ is a noisy function.

In summary, when the analytic Hessian is not readily available, approximating it by either finite-difference Algorithm A5.6.1 using analytic values of $\nabla f(x)$ or Algorithm A5.6.2 using only function values will usually not change the behavior of Newton's method, although Algorithm A5.6.1 is definitely preferred if $\nabla f(x)$ is analytically available. If $f(x)$ is quite noisy, Algorithm A5.6.2 may be unsatisfactory and a secant approximation to $\nabla^2 f(x)$ should probably be used instead (see Chapter 9). A secant approximation to the Hessian should certainly be used if the cost of the finite-difference approximation, n evaluations of $\nabla f(x)$ or $(n^2 + 3n)/2$ evaluations of $f(x)$ per iteration, is too expensive.

We now turn to the finite-difference approximation of the gradient in minimization algorithms. The reader can guess that this will be a stickier issue than the approximation of the Jacobian or Hessian, because the gradient is the quantity we are trying to make zero in our minimization algorithm. Therefore, a finite-difference approximation to the gradient will have to be particularly accurate.

Recall that we have two formulas for approximating $\nabla f(x_c)$ by finite differences: the forward difference formula

$$(g_c)_j = \frac{f(x_c + h_j e_j) - f(x_c)}{h_j}, \qquad j = 1, \ldots, n, \tag{5.6.5}$$

which is just a special case of Jacobian approximation and has error $O(h_j)$ in the jth component, and the central difference formula

$$(\hat{g}_c)_j = \frac{f(x_c + h_j e_j) - f(x_c - h_j e_j)}{2h_j}, \qquad j = 1, \ldots, n \tag{5.6.6}$$

which has error $O(h_j^2)$. Remember also that in the case of finite-difference Jacobians or Hessians, we have seen that if the stepsizes obey $h_j = O(\|e_c\|)$, where $e_c = x_c - x_*$, then under standard assumptions the q-quadratic convergence of Newton's method is retained. However, this cannot possibly be the case for the forward difference gradient approximation (5.6.5), because as x_c approaches a minimizer of $f(x)$, $\| \nabla f(x_c) \| \cong O(\| e_c \|)$ and so a stepsize of $O(\| e_c \|)$ in (5.6.5) would result in an approximation error $\| g_c - \nabla f(x_c) \|$ of the same order of magnitude as the quantity being approximated. Instead, the analysis of Theorem 5.4.1 can be expanded easily to show that if the Newton iteration for minimization is amended to

$$x_+ = x_c - A_c^{-1} g_c,$$

where g_c and A_c approximate $\nabla f(x_c)$ and $\nabla^2 f(x_c)$, respectively, then, under the analogous assumptions,

$$\| x_+ - x_* \| \leq \| A_c^{-1} \| \left[\| g_c - \nabla f(x_c) \| + \| A_c - \nabla^2 f(x_c) \| \, \| e_c \| + \frac{\gamma \| e_c^2 \|}{2} \right]$$

$$(5.6.7)$$

(see Exercise 17). Therefore, q-quadratic convergence requires $\| g_c - \nabla f(x_c) \| = O(\| e_c \|^2)$, which means $h_j = O(\| e_c \|^2)$ using forward differences or $h_j = O(\| e_c \|)$ using central differences. Furthermore, it is hard to estimate $\| e_c \|$, since $\nabla f(x_c)$ isn't analytically available.

In practice this analysis doesn't really apply, since we will just try to select h_j to minimize the combined effects of the errors due to the nonlinearity of f and to finite precision. For the forward difference formula (5.6.5) this just means using the same stepsize (5.6.2) as for a Jacobian approximation. For the central difference formula (5.6.6), the same analysis as in Section 5.4 shows that $h_j = O(\eta^{1/3})$ is optimal, and we suggest the same stepsize (5.6.4) as is used in approximating the Hessian with function values. In our experience, forward difference approximation of the gradient is usually quite sufficient, and since central differences are twice as expensive, the algorithm we recommend for gradient approximation is Algorithm A5.6.3, the Jacobian Algorithm A5.4.1 with m (the number of equations) $= 1$. One should realize, however, that the accuracy in the solution x_* will be limited by the accuracy in the gradient approximation. For this reason, and to cover the unusual case when forward differences are too inaccurate, a central difference gradient routine is provided in Algorithm A5.6.4 in the appendix. Some production codes decide automatically whether and when to switch from forward difference to central difference gradient approximation. [See, e.g., Stewart (1967).]

In conclusion, a user should be aware of the possible difficulties in using finite-difference gradients in minimization algorithms. Boggs and Dennis (1976) analyze the error inherent in using such algorithms to find approximate minimizers. Although their bounds are too pessimistic in practice, users are encouraged strongly to supply an analytic gradient whenever possible. A user who obtains bad results with forward difference gradients or requires increased accuracy might want to try central differences instead. Owing to the accuracy needed in the gradient approximation, we will see that there is no useful secant approximation to the gradient in minimization algorithms.

5.7 EXERCISES

1. The system of equations

$$2(x_1 + x_2)^2 + (x_1 - x_2)^2 - 8 = 0,$$

$$5x_1^2 + (x_2 - 3)^2 - 9 = 0,$$

has a solution $x_* = (1, 1)^T$. Carry out one iteration of Newton's method from $x_0 = (2, 0)^T$.

2. Carry out the second iteration of Newton's method on the system

$$\begin{bmatrix} e^{x_1} - 1 \\ e^{x_2} - 1 \end{bmatrix} = 0$$

starting from $x_0 = (-10, -10)^T$. (The first iteration is done in Section 5.1.) What will happen if Newton's method is continued on this problem, using finite-precision arithmetic?

Exercises 3 and 4 are meant to give insight into the region of convergence given by the convergence theorems for Newton's method.

3. For each of the functions $f(x) = x, f(x) = x^2 + x, f(x) = e^x - 1$, answer the following questions:
   (a) What is $f'(x)$ at the root $x_* = 0$?
   (b) What is a Lipschitz constant for $f'(x)$ in the interval $[-a, a]$; i.e., what is a bound on $|(f'(x) - f'(0))/x|$ in this interval?
   (c) What region of convergence of Newton's method on $f(x)$ is predicted by Theorem 2.4.3?
   (d) In what interval $[b, c]$, $b < 0 < c$, is Newton's method on $f(x)$ actually convergent to $x_* = 0$?

4. Using your answers to Exercise 3, consider applying Newton's method to

$$F(x) = \begin{bmatrix} x_1 \\ x_2^2 + x_2 \\ e^{x_3} - 1 \end{bmatrix}$$

   (a) What is $J(x)$ at the root $x_* = (0, 0, 0)^T$?
   (b) What is a Lipschitz constant on $J(x)$ in an interval of radius a around x_*?
   (c) What region of convergence for Newton's method on $F(x)$ is predicted by Theorem 5.2.1?
   (d) What would the region of convergence be if $(x_0)_3 = 0$? If $(x_0)_2 = (x_0)_3 = 0$?

5. Show that if the assumption $J \in \text{Lip}_\gamma(N(x_*, r))$ in Theorem 5.2.1 is replaced by the Hölder continuity assumption of Exercise 4.8, then Theorem 5.2.1 remains true if (5.2.1) is changed to

$$\| x_{k+1} - x_* \| \le \beta\gamma \| x_k - x_* \|^{1+\alpha}, \qquad k = 0, 1, \dots..$$

   [*Hint*: Change (5.2.2) to $\varepsilon = \min \{r, (\alpha/((1 + \alpha)\beta\gamma))^{1/\alpha}\}$, which implies that $\beta\gamma\varepsilon^\alpha \le \alpha/(1 + \alpha)$.]

Exercises 6 through 8 develop the proof of the Kantorovich theorem.

6. Prove that the conditions and conclusions of the Kantorovich theorem are satisfied for the quadratic

$$f(t) = \frac{\gamma}{2} t^2 - \frac{t}{\beta} + \frac{\eta}{\beta}, \qquad t_0 = 0.$$

[*Hint*: Prove (5.3.1) by proving the stronger inequality

$$|t_k - t_*| \le 2^{-k}(1 - \sqrt{1 - 2\alpha})^{2k} \frac{\eta}{\alpha}, \qquad k = 0, 1, \ldots.$$

To prove this, show inductively that

$$|f'(t_{k+1})| \ge \left| \frac{f'(t_k)}{2} \right|$$

and, using this, that

$$|t_{k+1} - t_*| \le 2^{k-1}\beta\gamma |t_k - t_*|^2.]$$

7. Let $F: \mathbb{R}^n \longrightarrow \mathbb{R}^n$, $x_0 \in \mathbb{R}^n$, satisfy the assumptions of the Kantorovich theorem and let $f(t)$, t_0 be defined as in Exercise 6. Let $\{x_k\}$, $\{t_k\}$ be the sequences generated by applying Newton's method to $F(x)$ and $f(t)$ from x_0 and t_0, respectively. Prove by induction that

$$\| J(x_k)^{-1} \| \le |f'(t_k)^{-1}|,$$

$$\| x_{k+1} - x_k \| \le |t_{k+1} - t_k|,$$

$k = 0, 1, \ldots.$ [*Hint*: At the induction step, use (3.1.20) to help prove the first relation. To prove the second, first use the equation

$$F(x_k) = F(x_k) - F(x_{k-1}) - J(x_{k-1})(x_k - x_{k-1})$$

to show that

$$\| F(x_k) \| \le \frac{\gamma}{2} \| x_k - x_{k-1} \|^2.$$

Similarly, show that

$$|f(t_k)| = \frac{\gamma}{2} (t_k - t_{k-1})^2.$$

Then use these two equations and the first induction relation to prove the second.]

8. Use Exercises 6 and 7 to prove the Kantorovich theorem.

9. Read and enjoy Bryan (1968) for a beautiful application of the Kantorovich theory.

10. Prove the contractive mapping theorem by showing that, under the assumptions of Theorem 5.3.2:
    (a) for any $j \ge 0$,

$$\sum_{i=0}^{j} \| x_{i+1} - x_i \| \le \frac{\| x_1 - x_0 \|}{1 - \alpha}.$$

[*Hint*: use equations (5.3.3).]

(b) using (a), that the sequence $\{x_i\}$ has a limit x_*, because $\{x_i\}$ is a Cauchy sequence, and that $G(x_*) = x_*$.

(c) using (b) and equation (5.3.3), that x_* is the unique fixed point of G in D and that the q-linear convergence and error bounds given by Theorem 5.3.2 hold.

(d) Use (a) to prove another version of the contractive mapping theorem without the assumption that $F(D) \subset D$.

11. Let $f(x) = x^2 - 1$, $a \in \mathbb{R}$, $a > 1$. Prove that for some $\delta > 0$ (δ dependent on a), the sequence of points $\{x_i\}$ generated from any $x_0 \in (1 - \delta, 1 + \delta)$ by

$$x_{i+1} = x_i - \frac{f(x_i)}{a}$$

converges q-linearly to $x_* = 1$, but that

$$\lim_{i \to \infty} \left| \frac{x_{i+1} - x_*}{x_i - x_*} \right| > 0 \qquad \text{unless } a = 2.$$

12. Complete the proof of Theorem 5.4.1.

13. Let $F: \mathbb{R}^2 \longrightarrow \mathbb{R}^2$,

$$F(x) = \begin{pmatrix} x_1^2 \\ x_2^2 \end{pmatrix},$$

let $x_k = (10^7, 10^{-7})^T$, and suppose we are using a computer with 14 base-10 digits. What will happen if we compute an approximation to $J(x_*)$ by equation (5.4.1) with $h_k = 1$? with $h_k = 10^{-14}$? Is there any choice of h_k that is good for approximating both $J(x_k)_{11}$ and $J(x_k)_{22}$ in this case?

14. Let $p \in \mathbb{R}$, $p \neq 0$ or $1, f_1(x) = x^p, f_2(x) = e^{px}$. Letting $\hat{F} = f(x)$ and $\gamma = f''(x)$, what is the optimal step size given by (5.4.13) for $f_1(x)$? for $f_2(x)$? How do these compare with step size (5.4.10)? What special problems occur in the finite-difference approximation of $f'_2(x)$ when x is a very large positive number?

15. Let $f(x_1, x_2) = \frac{1}{2}(x_1^2 - x_2)^2 + \frac{1}{2}(1 - x_1)^2$. What is the minimizer of $f(x)$? Compute one iteration of Newton's method for minimizing $f(x)$ starting from $x_0 = (2, 2)^T$. Is this a good step? Before you decide, compute $f(x_0)$ and $f(x_1)$.

16. Consider applying Newton's method to find a minimizer of $f(x) = \sin x$ from $x_0 \in [-\pi, \pi]$. The desired answer is $x_* = -\pi/2$. Let ε be a small positive number. Show that if $x_0 = -\varepsilon$, Newton's method gives $x_1 \cong -1/\varepsilon$. Similarly, what happens if $x_0 = +\varepsilon$, but $f''(x_0)$ is modified to make it a small positive number?

17. Let $f_1(x) = -x_1^2 - x_2^2$, $f_2(x) = x_1^2 - x_2^2$. Show that Algorithm 5.5.3 will not converge to the maximizer $x_* = (0, 0)^T$ of $f_1(x)$ if $x_0 \neq x_*$. Also show that Algorithm 5.5.3 will not converge to the saddle point $x_* = (0, 0)^T$ of $f_2(x)$ unless $(x_0)_2 = 0$.

18. On a computer of your choice, compute the finite-difference approximation to the Hessian of $f(x)$ given in Exercise 15 at $x_0 = (2, 2)^T$, using equation (5.6.3) with $h_j = (\text{macheps})^{1/2} x_j$, and again with $h_j = (\text{macheps})^{1/3} x_j$. Compare your results with the analytic value of $\nabla^2 f(x_0)$.

19. Let $f: \mathbb{R}^n \longrightarrow \mathbb{R}$ obey $\nabla^2 f(x) \in \text{Lip}_\gamma D$ for some $D \subset \mathbb{R}^n$ and assume there exists $x_* \in D$ for which $\nabla f(x_*) = 0$. Show that if $x_c \in D$, $g_c \in \mathbb{R}^n$, and $A_c \in \mathbb{R}^{n \times n}$ is nonsingular, then the x_+ generated by $x_+ := x_c - A_c^{-1} g_c$ obeys (5.6.7).

20. Show how to efficiently combine the central difference algorithm for computing $\nabla f(x)$ (Algorithm A5.6.4) and the forward difference algorithm for computing $\nabla^2 f(x)$ (Algorithm A5.6.2) into one routine. How does the number of function evaluations required compare with: (a) using the two routines separately, and (b) using forward difference algorithms A5.6.3 for the gradient and A5.6.2 for the Hessian? What disadvantages would there be in using your new routine in practice?

6

Globally Convergent Modifications of Newton's Method

In the last chapter, Newton's method was shown to be locally q-quadratically convergent. This means that when the current solution approximation is good enough, it will be improved rapidly and with relative ease. Unfortunately, it is not unusual to expend significant computational effort in getting close enough. In addition, the strategies for getting close constitute the major part of the program and the programming effort, and they can be sensitive to small differences in implementation.

This chapter will be devoted to the two major ideas for proceeding when the Newton step is unsatisfactory. Section 6.1 will set the framework for the class of algorithms we want to consider, and Section 6.2 will reintroduce the concept of a descent direction that we saw briefly in the proof of Lemma 4.3.1. Section 6.3 discusses the first major global approach, modern versions of the traditional idea of backtracking along the Newton direction if a full Newton step is unsatisfactory. Section 6.4 discusses the second major approach, based on estimating the region in which the local model, underlying Newton's method, can be trusted to adequately represent the function, and taking a step to approximately minimize the model in this region. Both approaches are derived initially for the unconstrained minimization problem; their application to solving systems of nonlinear equations is the topic of Section 6.5.

---

* For our definition of "globally convergent," see the last paragraph of Section 1.1.

111

6.1   THE QUASI-NEWTON FRAMEWORK

The basic idea in forming a successful nonlinear algorithm is to combine a globally convergent strategy with a fast local strategy in a way that derives the benefits of both. The framework for doing this, a slight expansion of the hybrid algorithm discussed in Chapter 2 (Algorithm 2.5.1), is outlined in Algorithm 6.1.1 below. The most important point is to try Newton's method, or some modification of it, first at each iteration. If it seems to be taking a reasonable step—for example, if f decreases in a minimization application—use it. If not, fall back on a step dictated by a global method. Such a strategy will always end up using Newton's method close to the solution and thus retain its fast local convergence rate. If the global method is chosen and incorporated properly, the algorithm will also be globally convergent. We will call an algorithm that takes this approach *quasi-Newton*.

ALGORITHM 6.1.1   QUASI-NEWTON   ALGORITHM   FOR   NONLINEAR EQUATIONS OR UNCONSTRAINED OPTIMIZATION [for optimization, replace $F(x_k)$ by $\nabla f(x_k)$ and J_k by H_k]

Given $F: \mathbb{R}^n \longrightarrow \mathbb{R}^n$ continuously differentiable, and $x_0 \in \mathbb{R}^n$. At each iteration k:

1. Compute $F(x_k)$, if it is not already done, and decide whether to stop or continue.
2. Compute J_k to be $J(x_k)$, or an approximation to it.
3. Apply a factorization technique to J_k and estimate its condition number. If J_k is ill-conditioned, perturb it in an appropriate manner.
4. Solve $J_k s_k^N = -F(x_k)$.
5. Decide whether to take a Newton step, $x_{k+1} = x_k + s_k^N$, or to choose x_{k+1} by a global strategy. This step often furnishes $F(x_k)$ to step 1.

At this point we have covered steps 2, 3, and 4; step 5 is the topic of this chapter. At the end of this chapter, then, the reader could construct a complete quasi-Newton algorithm, with the exception of the stopping criteria, which are covered in Chapter 7. The secant updates of Chapters 8 and 9 are alternate versions of step 2.

We remind the reader that a modular system of algorithms for nonlinear equations and unconstrained minimization is provided in Appendix A, supplying a variety of completely detailed quasi-Newton algorithms for the two problems. Algorithm 6.1.1 is the basis of the driver for these algorithms, which in the appendix is Algorithm D6.1.1 for minimization and Algorithm D6.1.3 for nonlinear equations. The main difference between Algorithm 6.1.1 above and the two drivers in the appendix is that the actual drivers contain a separate initialization section preceeding the iteration section.

The purpose, organization, and use of the modular system of algorithms is discussed at the beginning of Appendix A. We recommend reading this discussion now if you have not previously done so, and then referring back to the appendix for details as the algorithms arise in the text.

6.2 DESCENT DIRECTIONS

We start our discussion of global methods by considering the unconstrained minimization problem

$$\min_{x \in \mathbb{R}^n} f: \mathbb{R}^n \longrightarrow \mathbb{R}, \tag{6.2.1}$$

because there is a natural global strategy for minimization problems; it is to make sure that each step decreases the value of f. There are corresponding strategies for systems of nonlinear equations, in particular, making sure each step decreases the value of some norm of $F: \mathbb{R}^n \longrightarrow \mathbb{R}^n$. However, by using a norm, one is really transforming the problem of solving the system of equations $F(x) = 0$ into the problem of minimizing a function, such as

$$f \triangleq \tfrac{1}{2} \| F \|_2^2 : \mathbb{R}^n \longrightarrow \mathbb{R}. \tag{6.2.2}$$

Therefore, for this section and the next two we will discuss global strategies for unconstrained minimization. In Section 6.5 we will apply these strategies to systems of nonlinear equations, using the transformation (6.2.2).

The basic idea of a global method for unconstrained minimization is geometrically obvious: take steps that lead "downhill" for the function f. More precisely, one chooses a direction p from the current point x_c in which f decreases initially, and a new point x_+ in this direction from x_c such that $f(x_+) < f(x_c)$. Such a direction is called a *descent direction*. Mathematically, p is a descent direction from x_c if the directional derivative of f at x_c in the direction p is negative—that is, from Section 4.1, if

$$\nabla f(x_c)^T p < 0. \tag{6.2.3}$$

If (6.2.3) holds, then it is guaranteed that for sufficiently small positive δ, $f(x_c + \delta p) < f(x_c)$. Descent directions form the basis of some global methods for minimization and are important to all of them, and therefore they are the topic of this section.

So far, the only direction we have considered for minimization is the Newton direction $s^N = -H_c^{-1} \nabla f(x_c)$, where H_c is either $\nabla^2 f(x_c)$ or an approximation to it. Therefore, it is natural to ask whether the Newton direction is a descent direction. By (6.2.3), it is if and only if

$$\nabla f(x_c)^T s^N = -\nabla f(x_c)^T H_c^{-1} \nabla f(x_c) < 0,$$

which is true if H_c^{-1} or, equivalently, H_c is positive definite. This is the second reason why, in Algorithm 5.4.2, we coerce H_c to be positive definite if it isn't already. It guarantees not only a quadratic model with a unique minimizer but also that the resultant Newton direction will be a descent direction for the actual problem. This in turn guarantees that for sufficiently small steps in the Newton direction, the function value will be decreased. In all our methods for unconstrained minimization, the model Hessian H_c will be formed or coerced to make it positive definite.

A second natural question is: as long as one is taking steps in descent directions, what is the direction p in which f decreases most rapidly from x? The notion of direction needs to be made more explicit before we can answer this question, since, for a given perturbation vector p, the directional derivative of f in the direction p is directly proportional to the length of p. A reasonable way to proceed is to choose a norm $\| \cdot \|$, and define the direction of any p as $p/\| p \|$. When we speak of a direction p, we will assume that this normalization has been carried out. We can now pose our question about a most rapid descent direction for a given norm as

$$\min_{p \in \mathbb{R}^n} \nabla f(x)^T p \qquad \text{subject to } \| p \| = 1,$$

which, in the l_2 norm, has solution $p = -\nabla f(x)/\| \nabla f(x) \|_2$. Thus the negative of the gradient direction is the steepest downhill direction from x in the l_2 norm, and it is referred to as the *steepest-descent direction*.

A classic minimization algorithm due to Cauchy is based solely on the steepest-descent direction. Its theoretical form is given below.

ALGORITHM 6.2.1   METHOD OF STEEPEST DESCENT

Given $f: \mathbb{R}^n \longrightarrow \mathbb{R}$ continuously differentiable, $x_0 \in \mathbb{R}^n$. At each iteration k:

find the lowest point of f in the direction $-\nabla f(x_k)$ from x_k—i.e., find λ_k that solves

$$\min_{\lambda_k > 0} f(x_k - \lambda_k \nabla f(x_k))$$

$$x_{k+1} := x_k - \lambda_k \nabla f(x_k)$$

This is not a computational method, because each step contains an exact one-dimensional minimization problem, but it can be implemented by doing an inexact minimization. In either case, under mild conditions it can be shown to converge to a local minimizer or saddle point of $f(x)$. [See, e.g., Goldstein (1967).] However, the convergence is only linear, and sometimes very slowly linear. Specifically, if Algorithm 6.2.1 converges to a local minimizer x_* where $\nabla^2 f(x_*)$ is positive definite, and $\text{ev}_{\max}$ and $\text{ev}_{\min}$ are the largest and smallest eigenvalues of $\nabla^2 f(x_*)$, then one can show that $\{x_k\}$ satisfies

$$\limsup_{k \to \infty} \frac{\|x_{k+1} - x_*\|}{\|x_k - x_*\|} \le c, \qquad c \triangleq \left(\frac{\text{ev}_{\max} - \text{ev}_{\min}}{\text{ev}_{\max} + \text{ev}_{\min}}\right) \qquad (6.2.4)$$

in a particular weighted l_2 norm, and that the bound on c is tight for some starting x_0. If $\nabla^2 f(x_*)$ is nicely scaled with $\text{ev}_{\max} \cong \text{ev}_{\min}$, then c will be very small and convergence will be fast, but if $\nabla^2 f(x_*)$ is even slightly poorly scaled—e.g., $\text{ev}_{\max} = 10^2 \, \text{ev}_{\min}$—then c will be almost 1 and convergence may be very slow (see Figure 6.2.1). An example, where $c = 0.8$ and $\|x_{k+1} - x_*\| = c \|x_k - x_*\|$ at each iteration, is given in Example 6.2.2. It is any easy exercise to generalize this example to make c arbitrarily close to 1.

EXAMPLE 6.2.2   Let $f(x_1, x_2) = \frac{1}{2}x_1^2 + \frac{9}{2}x_2^2$. This is a positive definite quadratic with minimizer at $x_* = (0, 0)^T$ and

$$\nabla^2 f(x) = \begin{bmatrix} 1 & 0 \\ 0 & 9 \end{bmatrix} \qquad \text{for all } x.$$

Thus, the c given by (6.2.4) is $(9 - 1)/(9 + 1) = 0.8$. Also, the reader can verify that if $f(x)$ is a positive definite quadratic, the steepest descent step of Algorithm 6.2.1 is given by

$$x_{k+1} := x_k - \left(\frac{g^T g}{g^T \nabla^2 f(x_k) g}\right) g, \qquad (6.2.5)$$

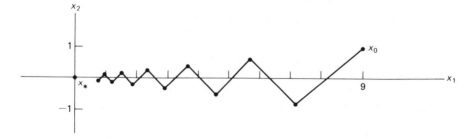

Figure 6.2.1   Method of steepest descent on a slightly poorly scaled quadratic

where $g = \nabla f(x_k)$. Using (6.2.5), it is easy to verify that the sequence of points generated by Algorithm 6.2.1 with the given $f(x)$ and $x_0 = (9, 1)^T$ is

$$x_k = \begin{pmatrix} 9 \\ (-1)^k \end{pmatrix} (0.8)^k, \qquad k = 1, 2, \ldots . \tag{6.2.6}$$

Therefore, $\| x_{k+1} - x_* \| / \| x_k - x_* \| = \| x_{k+1} \| / \| x_k \| = c$ in any l_p norm.
The sequence $\{x_k\}$ given in (6.2.6) is drawn in Figure 6.2.1.

Thus, the method of steepest descent should not be used as a computational algorithm in most circumstances, since a quasi-Newton algorithm will be far more efficient. However, when our global strategy must take steps much smaller than the Newton step, we will see in Section 6.4 that it may take steps in, or close to, the steepest descent direction.

We note finally that the steepest descent direction is very sensitive to changes in the scale of x, while the Newton direction is independent of such changes. For this reason, when we seem to use the steepest descent direction in our algorithms of Section 6.4, the corresponding implementations in the appendix will actually first premultiply it by a diagonal scaling matrix. We defer to Section 7.1 consideration of this important practical topic called scaling.

6.3 LINE SEARCHES

The first strategy we consider for proceeding from a solution estimate outside the convergence region of Newton's method is the method of line searches. This is really a very natural idea after the discussion of descent directions in Section 6.2.

The idea of a line-search algorithm is simple: given a descent direction p_k, we take a step in that direction that yields an "acceptable" x_{k+1}. That is,

at iteration k:

calculate a descent direction p_k,

set $x_{k+1} := x_k + \lambda_k p_k$ for some $\lambda_k > 0$ that makes x_{k+1}
an acceptable next iterate. $\hspace{2cm}$ (6.3.1)

Graphically, this means selecting x_{k+1} by considering the half of a one-dimensional cross section of $f(x)$ in which $f(x)$ decreases initially from x_k, as depicted in Figure 6.3.1. However, rather than setting p_k to the steepest descent direction $-\nabla f(x_k)$ as in Algorithm 6.2.1, we will use the quasi-Newton direction $-H_k^{-1} \nabla f(x_k)$ discussed in Section 5.5, where $H_k = \nabla^2 f(x_k) + \mu_k I$ is positive definite with $\mu_k = 0$ if $\nabla^2 f(x_k)$ is safely positive definite. This will allow us to retain fast local convergence.

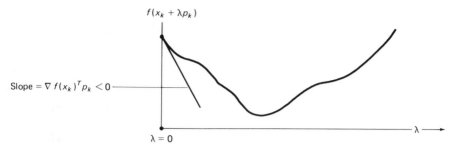

Figure 6.3.1   A cross section of $f(x)$: $\mathbb{R}^n \to \mathbb{R}$ from x_k in the direction p_k

The term "line search" refers to a procedure for choosing λ_k in (6.3.1). Until the mid 1960s the prevailing belief was that λ_k should be chosen to solve the one-dimensional minimization problem accurately. More careful computational testing has led to a complete turnabout, and in this section we will give weak acceptance criteria for $\{\lambda_k\}$ that lead to methods that perform as well in theory and better in practice.

The common procedure now is to try the full quasi-Newton step first and, if $\lambda_k = 1$ fails to satisfy the criterion in use, to backtrack in a systematic way along the direction defined by that step. *Computational experience has shown the importance of taking a full quasi-Newton step whenever possible.* Failure to do so leads to forfeiture of the advantage of Newton's method near the solution. Therefore, it is important that our procedure for choosing λ_k allow $\lambda_k = 1$ near the solution.

In the remainder of this section we discuss the choice of λ_k in theory and in practice. Since there is no compelling reason to confine our discussion to line searches in the quasi-Newton direction, we consider the search for $x_+ = x_c + \lambda_c p$ along a general descent direction p from the current solution estimate x_c.

While no step-acceptance rule will always be optimal, it does seem to be common sense to require that

$$f(x_{k+1}) < f(x_k). \tag{6.3.2}$$

It comes as no great surprise that this simple condition does not gurantee that $\{x_k\}$ will converge to a minimizer of f. We look now at two simple one-dimensional examples that show two ways that a sequence of iterates can satisfy (6.3.2) but fail to converge to a minimizer. These examples will guide us to useful step-acceptance criteria.

First, let $f(x) = x^2, x_0 = 2$. If we choose $\{p_k\} = \{(-1)^{k+1}\}$, $\{\lambda_k\} = \{2 + 3(2^{-(k+1)})\}$, then $\{x_k\} = \{2, -\frac{3}{2}, \frac{5}{4}, -\frac{9}{8}, \ldots\} = \{(-1)^k(1 + 2^{-k})\}$, each p_k is a descent direction from x_k, and $f(x_k)$ is monotonically decreasing with

$$\lim_{k \to \infty} f(x_k) = 1$$

[see Figure 6.3.2(a)]. Of course this is not a minimum of any sort for f, and furthermore $\{x_k\}$ has limit points ± 1, so it does not converge.

Now consider the same function with the same initial estimate, and let us take $\{p_k\} = \{-1\}$, $\{\lambda_k\} = \{2^{-k+1}\}$. Then $\{x_k\} = \{2, \frac{3}{2}, \frac{5}{4}, \frac{9}{8}, \ldots\} = \{1 + 2^{-k}\}$, each p_k is again a descent direction, $f(x_k)$ decreases monotonically, and $\lim_{k \to \infty} x_k = 1$, which again is not a minimizer of f [see Figure 6.3.2(b)].

The problem in the first case is that we achieved very small decreases in f values relative to the lengths of the steps. We can fix this by requiring that the average rate of decrease from $f(x_c)$ to $f(x_+)$ be at least some prescribed fraction of the initial rate of decrease in that direction; that is, we pick an $\alpha \in (0, 1)$ and choose λ_c from among those $\lambda > 0$ that satisfy

$$f(x_c + \lambda p) \le f(x_c) + \alpha\lambda \, \nabla f(x_c)^T p \qquad (6.3.3a)$$

(see Figure 6.3.3). Equivalently, λ_c must be chosen so that

$$f(x_+) \le f(x_c) + \alpha \, \nabla f(x_c)^T(x_+ - x_c). \qquad (6.3.3b)$$

It can be verified that this precludes the first unsuccessful choice of points but not the second.

The problem in the second example is that the steps are too small, relative to the initial rate of decrease of f. There are various conditions that ensure sufficiently large steps; we will present one that will also be useful to us in Chapter 9. We will require that the rate of decrease of f in the direction p at x_+ be larger than some prescribed fraction of the rate of decrease in the direction p at x_c—that is,

$$\nabla f(x_+)^T p \triangleq \nabla f(x_c + \lambda_c p)^T p \ge \beta \, \nabla f(x_c)^T p, \qquad (6.3.4a)$$

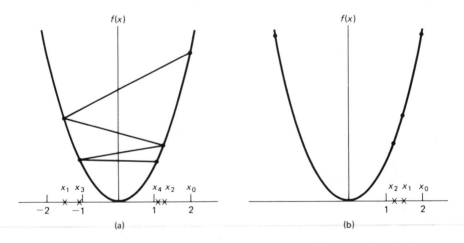

Figure 6.3.2 Monotonically decreasing sequences of iterates that don't converge to the minimizer

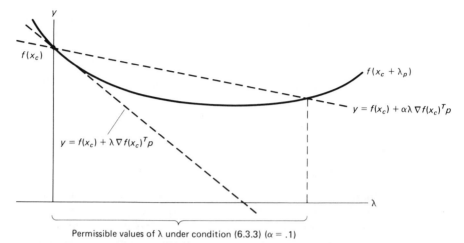

Permissible values of λ under condition (6.3.3) (α = .1)

Figure 6.3.3   Permissible values of λ under condition (6.3.3) ($\alpha = 0.1$)

or equivalently,

$$\nabla f(x_+)^T(x_+ - x_c) \geq \beta \; \nabla f(x_c)^T(x_+ - x_c) \tag{6.3.4b}$$

for some fixed constant $\beta \in (\alpha, 1)$ (see Figure 6.3.4). The condition $\beta > \alpha$ guarantees that (6.3.3) and (6.3.4) can be satisfied simultaneously. In practice, (6.3.4) generally is not needed because the use of a backtracking strategy avoids excessively small steps. An alternative condition to (6.3.4) is given in Exercise 7.

Conditions (6.3.3) and (6.3.4) are based on work of Armijo (1966) and Goldstein (1967). Example 6.3.1 demonstrates the effect of these conditions on a simple function. In Subsection 6.3.1 we prove powerful convergence results for any algorithm obeying these conditions. Subsection 6.3.2 discusses a practical line-search algorithm.

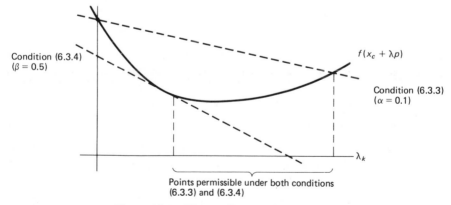

Points permissible under both conditions
(6.3.3) and (6.3.4)

Figure 6.3.4   The two line search conditions

EXAMPLE 6.3.1   Let $f(x_1, x_2) = x_1^4 + x_1^2 + x_2^2$, $x_c = (1, 1)^T$, $p_c = (-3, -1)^T$, and let $\alpha = 0.1$ in (6.3.3), $\beta = 0.5$ in (6.3.4). Since

$$\nabla f(x_c)^T p_c = (6, 2)(-3, -1)^T = -20,$$

p_c is a descent direction for $f(x)$ from x_c. Now consider $x(\lambda) = x_c + \lambda p_c$. If $x_+ = x(1) = (-2, 0)^T$,

$$\nabla f(x_+)^T p_c = (-36, 0)(-3, -1)^T = 108 > -10 = \beta \, \nabla f(x_c)^T p_c,$$

so that x_+ satisfies (6.3.4), but

$$f(x_+) = 20 > 1 = f(x_c) + \alpha \, \nabla f(x_c)^T p_c,$$

so that x_+ doesn't satisfy (6.3.3). Similarly, the reader can verify that $x_+ = x(0.1) = (0.7, 0.9)^T$ satisfies (6.3.3) but not (6.3.4), and that $x_+ = x(0.5) = (-0.5, 0.5)^T$ satisfies both (6.3.3) and (6.3.4). These three points correspond to points to the right of, to the left of, and in the "permissible" region in Figure 6.3.4, respectively.

6.3.1   CONVERGENCE RESULTS FOR PROPERLY CHOSEN STEPS

Using conditions (6.3.3) and (6.3.4), we can prove some amazingly powerful results: that given any direction p_k such that $\nabla f(x_k)^T p_k < 0$, there exist $\lambda_k > 0$ satisfying (6.3.3) and (6.3.4); that any method that generates a sequence $\{x_k\}$ obeying $\nabla f(x_k)^T (x_{k+1} - x_k) < 0$, (6.3.3), and (6.3.4) at each iteration is essentially globally convergent; and that close to a minimizer of f where $\nabla^2 f$ is positive definite, Newton steps satisfy (6.3.3) and (6.3.4). This means we can readily create algorithms that are globally convergent and, by always trying the Newton step first, are also quickly locally convergent on most functions.

Theorem 6.3.2, due to Wolfe (1969, 1971), shows that given any descent direction p_k, there exist points $x_k + \lambda_k p_k$ satisfying (6.3.3) and (6.3.4).

THEOREM 6.3.2   Let $f: \mathbb{R}^n \longrightarrow \mathbb{R}$ be continuously differentiable on $\mathbb{R}^n$. Let $x_k \in \mathbb{R}^n$, $p_k \in \mathbb{R}^n$ obey $\nabla f(x_k)^T p_k < 0$, and assume $\{f(x_k + \lambda p_k) \,|\, \lambda > 0\}$ is bounded below. Then if $0 < \alpha < \beta < 1$, there exist $\lambda_u > \lambda_l > 0$ such that $x_{k+1} = x_k + \lambda_k p_k$ satisfies (6.3.3) and (6.3.4) if $\lambda_k \in (\lambda_l, \lambda_u)$.

Proof.   Since $\alpha < 1$, $f(x_k + \lambda p_k) < f(x_k) + \alpha\lambda \, \nabla f(x_k)^T p_k$ for all $\lambda > 0$ sufficiently small. Since $f(x_k + \lambda p_k)$ is also bounded below, there exists some smallest positive $\hat{\lambda}$ such that

$$f(\hat{x}) = f(x_k + \hat{\lambda} p_k) = f(x_k) + \alpha\hat{\lambda} \, \nabla f(x_k)^T p_k. \tag{6.3.5}$$

Thus any $x \in (x_k, \hat{x})$ satisfies (6.3.3). By the mean value theorem, there exists $\bar{\lambda} \in (0, \hat{\lambda})$ such that

$$f(\hat{x}) - f(x_k) = \nabla f(x_k + \bar{\lambda} p_k)^T (\hat{x} - x_k)$$
$$= \nabla f(x_k + \bar{\lambda} p_k)^T \hat{\lambda} p_k, \tag{6.3.6}$$

and so from (6.3.5) and (6.3.6),

$$\nabla f(x_k + \bar{\lambda} p_k)^T p_k = \alpha \, \nabla f(x_k)^T p_k > \beta \, \nabla f(x_k)^T p_k \tag{6.3.7}$$

since $\alpha < \beta$ and $\nabla f(x_k)^T p_k < 0$. By the continuity of ∇f, (6.3.7) still holds for λ in some interval (λ_l, λ_u) about $\bar{\lambda}$. Therefore, if we restrict (λ_l, λ_u) to be in $(0, \hat{\lambda})$, $x_{k+1} = x_k + \lambda_k p_k$ satisfies (6.3.3) and (6.3.4) for any $\lambda_k \in (\lambda_l, \lambda_u)$. $\square$

Theorem 6.3.3 also is due to Wolfe (1969, 1971). It shows that if any sequence $\{x_k\}$ is generated to obey $\nabla f(x_k)^T (x_{k+1} - x_k) < 0$, (6.3.3), and (6.3.4), then, unless the angle between $\nabla f(x_k)$ and $(x_{k+1} - x_k)$ converges to $90°$ as $k \longrightarrow \infty$, either the gradient converges to 0, or f is unbounded below. Of course, both may be true. We comment below that the case where the angle between $\nabla f(x_k)$ and $(x_{k+1} - x_k)$ approaches $90°$ can be prevented by the algorithm.

THEOREM 6.3.3   Let $f: \mathbb{R}^n \longrightarrow \mathbb{R}$ be continuously differentiable on $\mathbb{R}^n$, and assume there exists $\gamma \geq 0$ such that

$$\| \nabla f(z) - \nabla f(x) \|_2 \leq \gamma \| z - x \|_2 \tag{6.3.8}$$

for every $x, z \in \mathbb{R}^n$. Then, given any $x_0 \in \mathbb{R}^n$, either f is unbounded below, or there exists a sequence $\{x_k\}$, $k = 0, 1, \ldots$, obeying (6.3.3), (6.3.4), and either

$$\nabla f(x_k)^T s_k < 0 \tag{6.3.9}$$

or

$$\nabla f(x_k) = 0 \quad \text{and} \quad s_k = 0,$$

for each $k \geq 0$, where

$$s_k \triangleq x_{k+1} - x_k.$$

Furthermore, for any such sequence, either

(a) $\nabla f(x_k) = 0$ for some $k \geq 0$, or

(b) $\lim\limits_{k \to \infty} f(x_k) = -\infty$, or

(c) $\lim\limits_{k \to \infty} \dfrac{\nabla f(x_k)^T s_k}{\| s_k \|_2} = 0.$

Proof. For each k, if $\nabla f(x_k) = 0$, then (a) holds and the sequence is constant subsequently. If $\nabla f(x_k) \neq 0$, then there exists p_k—e.g., $p_k = -\nabla f(x_k)$—such that $f(x_k)^T p_k < 0$. By Theorem 6.3.2, either f is unbounded below, or there exists $\lambda_k > 0$ such that $x_{k+1} = x_k + \lambda_k p_k$ satisfies (6.3.3) and (6.3.4). In order to simplify notation, let us assume that $\| p_k \|_2 = 1$, so $\lambda_k = \| s_k \|_2$. This constitutes no loss of generality.

So far we see that either f is unbounded below, or $\{x_k\}$ exists, and either $\{x_k\}$ satisfies (a) or else $s_k \neq 0$ for every k. We must now show that if no term of $\{s_k\}$ is zero, then (b) or (c) must hold. It will be useful to have

$$\sigma_k \triangleq \frac{\nabla f(x_k)^T s_k}{\| s_k \|_2}.$$

By (6.3.3) and $\lambda_k \sigma_k < 0$ for every k, for any $j > 0$,

$$f(x_j) - f(x_0) = \sum_{k=0}^{j-1} f(x_{k+1}) - f(x_k)$$

$$\leq \sum_{k=0}^{j-1} \alpha \, \nabla f(x_k)^T s_k$$

$$= \alpha \sum_{k=0}^{j-1} \lambda_k \sigma_k < 0.$$

Hence, either

$$\lim_{j \to \infty} f(x_j) = -\infty \quad \text{or} \quad -\sum_{k=0}^{\infty} \lambda_k \sigma_k < \infty$$

is convergent.

In the first case (b) is true and we are finished, so we consider the second. In particular, we deduce that $\lim_{k \to \infty} \lambda_k \sigma_k = 0$. Now we want to conclude that $\lim_{k \to \infty} \sigma_k = 0$, and so we need to use condition (6.3.4), since it was imposed to ensure that the steps don't get too small.

We have for each k that

$$\nabla f(x_{k+1})^T s_k \geq \beta \, \nabla f(x_k)^T s_k$$

and so

$$[\nabla f(x_{k+1}) - \nabla f(x_k)]^T s_k \geq (\beta - 1) \, \nabla f(x_k)^T s_k > 0,$$

by (6.3.9) and $\beta < 1$. Applying the Cauchy-Schwarz inequality and (6.3.8) to the left side of the last equation, and using the definition of σ_k, gives us

$$0 < (\beta - 1)\lambda_k \sigma_k \leq \gamma \, \| s_k \|^2 = \gamma \lambda_k^2,$$

so

$$\lambda_k \geq \frac{\beta - 1}{\gamma} \sigma_k > 0$$

and

$$\lambda_k \sigma_k \leq \frac{\beta - 1}{\gamma} \sigma_k^2 < 0.$$

Thus

$$0 = \lim_{k \to \infty} \lambda_k \sigma_k \leq \frac{\beta - 1}{\gamma} \lim_{k \to \infty} \sigma_k^2 \leq 0,$$

which shows that

$$\lim_{k \to \infty} \sigma_k = 0$$

[i.e., (c) is true] and completes the proof.   □

Note that while Theorem 6.3.3 applies readily to any line-search algorithm, it is completely independent of the method for selecting the descent directions or the step lengths. Therefore, this theorem gives sufficient conditions for global convergence, in a weak sense, of any optimization algorithm, including the model-trust region algorithms of Section 6.4. Furthermore, while the Lipschitz condition (6.3.8) is assumed on all of $\mathbb{R}^n$, it is used only in a neighborhood of the solution x_*. Finally, although $\{\sigma_k\} \longrightarrow 0$ in Theorem 6.3.3 does not necessarily imply $\{\nabla f(x_k)\} \longrightarrow 0$, it does as long as the angle between $\nabla f(x_k)$ and s_k is bounded away from $90°$. This can easily be achieved in practice. For example, in a quasi-Newton line-search algorithm where $p_k = -H_k^{-1} \nabla f(x_k)$ and H_k is positive definite, all that is needed is that the condition numbers of $\{H_k\}$ are uniformly bounded above. Thus, Theorem 6.3.3 can be viewed as implying global covergence toward $f = -\infty$ or $\nabla f = 0$, although the conditions are too weak to imply that $\{x_k\}$ converges.

Theorem 6.3.4, due to Dennis and Moré (1974), shows that our global strategy will permit full quasi-Newton steps $x_{k+1} := x_k - H_k^{-1} \nabla f(x_k)$ close to a minimizer of f as long as $-H_k^{-1} \nabla f(x_k)$ is close enough to the Newton step.

THEOREM 6.3.4   Let $f: \mathbb{R}^n \longrightarrow \mathbb{R}$ be twice continuously differentiable in an open convex set D, and assume that $\nabla^2 f \in \text{Lip}_\gamma(D)$. Consider a sequence $\{x_k\}$ generated by $x_{k+1} := x_k + \lambda_k p_k$, where $\nabla f(x_k)^T p_k < 0$ for all k and λ_k is chosen to satisfy (6.3.3) with an $\alpha < \frac{1}{2}$, and (6.3.4). If $\{x_k\}$ converges to a point $x_* \in D$ at which $\nabla^2 f(x_*)$ is positive definite, and if

$$\lim_{k \to \infty} \frac{\| \nabla f(x_k) + \nabla^2 f(x_k) p_k \|_2}{\| p_k \|_2} = 0, \tag{6.3.10}$$

then there is an index $k_0 \geq 0$ such that for all $k \geq k_0$, $\lambda_k = 1$ is admissible. Furthermore, $\nabla f(x_*) = 0$, and if $\lambda_k = 1$ for all $k \geq k_0$, then $\{x_k\}$ converges q-superlinearly to x_*.

Proof. The proof is really just a generalization of the easy exercise that if $f(x)$ is a positive definite quadratic and $p_k = -\nabla^2 f(x_k)^{-1} \nabla f(x_k)$, then $\nabla f(x_k)^T p_k < 0$ and $x_{k+1} = x_k + p_k$ satisfies (6.3.3) for any $\alpha \leq \frac{1}{2}$, and (6.3.4) for any $\beta \geq 0$. Some readers may wish to skim or skip it, as the details are not too illuminating. We set

$$\rho_k = \frac{\| \nabla f(x_k) + \nabla^2 f(x_k)p_k \|}{\| p_k \|}, \tag{6.3.11}$$

where throughout this proof, $\| \cdot \|$ means $\| \cdot \|_2$. First we show that

$$\lim_{k \to \infty} \| p_k \| = 0.$$

By a now familiar argument, if $\bar{x}$ is near enough x_*, then $\nabla^2 f(\bar{x})^{-1}$ exists and is positive definite, and if $\mu^{-1} = \| \nabla^2 f(x_*)^{-1} \|$, then $\frac{1}{2}\mu^{-1} \leq \| \nabla^2 f(\bar{x})^{-1} \| \leq 2\mu^{-1}$. Therefore, since

$$- \frac{\nabla f(x_k)^T p_k}{\| p_k \|} = \frac{p_k^T \nabla^2 f(x_k)p_k}{\| p_k \|} - \frac{(\nabla f(x_k) + \nabla^2 f(x_k)p_k)^T p_k}{\| p_k \|}$$

and $p_k^T \nabla^2 f(x_k)p_k > \frac{1}{2}\mu \| p_k \|^2$, (6.3.11) and (6.3.10) show that for k sufficiently large,

$$- \frac{\nabla f(x_k)^T p_k}{\| p_k \|} \geq (\tfrac{1}{2}\mu - \rho_k)\| p_k \| \geq \tfrac{1}{4}\mu \| p_k \|. \tag{6.3.12}$$

Since from Theorem 6.3.3 we have

$$\lim_{k \to \infty} \frac{\nabla f(x_k)^T p_k}{\| p_k \|} = 0,$$

(6.3.12) implies

$$\lim_{k \to \infty} \| p_k \| = 0.$$

We now show that $\lambda_k = 1$ satisfies (6.3.3), for all k greater than or equal to some k_0. From the mean value theorem, for some $\bar{x}_k \in [x_k, x_k + p_k]$,

$$f(x_k + p_k) - f(x_k) = \nabla f(x_k)^T p_k + \tfrac{1}{2}p_k^T \nabla^2 f(\bar{x}_k)p_k,$$

or

$$
\begin{aligned}
f(x_k + p_k) &- f(x_k) - \tfrac{1}{2} \nabla f(x_k)^T p_k \\
&= \tfrac{1}{2}(\nabla f(x_k) + \nabla^2 f(\bar{x}_k)p_k)^T p_k \\
&= \tfrac{1}{2}(\nabla f(x_k) + \nabla^2 f(x_k)p_k)^T p_k + \tfrac{1}{2} p_k^T [\nabla^2 f(\bar{x}_k) - \nabla^2 f(x_k)]p_k \\
&\leq \tfrac{1}{2}(\rho_k + \gamma \| p_k \|)\| p_k \|^2
\end{aligned}
\tag{6.3.13}
$$

because of (6.3.11) and the Lipschitz continuity of $\nabla^2 f$. Now choose k_0 so that for $k \geq k_0$, (6.3.12) holds and

$$\rho_k + \gamma \| p_k \| \leq \tfrac{1}{4}\mu \cdot \min \, (\beta, \, 1 - 2\alpha). \tag{6.3.14}$$

If $k \geq k_0$, $\lambda_k = 1$ satisfies (6.3.3) because from (6.3.13), (6.3.12), and (6.3.14)

$$
\begin{aligned}
f(x_k + p_k) - f(x_k) &\leq \tfrac{1}{2}\, \nabla f(x_k)^T p_k + \tfrac{1}{8}\mu \cdot (1 - 2\alpha) \| p_k \|^2 \\
&\leq \tfrac{1}{2}(1 - (1 - 2\alpha)) \, \nabla f(x_k)^T p_k \\
&= \alpha \, \nabla f(x_k)^T p_k.
\end{aligned}
$$

To show that (6.3.4) is satisfied by $\lambda_k = 1$ for $k \geq k_0$, we use the mean value theorem again to get, for some $z_k \in (x_k, \, x_k + p_k)$, that

$$
\begin{aligned}
\nabla f(x_k + p_k)^T p_k &= (\nabla f(x_k) + \nabla^2 f(z_k) p_k)^T p_k \\
&= (\nabla f(x_k) + \nabla^2 f(x_k) p_k)^T p_k \\
&\quad + p_k^T (\nabla^2 f(z_k) - \nabla^2 f(x_k)) p_k
\end{aligned}
$$

so

$$
\begin{aligned}
| \nabla f(x_k + p_k)^T p_k | &\leq \rho_k \| p_k \|^2 + \gamma \| p_k \|^3 \\
&\leq \frac{\mu \beta}{4} \| p_k \|^2 \\
&\leq -\beta \, \nabla f(x_k)^T p_k
\end{aligned}
$$

by (6.3.11), the Lipschitz continuity of $\nabla^2 f$, (6.3.14), and (6.3.12). This yields $\nabla f(x_k + p_k)^T p_k \geq \beta \nabla f(x_k)^T p_k$ for $k \geq k_0$. Thus $\lambda_k = 1$ eventually satisfies (6.3.3, 6.3.4), and so it is admissible. It is an easy consequence of (6.3.10) and $\lim_{k \to \infty} p_k = 0$ that $\nabla f(x_*) = 0$. This leaves only the proof of q-superlinear convergence if $\lambda_k = 1$ is chosen for all but finitely many terms. We postpone this until it follows from a much more general result in Chapter 8. $\square$

Taken together, the conclusions of Theorems 6.3.3 and 6.3.4 are quite remarkable. They say that if f is bounded below, then the sequence $\{x_k\}$ generated by any algorithm that takes descent steps whose angles with the gradients are bounded away from $90°$, and that satisfy conditions (6.3.3) and (6.3.4), will obey

$$\lim_{k \to \infty} \nabla f(x_k) = 0.$$

Furthermore, if any such algorithm tries a Newton or quasi-Newton step first at each iteration, then $\{x_k\}$ will also converge q-quadratically or q-superlinearly to a local minimizer x_* if any x_k is sufficiently close to x_*, and if local convergence assumptions for the Newton or quasi-Newton method are met. These results are all we will need for the global convergence of the algorithms we discuss in the remainder of this book. (See also Exercise 25.)

6.3.2 STEP SELECTION
BY BACKTRACKING

We now specify how our line-search algorithm will choose λ_k. As we have stated, the modern strategy is to start with $\lambda_k = 1$, and then, if $x_k + p_k$ is not acceptable, "backtrack" (reduce λ_k) until an acceptable $x_k + \lambda_k p_k$ is found. The framework of such an algorithm is given below. Recall that condition (6.3.4) is not implemented because the backtracking strategy avoids excessively small steps. (See also Exercise 25.)

ALGORITHM 6.3.5   BACKTRACKING LINE-SEARCH FRAMEWORK

Given $\alpha \in (0, \frac{1}{2})$, $0 < l < u < 1$

$\lambda_k = 1$;

while $f(x_k + \lambda_k p_k) > f(x_k) + \alpha \lambda_k \nabla f(x_k)^T p_k$, do

$\quad \lambda_k := \rho \lambda_k$ for some $\rho \in [l, u]$;

$\quad$ (*ρ is chosen anew each time by the line search*)

$x_{k+1} := x_k + \lambda_k p_k$;

In practice, α is set quite small, so that hardly more than a decrease in function value is required. Our algorithm uses $\alpha = 10^{-4}$. Now just the strategy for reducing λ_k (choosing ρ) remains. Let us define

$$\hat{f}(\lambda) \triangleq f(x_k + \lambda p_k),$$

the one-dimensional restriction of f to the line through x_k in the direction p_k. If we need to backtrack, we will use our most current information about $\hat{f}$ to model it, and then take the value of λ that minimizes this model as our next value of λ_k in Algorithm 6.3.5.

Initially, we have two pieces of information about $\hat{f}(\lambda)$,

$$\hat{f}(0) = f(x_k) \quad \text{and} \quad \hat{f}'(0) = \nabla f(x_k)^T p_k. \tag{6.3.15}$$

After calculating $f(x_k + p_k)$, we also know that

$$\hat{f}(1) = f(x_k + p_k), \tag{6.3.16}$$

so if $f(x_k + p_k)$ does not satisfy (6.3.3) [i.e., $\hat{f}(1) > \hat{f}(0) + \alpha \hat{f}'(0)$], we model $\hat{f}(\lambda)$ by the one-dimensional quadratic satisfying (6.3.15) and (6.3.16),

$$\hat{m}_q(\lambda) = [\hat{f}(1) - \hat{f}(0) - \hat{f}'(0)]\lambda^2 + \hat{f}'(0)\lambda + \hat{f}(0),$$

and calculate the point

$$\hat{\lambda} = \frac{-\hat{f}'(0)}{2[\hat{f}(1) - \hat{f}(0) - \hat{f}'(0)]} \tag{6.3.17}$$

for which $\hat{m}_q'(\hat{\lambda}) = 0$. Now

$$m_q''(\lambda) = 2[\hat{f}(1) - \hat{f}(0) - \hat{f}'(0)] > 0,$$

since $\hat{f}(1) > \hat{f}(0) + \alpha\hat{f}'(0) > \hat{f}(0) + \hat{f}'(0)$. Thus $\hat{\lambda}$ minimizes $\hat{m}_q(\lambda)$. Also $\hat{\lambda} > 0$, because $\hat{f}'(0) < 0$. Therefore we take $\hat{\lambda}$ as our new value of λ_k in Algorithm 6.3.5 (see Figure 6.3.5). Note that since $\hat{f}(1) > \hat{f}(0) + \alpha\hat{f}'(0)$, we have

$$\hat{\lambda} < \frac{1}{2(1 - \alpha)}.$$

In fact, if $\hat{f}(1) \geq \hat{f}(0)$, then $\hat{\lambda} \leq \frac{1}{2}$. Thus, (6.3.17) gives a useful implicit upper bound of $u \approx \frac{1}{2}$ on the first value of ρ in Algorithm 6.3.5. On the other hand, if $\hat{f}(1)$ is much larger than $\hat{f}(0)$, $\hat{\lambda}$ can be very small. We probably do not want to decrease λ_k too much based on this information, since probably it indicates that $\hat{f}(\lambda)$ is poorly modeled by a quadratic in this region, so we impose a lower bound of $l = \frac{1}{10}$ in Algorithm 6.3.5. This means that at the first backtrack at each iteration, if $\hat{\lambda} \leq 0.1$, then we next try $\lambda_k = \frac{1}{10}$.

EXAMPLE 6.3.6   Let $f: \mathbb{R}^n \longrightarrow \mathbb{R}$, and x_c and p be given by Example 6.3.1. Since $f(x_c) = 3$ and $f(x_c + p) = 20$, $x_c + p$ is not acceptable and a backtrack is needed. Then $\hat{m}_q(0) = f(x_c) = 3$, $\hat{m}_q'(0) = \nabla f(x_c)^T p = -20$, $\hat{m}_q(1) = f(x_c + p) = 20$, and (6.3.17) gives $\hat{\lambda} = \frac{10}{37} \cong 0.270$. Now $x_c + \hat{\lambda}p \cong (0.189, 0.730)^T$ satisfies condition (6.3.3) with $\alpha = 10^{-4}$, since $f(x_c + \hat{\lambda}p) \cong 0.570 \leq 2.99946 = f(x_c) + \alpha\lambda \ \nabla f(x_c)^T p$. Therefore $x_+ := x_c + \hat{\lambda}p$. Incidentally, the minimum of $f(x_c + \lambda p)$ occurs at $\lambda_* \cong 0.40$.

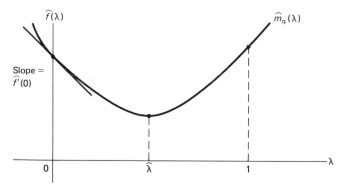

Figure 6.3.5   Backtracking at the first iteration, using a quadratic model

Now suppose $\hat{f}(\lambda_k) = f(x_k + \lambda_k p_k)$ doesn't satisfy (6.3.3). In this case we need to backtrack again. Although we could use a quadratic model as we did on the first backtrack, we now have four pieces of information about $\hat{f}(\lambda)$. So at this and any subsequent backtrack during the current iteration, we use a cubic model of $\hat{f}$, fit $\hat{m}_{cu}(\lambda)$ to $\hat{f}(0)$, $\hat{f}'(0)$, and the last two values of $\hat{f}(\lambda)$, and, subject to the same sort of upper and lower limits as before, set λ_k to the value of λ at which $\hat{m}_{cu}(\lambda)$ has its local minimizer (see Figure 6.3.6). The reason for using a cubic is that it can more accurately model situations where f has negative curvature, which are likely when (6.3.3) has failed for two positive values of λ. Furthermore, such a cubic has a unique local minimizer, as illustrated in Figure 6.3.6.

The calculation of λ proceeds as follows. Let λprev and λ2prev be the last two previous values of λ_k. Then the cubic that fits $\hat{f}(0)$, $\hat{f}'(0)$, $\hat{f}(\lambda$prev$)$, and $\hat{f}(\lambda$2prev$)$ is

$$\hat{m}_{cu}(\lambda) = a\lambda^3 + b\lambda^2 + \hat{f}'(0)\lambda + \hat{f}(0),$$

where

$$
\begin{bmatrix} a \\ b \end{bmatrix} = \frac{1}{\lambda\text{prev} - \lambda2\text{prev}}
$$

$$
\times \begin{bmatrix} \dfrac{1}{\lambda\text{prev}^2} & \dfrac{-1}{\lambda2\text{prev}^2} \\[2mm] \dfrac{-\lambda2\text{prev}}{\lambda\text{prev}^2} & \dfrac{\lambda\text{prev}}{\lambda2\text{prev}^2} \end{bmatrix} \begin{bmatrix} \hat{f}(\lambda\text{prev}) - \hat{f}(0) - \hat{f}'(0)\lambda\text{prev} \\[2mm] \hat{f}(\lambda2\text{prev}) - \hat{f}(0) - \hat{f}'(0)\lambda2\text{prev} \end{bmatrix}.
$$

Its local minimizing point $\hat{\lambda}$ is

$$\frac{-b + \sqrt{b^2 - 3a\hat{f}'(0)}}{3a}. \tag{6.3.18}$$

It can be shown that if $\hat{f}(\lambda$prev$) \geq \hat{f}(0)$, then $\hat{\lambda} < \frac{2}{3}\lambda$prev, but this reduction can be achieved in practice and is considered too small (see Example 6.5.1). Therefore the upper bound $u = 0.5$ is imposed, which means that if $\hat{\lambda} > \frac{1}{2}\lambda$prev, we

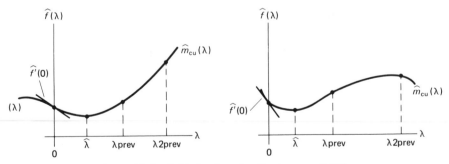

Figure 6.3.6   Cubic backtrack—the two possibilities

set the new $\lambda_k = \frac{1}{2}\lambda$prev. Also, since $\hat{\lambda}$ can be an arbitrarily small fraction of λprev, the lower bound $l = \frac{1}{10}$ is used again (i.e., if $\lambda < \frac{1}{10}\lambda$prev, we set the new $\lambda_k = \frac{1}{10}\lambda$prev). It can be shown that (6.3.18) is never imaginary if α in Algorithm 6.3.5 is less than $\frac{1}{4}$.

There is one important case for which backtracking based on cubic interpolation can be used at every step of Algorithm 6.3.5. Sometimes it is possible to obtain $\nabla f(x_k + \lambda_k p_k)$ at very little additional cost while computing $f(x_k + \lambda_k p_k)$. In this case, $\hat{m}_{cu}(\lambda)$ can be fitted to $\hat{f}(0), \hat{f}'(0), \hat{f}(\lambda$prev$), \hat{f}'(\lambda$prev$)$ at each backtrack. It is a valuable exercise for the student to work out the resulting algorithm in detail.

Algorithm A6.3.1 in the appendix contains our backtracking line-search procedure. It has two additional features. A minimum step length is imposed called *minstep*; it is the quantity used to test for convergence in the upcoming Algorithm A7.2.1. If condition (6.3.3) is not satisfied, but $\| \lambda_k p_k \|_2$ is smaller than *minstep*, then the line search is terminated, since convergence to x_k will be detected at the end of the current iteration anyway. This criterion prevents the line search from looping forever if p_k is not a descent direction. (This sometimes occurs at the final iteration of minimization algorithms, owing to finite-precision errors, especially if the gradient is calculated by finite differences.) Because this condition could also indicate convergence to a false solution, a warning message should be printed. A maximum allowable step length also is declared by the user, because excessively long steps could occur in practice when $p_k = s_k^N = -H_k^{-1} \nabla f(x_k)$ and H_k is nearly singular. The maximum step length is imposed precisely to prevent taking steps that would result in the algorithm's leaving the domain of interest and possibly cause an overflow in the computation of $\hat{f}(1)$.

When we use a line search in conjunction with the secant algorithms of Chapter 9, we will see that we also need to assume that condition (6.3.4) is satisfied at each iteration. Algorithm A6.3.1 mod is the modification of Algorithm A6.3.1 that explicitly enforces this condition. Usually it results in the same steps as Algorithm A6.3.1.

6.4 THE MODEL-TRUST REGION APPROACH

The last section dealt exclusively with the problem of finding an acceptable step length in a given direction of search. The underlying assumptions were that the direction would be the quasi-Newton direction, and that the full quasi-Newton step would always be the first trial step. The resulting back-tracking algorithm incorporated these assumptions to attempt global convergence without sacrificing the local convergence properties of the quasi-Newton method, clearly the goal of any global strategy. In this section we seek the same goal, but we drop the assumption that shortened steps must be in the quasi-Newton direction.

Suppose that the full quasi-Newton step is unsatisfactory. This indicates that our quadratic model does not adequately model f in a region containing the full quasi-Newton step. In line-search algorithms we retain the same *step direction* and choose a shorter *step length*. This new length is determined by building a new one-dimensional quadratic or cubic model, based only on function and gradient information in the quasi-Newton direction. While this strategy is successful, it has the disadvantage that it makes no further use of the n-dimensional quadratic model, including the model Hessian. In this section, when we need to take a shorter step, we first will choose a shorter *step length*, and then use the full n-dimensional quadratic model to choose the *step direction*.

Before we begin to consider ways to choose such directions, it will be useful to make a few preliminary remarks about prespecifying step length. It seems reasonable that after the first iteration we would begin to get at least a rough idea about the length of step we can reasonably expect to make. For example, we might deduce from the length of the step we took at iteration k an upper bound on the length of step likely to be successful at iteration $k + 1$. In fact, in Subsection 6.4.3 we will see how to use information about f, gained as the iteration proceeds, to better estimate the length of step likely to be successful. Given this estimate, we might want to start our next iteration with a step of about this length, and not waste possibly expensive function evaluations on longer steps that are unlikely to be successful. Of course, this means that in order to save function evaluations, we might not try a full quasi-Newton step at some iteration when it might be satisfactory. In Subsection 6.4.3 we give some heuristics that are intended to minimize the chances of this occurrence.

Now suppose that we have x_c and some estimate δ_c of the maximum length of a successful step we are likely to be able to take from x_c. This raises the question: how can we best select a step of maximal length δ_c from x_c? A natural answer exists if we return to our idea of a quadratic model. From the beginning we have taken the view that the quasi-Newton step s_c^N is reasonable because it is the step from x_c to the global minimizer of the local quadratic model m_c (if the model Hessian H_c is positive definite). If we add the idea of bounding the maximal step length by $\delta_c > 0$, then the corresponding answer to our question is to try the step s_c that solves

$$\min\ m_c(x_c + s) = f(x_c) + \nabla f(x_c)^T s + \tfrac{1}{2} s^T H_c\, s,$$

$$\text{subject to} \quad \|s\|_2 \le \delta_c. \tag{6.4.1}$$

Problem (6.4.1) is the basis of the "model-trust region" approach to minimization. Its solution is given in Lemma 6.4.1 below. The name comes from viewing δ_c as providing a *region* in which we can *trust* m_c to adequately *model f*.

In the next chapter we will see the merit of using a scaled version of the l_2 norm in the step-length bound, but to do so now would only add clutter.

LEMMA 6.4.1   Let $f: \mathbb{R}^n \longrightarrow \mathbb{R}$ be twice continuously differentiable, $H_c \in \mathbb{R}^{n \times n}$ be symmetric and positive definite, and let $\| \cdot \|$ designate the l_2 norm. Then problem 6.4.1 is solved by

$$s(\mu) \triangleq -(H_c + \mu I)^{-1} \nabla f(x_c) \qquad (6.4.2)$$

for the unique $\mu \geq 0$ such that $\| s(\mu) \| = \delta_c$, unless $\| s(0) \| \leq \delta_c$, in which case $s(0) = s_c^N$ is the solution. For any $\mu \geq 0$, $s(\mu)$ defines a descent direction for f from x_c.

Proof. (6.4.2) is straightforward from the necessary and sufficient conditions for constrained optimization, but we do not need this generality elsewhere in the book. While reading the proof, the reader may want to refer to Figure 6.4.1. It shows x_c, a positive definite quadratic model m_c with minimum at x^N surrounded by contours at some increasing values of m_c, and a step bound δ_c.

Let us call the solution to (6.4.1) s_*, and let $x_* = x_c + s_*$. Since the Newton point $x_c + s_c^N$ is the global minimizer of m_c, it is clear that if $\|s_c^N\| \leq \delta_c$, then $s_c^N = s_*$. Now consider the case when $x_c + s_c^N$ is outside the step bound as in Figure 6.4.1. Let $\bar{x}$ be any point interior to the constraint region—i.e., $\| \bar{x} - x_c \| < \delta_c$. Then $\nabla m_c(\bar{x}) \neq 0$, so it is possible to decrease m_c from $\bar{x}$ while staying inside the constrained region by

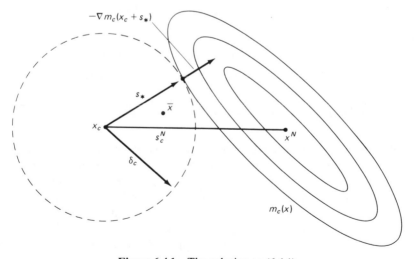

Figure 6.4.1   The solution to (6.4.1)

considering points of the form $\bar{x} - \lambda \, \nabla m_c(\bar{x})$. This implies that $\bar{s}$ cannot be the solution to (6.4.1), so s_* must satisfy $\| s_* \| = \delta_c$ unless $\| s(0) \| < \delta_c$. Also, since the constrained region in problem (6.4.1) is closed and compact, some s for which $\| s \| = \delta_c$ must be the solution.

Now consider some s such that $\| s \| = \delta_c$. For s to be the solution to (6.4.1), it is necessary that if we move an arbitrarily small distance from $x_c + s$ in any descent direction for m_c, we increase the distance from x_c. A descent direction for m_c from $x_c + s$ is any vector p for which

$$0 > p^T \, \nabla m_c(x_c + s) = p^T[H_c \, s + \nabla f(x_c)]. \tag{6.4.3}$$

Similarly a direction $\bar{p}$ from $x_c + s$ increases distance from x_c if and only if

$$\bar{p}^T s > 0. \tag{6.4.4}$$

What we are saying is that for s to solve (6.4.1), any p that satisfies (6.4.3) must also satisfy (6.4.4). Since we know that $\nabla m_c(x_c + s) \neq 0$, this can occur only if $\nabla m_c(x_c + s)$ and $-s$ point in the same direction, or in other words, if for some $\mu > 0$,

$$\mu s = -\nabla m_c(x_c + s) = -(H_c \, s + \nabla f(x_c))$$

which is just $(H_c + \mu I)s = -\nabla f(x_c)$. Thus $s_* = s(\mu)$ for some $\mu > 0$, when $\| s_c^N \| > \delta_c$. Since $\mu \geq 0$ and H_c is symmetric and positive definite, $(H_c + \mu I)$ is symmetric and positive definite, so $s(\mu)$ is a descent direction for f from x_c. In order to finish the proof, we need to show that s_* is unique, which is implied by the stronger result that if $\mu_1 > \mu_2 \geq 0$, then $\| s(\mu_1) \| < \| s(\mu_2) \|$. This shows that $\| s(\mu) \| = \delta_c$ has a unique solution that must be the solution to (6.4.1). The proof is straightforward, because if $\mu \geq 0$ and

$$\eta(\mu) \triangleq \| s(\mu) \|_2 = \| (H_c + \mu I)^{-1} \, \nabla f(x_c) \|,$$

then

$$\eta'(\mu) = \frac{-\nabla f(x_c)^T(\mu I + H_c)^{-3} \, \nabla f(x_c)}{\| s(\mu) \|} = \frac{-s(\mu)^T(H_c + \mu I)^{-1}s(\mu)}{\| s(\mu) \|}$$

and so $\eta'(\mu) < 0$ as long as $\nabla f(x_c) \neq 0$. Thus η is a monotonically decreasing function of μ. □

The trouble with using Lemma 6.4.1 as the basis of a step in a minimization algorithm is that there is no finite method of determining the $\mu_c > 0$ such that $\| s(\mu_c) \|_2 = \delta_c$, when $\delta_c < \| H_c^{-1} \, \nabla f(x_c) \|_2$. Therefore, in the next two subsections we will describe two computational methods for approximately solving problem (6.4.1). The first, the locally constrained optimal (or "hook") step, finds a μ_c such that $\| s(\mu_c) \|_2 \cong \delta_c$, and takes $x_+ = x_c + s(\mu_c)$. The second, the dogleg step, makes a piecewise linear approximation to the curve $s(\mu)$, and takes x_+ as the point on this approximation for which $\| x_+ - x_c \| = \delta_c$.

There is no guarantee that the x_+ that approximately or exactly solves (6.4.1) will be an acceptable next point, although we hope it will be if δ_c is a good step bound. Therefore, a complete step of a trust-region algorithm will have the following form:

ALGORITHM 6.4.2   A GLOBAL STEP BY THE MODEL-TRUST REGION APPROACH

Given $f: \mathbb{R}^n \longrightarrow \mathbb{R}$, $\delta_c > 0$, $x_c \in \mathbb{R}^n$, $H_c \in \mathbb{R}^{n \times n}$ symmetric and positive definite:

repeat
  (1) $s_c :=$ approximate solution to (6.4.1),
      $x_+ := x_c + s_c$,
  (2) decide whether x_+ is acceptable, and calculate a new value of δ_c

until x_+ is an acceptable next point;

$\delta_+ := \delta_c$

To complete the model-trust region algorithm, we discuss the above Step 2 in Subsection 6.4.3.

Finally, some observations are in order. First, Gay (1981) shows that even if H_c has negative eigenvalues, the solution to (6.4.1) is still an s_* satisfying $(H_c + \mu I)s_* = -\nabla f(x_c)$ for some $\mu > 0$ such that $H_c + \mu I$ is at least positive semidefinite. This leads to another justification for our strategy from Section 5.5 of perturbing the model Hessian to a positive definite $H_c = \nabla^2 f(x_c) + \mu I$ when $\nabla^2 f(x_c)$ is not positive definite; the resultant quasi-Newton step $-H_c^{-1} \nabla f(x_c)$ is to the minimizer of the original (non positive definite) quadratic model in some spherical region around x_c. For the remainder of Section 6.4 we will assume that H_c is positive definite.

Second, it is important to note that $s(\mu)$ runs smoothly from the quasi-Newton step $-H_c^{-1} \nabla f(x_c)$, when $\mu = 0$, to $s(\mu) \cong - (1/\mu) \nabla f(x_c)$, when μ gets large. Thus when δ_c is very small, the solution to (6.4.1) is a step of length δ_c in approximately the steepest-descent direction. Figure 6.4.2 traces the curve $s(\mu)$, $0 \le \mu < \infty$, for the same quadratic model as in Figure 6.4.1. We note that in general when $n > 2$, this curve does not lie in the subspace spanned by $\nabla f(x_c)$ and $H_c^{-1} \nabla f(x_c)$ (see Exercise 6.10).

Fletcher (1980) calls these algorithms *restricted step methods*. This term emphasizes the procedure to be covered in Subsection 6.4.3 for updating δ_c, which could also be used in line-search algorithms. Some important groundwork for the algorithms of this section was provided by Levenberg (1944), Marquardt (1963), and Goldfeldt, Quandt, and Trotter (1966). More recent references are cited in the following pages.

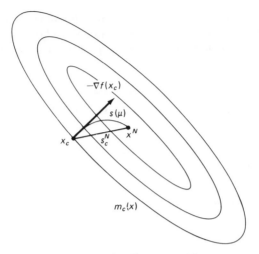

Figure 6.4.2   The curve $s(\mu)$

6.4.1   THE LOCALLY CONSTRAINED OPTIMAL ("HOOK") STEP

Our first approach for calculating the step in model-trust region Algorithm 6.4.2, when $\| s(0) \|_2 > \delta_c$, is to find an $s(\mu) \triangleq -(H_c + \mu I)^{-1} \nabla f(x_c)$ such that $\| s(\mu) \|_2 \cong \delta_c$, and then make a trial step to $x_+ := x_c + s(\mu)$. In this section we discuss an algorithm for finding an approximate solution μ_c to the scalar equation

$$\Phi(\mu) \triangleq \| s(\mu) \|_2 - \delta_c = 0. \tag{6.4.5}$$

In practice we will not require a very exact solution at all, but we defer that consideration to the end of this subsection.

An application of the ideas of Chapter 2 would cause us to use Newton's method to solve (6.4.5). Although this would work, it can be shown that for all $\mu \geq 0$, $\Phi'(\mu) < 0$ and $\Phi''(\mu) > 0$, so Newton's method will always underestimate the exact solution μ_* (see Figure 6.4.3). Therefore, we will construct a different method geared specifically to (6.4.5).

The proper idea to bring from Chapter 2 is the device of a local model of the problem to be solved. In this case, the one-dimensional version

$$\Phi(\mu) = | -(\mu + h_c)^{-1} g_c | - \delta_c$$

suggests a local model of the form

$$m_c(\mu) = \frac{\alpha_c}{\beta_c + \mu} - \delta_c$$

with two free parameters α, β. Before we discuss α, β, notice that we have slipped into a new notation for dealing with the iteration on μ, which comes from the model above, takes place inside the main x-iteration, and whose purpose is to find the μ_c satisfying (6.4.5). We will use boldface for the current values of the quantities $\boldsymbol{\alpha}_c$, $\boldsymbol{\beta}_c$, $\boldsymbol{\mu}_c$ that are changing in the inner iteration on μ, and normal type for the current values of δ_c, x_c, $\nabla f(x_c)$, H_c that come from the outer iteration and are unchanged during the solution of (6.4.5). Thus, we obtain μ_c as the inner iteration's last approximation $\boldsymbol{\mu}_c$ to $\boldsymbol{\mu}_*$, the exact solution of (6.4.5); and we use x_c and μ_c to define $x_+ = x_c + s(\mu_c)$.

The model $m_c(\mu)$ has two free parameters, $\boldsymbol{\alpha}_c$ and $\boldsymbol{\beta}_c$, and so as in the derivation of Newton's method, it is reasonable to choose them to satisfy the two conditions:

$$m_c(\boldsymbol{\mu}_c) = \frac{\boldsymbol{\alpha}_c}{\boldsymbol{\beta}_c + \boldsymbol{\mu}_c} - \delta_c = \Phi(\boldsymbol{\mu}_c) = \| s(\boldsymbol{\mu}_c) \|_2 - \delta_c$$

and

$$m_c'(\boldsymbol{\mu}_c) = -\frac{\boldsymbol{\alpha}_c}{(\boldsymbol{\beta}_c + \boldsymbol{\mu}_c)^2} = \Phi'(\boldsymbol{\mu}_c) = -\frac{s(\boldsymbol{\mu}_c)^T (H_c + \boldsymbol{\mu}_c I)^{-1} s(\boldsymbol{\mu}_c)}{\| s(\boldsymbol{\mu}_c) \|_2}$$

This gives

$$\boldsymbol{\alpha}_c = -\frac{(\Phi(\boldsymbol{\mu}_c) + \delta_c)^2}{\Phi'(\boldsymbol{\mu}_c)}, \tag{6.4.6}$$

$$\boldsymbol{\beta}_c = -\frac{(\Phi(\boldsymbol{\mu}_c) + \delta_c)}{\Phi'(\boldsymbol{\mu}_c)} - \boldsymbol{\mu}_c. \tag{6.4.7}$$

Notice that the computation of $s(\boldsymbol{\mu}_c)$ requires a factorization of $H_c + \boldsymbol{\mu}_c I$, and so it will require $O(n^3)$ arithmetic operations. Once we have this factorization, the computation of $(H_c + \boldsymbol{\mu}_c I)^{-1} s(\boldsymbol{\mu}_c)$ costs only $O(n^2)$ operations, so $\Phi'(\boldsymbol{\mu}_c)$ is relatively cheap.

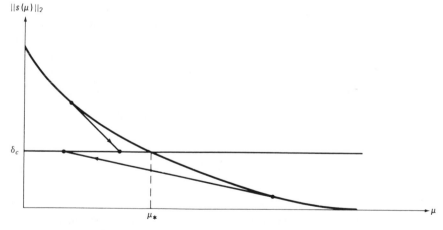

Figure 6.4.3   Newton's method for finding $\Phi(\mu) = 0$

Now that we have our model $\mathbf{m}_c(\mu)$, we naturally choose μ_+ such that $\mathbf{m}_c(\mu_+) = 0$—that is

$$\mu_+ = \frac{\alpha_c}{\delta_c} - \beta_c .$$

The reader can verify that substituting (6.4.6), (6.4.7), and (6.4.5) into the above gives

$$\mu_+ = \mu_c - \left[\frac{\Phi(\mu_c) + \delta_c}{\delta_c} \right] \left[\frac{\Phi(\mu_c)}{\Phi'(\mu_c)} \right]$$

$$= \mu_c - \frac{\| s(\mu_c) \|}{\delta_c} \left[\frac{\Phi(\mu_c)}{\Phi'(\mu_c)} \right] \qquad (6.4.8)$$

as our iteration for solving $\Phi(\mu) = 0$. The form of (6.4.8) shows that when $\mu_c < \mu_*$, we are taking a larger correction than Newton's method would give, but that when $\mu_c > \mu_*$, the step is smaller than the Newton step. As $\mu_c \longrightarrow \mu_*$, (6.4.8) becomes more and more like Newton's method; in fact, (6.4.8) is locally q-quadratically convergent to μ_*.

Several issues remain in transforming (6.4.8) into a computational algorithm. The following discussion, and our algorithm, is based on Hebden (1973) and Moré (1977). Another comprehensive reference is Gander (1978). Readers not interested in implementation details may wish to skip the next three paragraphs.

The first consideration is a starting value for μ in solving (6.4.5). Reinsch (1971) has shown that if (6.4.8) is started with $\mu_0 = 0$, then the μ iteration converges monotonically up to μ_*, but we would like a closer start because each iteration of (6.4.8) involves solving a linear system. Moré's implementation uses the approximate solution μ_- to the last instance of (6.4.5) to generate an initial guess for the current instance. The rule Moré uses is simple: if the current step bound δ_c is p times the last value of the step bound δ_-, then $\mu_0 = \mu_-/p$ is used to start (6.4.8). We prefer a different strategy. Remember where we are in the iteration: we have just made an x-step $s(\mu_-)$ from x_- to x_c, and we now want x_+; we have obtained δ_c from δ_-. Before we get H_c, we still have $(H_- + \mu_- I)$ in factored form, and so it costs only $O(n^2)$ to compute

$$\Phi \triangleq \| s(\mu_-) \| - \delta_c \quad \text{and} \quad \Phi' \triangleq \frac{s(\mu_-)^T (H_- + \mu_- I)^{-1} s(\mu_-)}{\| s(\mu_-) \|} .$$

In analogy with (6.4.8), we use

$$\mu_0 = \mu_- - \frac{\| s(\mu_-) \|}{\delta_c} \frac{\Phi}{\Phi'} , \qquad (6.4.9)$$

the value of μ gotten from μ_-, the previous model, and the new trust radius δ_c. If $\delta_c = \delta_-$, then μ_0 is exactly the value we would have gotten had we taken one more μ-iteration on the previous model. On the other hand, if the previous iteration took the Newton step, then we find μ_0 by a different technique mentioned later.

Another computational detail, important to the performance of the algorithm for solving (6.4.5), is the generation and updating of lower and upper bounds u_+ and l_+ on μ_+. These bounds are used with (6.4.8) in the same way that bounds were used to safeguard the backtrack steps in Algorithm 6.3.5; that is, we restrict μ_+ to be in $[l_+, u_+]$. Since the Newton iteration applied to (6.4.5) always undershoots μ_*, we take $l_0 = -\Phi(0)/\Phi'(0)$. We then calculate

$$\mu_+^N = \mu_c - \Phi(\mu_c)/\Phi'(\mu_c)$$

along with each calculation of (6.4.8), and update the lower bound to $l_+ = \max\{l_c, \mu_+^N\}$, where l_c is the current lower bound. Also, since

$$\delta_c = \|(H_c + \mu_* I)^{-1} \nabla f(x_c)\|_2 < \frac{\|\nabla f(x_c)\|_2}{\mu_*}$$

because H_c is positive definite and $\mu_* > 0$, we take $\|\nabla f(x_c)\|_2/\delta_c$ as our initial upper bound u_0 on μ_*. Then, at each iteration, if $\Phi(\mu_c) < 0$ we update the upper bound to $u_+ = \min\{u_c, \mu_c\}$, where u_c is the current upper bound.

If, at any iteration, μ_+ is not in $[l_+, u_+]$, we follow Moré in choosing μ_+ by

$$\mu_+ = \max\{(l_+ \cdot u_+)^{1/2}, 10^{-3} u_+\}, \tag{6.4.10}$$

the second term being a safeguard against near-zero values of l_+. In practice, these bounds are invoked most frequently in calculating μ_0. In particular, (6.4.10) is used to define μ_0 whenever (6.4.9) cannot be used because the previous iteration used the Newton step.

Finally, we do not solve (6.4.5) to any great accuracy, settling instead for

$$\|s(\mu)\|_2 \in \left[\frac{3\delta_c}{4}, \frac{3\delta_c}{2}\right]$$

The reason is that, as will be seen in Subsection 6.4.3, the trust region is never increased or decreased by a factor smaller than 2. So if the current trust radius is δ_c, the previous one was either greater than or equal to $2\delta_c$, or less than or equal to $\delta_c/2$. Thus we consider the actual value of δ_c to be rather arbitrary within the range $[3\delta_c/4, 3\delta_c/2]$, which splits the difference in either direction, and it seems reasonable to allow $\|s(\mu)\|_2$ to have any value in this range. Some other implementations, for example Moré (1977), require $\|s(\mu)\|_2 \in [0.9\delta_c, 1.1\delta_c]$, and it is not clear whether either is better. Experience shows that either choice is satisfied in practice with an average of between 1 and 2 values of μ per x-iteration.

We have now completed our discussion of an algorithm for approximately solving (6.4.5); a simple example is given below.

EXAMPLE 6.4.3   Let $f(x_1, x_2) = x_1^4 + x_1^2 + x_2^2$, $x_c = (1, 1)^T$, $H_c = \nabla^2 f(x_c)$, $\delta_c = \frac{1}{2}$, $\mu_- = 0$. Then

$$\nabla f(x_c) = (6, 2)^T,$$

$$H_c = \begin{bmatrix} 14 & 0 \\ 0 & 2 \end{bmatrix},$$

$$s_c^N = -H_c^{-1} \nabla f(x_c) = (-\tfrac{3}{7}, -1)^T,$$

$$\| s_c^N \|_2 \cong 1.088.$$

Since $\| s_c^N \|_2 > \frac{3}{2}\delta_c$, the Newton step is too long, and we seek some $\mu > 0$ such that

$$\|(H_c + \mu I)^{-1} \nabla f(x_c)\|_2 \in \left[\frac{3\delta_c}{4}, \frac{3\delta_c}{2} \right] = [0.375, 0.75]$$

(see Figure 6.4.4). Since $\mu_- = 0$, we calculate

$$l_0 = 0 - \frac{\Phi(0)}{\Phi'(0)} \cong \frac{0.588}{0.472} \cong 1.25,$$

$$u_0 = \frac{\|\nabla f(x_c)\|_2}{\delta_c} \cong \frac{6.325}{0.5} \cong 12.6$$

$$\mu_0 = (l_0 \cdot u_0)^{1/2} \cong 3.97.$$

Next we calculate $s(\mu_0) = -(H_c + \mu_0 I)^{-1} \nabla f(x_c) \cong (-0.334, -0.335)^T$. Since

$$\| s(\mu_0) \|_2 \cong 0.473 \in [0.375, 0.75],$$

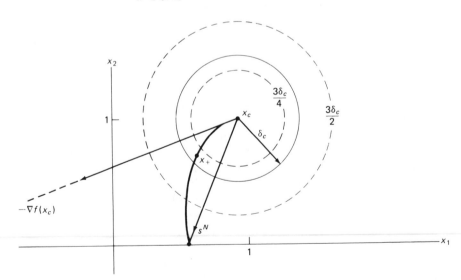

Figure 6.4.4   "Hook" step for Example 6.4.3

we take $\mu_c = \mu_0$ and $x_+ = x_c + s(\mu_c) \cong (0.666, 0.665)^T$ as our approximate solution to (6.4.1). Note that in this case we have not used iteration (6.4.8). For illustration, the reader can verify that one application of (6.4.8) would result in

$$\mu_1 = \mu_0 - \frac{\Phi(\mu_0)}{\Phi'(\mu_0)} \cdot \frac{\| s(\mu_0) \|_2}{\delta_c} = 3.97 - \frac{-0.0271}{-0.0528} \cdot \frac{+0.473}{+0.5} = 3.49$$

and that $s(\mu_1) = (-0.343, -0.365)^T$, $\| s(\mu_1) \|_2 = 0.5006$. Incidentally, (6.4.5) is solved exactly by $\mu_* \cong 3.496$.

Our algorithm for computing the locally constrained optimal ("hook") step by the techniques described in this section is given in Algorithm A6.4.2 in the appendix. It is preceded by the driver for a complete global step using the locally constrained optimal approach in Algorithm A6.4.1. The complete step involves selecting a new point x_+ by Algorithm A6.4.2, checking if x_+ is satisfactory and updating the trust radius (Algorithm A6.4.5), and repeating this process if necessary. All the algorithms use a diagonal scaling matrix on the variable space that will be discussed in Section 7.1.

Some recent research [Gay (1981), Sorensen (1982)] has centered on approximately solving the locally constrained model problem (6.4.1) by the techniques of this section expanded to cover the case when H_c is not positive definite. These techniques may become an important addition to the algorithms of this section.

6.4.2 THE DOUBLE DOGLEG STEP

The other implementation of the model-trust region approach that we discuss is a modification of the trust region algorithm introduced by Powell (1970a). It also finds an approximate solution to problem (6.4.1). However, rather than finding a point $x_+ = x_c + s(\mu_c)$ on the $s(\mu)$ curve such that $\| x_+ - x_c \|_2 \cong \delta_c$, it approximates this curve by a piecewise linear function connecting the "Cauchy point" C.P., the minimizer of the quadratic model m_c in the steepest-descent direction, to the Newton direction for m_c, as indicated in Figure 6.4.5. Then it chooses x_+ to be the point on this polygonal arc such that $\| x_+ - x_c \|_2 = \delta_c$, unless $\| H_c^{-1} \nabla f(x_c) \|_2 < \delta_c$, in which case x_+ is the Newton point. This strategy can alternatively be viewed as a simple strategy for looking in the steepest-descent direction when δ_c is small, and more and more toward the quasi-Newton direction as δ_c increases.

The specific way of choosing the double dogleg curve makes it have two important properties. First, as one proceeds along the piecewise linear curve from x_c to C.P. to $\hat{N}$ to x_+^N, the distance from x_c increases monotonically. Thus for any $\delta \leq \| H_c^{-1} \nabla f(x_c) \|$, there is a unique point x_+ on the curve such that $\| x_+ - x_c \|_2 = \delta$. This just makes the process well defined. Second, the value of the quadratic model $m_c(x_c + s)$ decreases monotonically as s goes

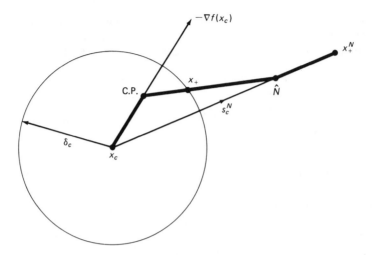

Figure 6.4.5 The double dogleg curve, $x_c \to$ C.P. $\to \hat{N} \to x_+^N$

along the curve from x_c to C.P. to $\hat{N}$ to x_+^N. This makes the process reasonable.

Point C.P. in Figure 6.4.5 is found by solving

$$\min_{\lambda \in \mathbb{R}} m_c(x - \lambda \nabla f(x_c)) = f(x_c) - \lambda \|\nabla f(x_c)\|_2^2 + \tfrac{1}{2}\lambda^2 \nabla f(x_c)^T H_c \nabla f(x_c),$$

which has the unique solution

$$\lambda_* = \frac{\|\nabla f(x_c)\|_2^2}{\nabla f(x_c)^T H_c \nabla f(x_c)}.$$

Therefore,

$$\text{C.P.} = x_c - \lambda_* \nabla f(x_c),$$

and if

$$\delta_c \leq \lambda_* \|\nabla f(x_c)\|_2 = \|\nabla f(x_c)\|_2^3 / \nabla f(x_c)^T H_c \nabla f(x_c),$$

the algorithm takes a step of length δ_c in the steepest-descent direction:

$$x_+ = x_c - \frac{\delta_c}{\|\nabla f(x_c)\|_2} \nabla f(x_c).$$

In order for the double dogleg curve to satisfy the first property stated above, it must be shown that the Cauchy point C.P. is no farther from x_c than the Newton point x_+^N. Let

$$s^{\text{C.P.}} = -\lambda_* \nabla f(x_c), \quad s^N = -H_c^{-1} \nabla f(x_c).$$

Then

$$\| s^{C.P.} \|_2 = \frac{\| \nabla f(x_c) \|_2^3}{\nabla f(x_c)^T H_c \, \nabla f(x_c)}$$

$$\leq \frac{\| \nabla f(x_c) \|_2^3}{\nabla f(x_c)^T H_c \, \nabla f(x_c)} \frac{\| \nabla f(x_c) \|_2 \, \| H_c^{-1} \, \nabla f(x_c) \|_2}{\nabla f(x_c)^T H_c^{-1} \, \nabla f(x_c)}$$

$$= \frac{\| \nabla f(x_c) \|_2^4}{(\nabla f(x_c)^T H_c \, \nabla f(x_c))(\nabla f(x_c)^T H_c^{-1} \, \nabla f(x_c))} \| s^N \|_2 \triangleq \gamma \| s^N \|_2. \qquad (6.4.11)$$

It is an exercise to prove that $\gamma \leq 1$ for any positive definite H_c, with $\gamma = 1$ only if $S^{C.P.} = S^N$, a case which we exclude for the remainder of this section. Thus,

$$\| s^{C.P.} \|_2 \leq \gamma \| s^N \|_2 \leq \| s^N \|_2.$$

The point $\hat{N}$ on the double dogleg curve is now chosen to have the form

$$\hat{N} = x_c - \eta H_c^{-1} \, \nabla f(x_c)$$

for some η such that

$$\gamma \leq \eta \leq 1$$

and

$$m_c(x) \text{ decreases monotonically along the line from C.P. to } \hat{N}. \qquad (6.4.12)\,'$$

Since we know that $m_c(x)$ decreases monotonically from x_c to C.P. and from $\hat{N}$ to x_+^N, (6.4.12) will guarantee that $m_c(x)$ decreases monotonically along the entire double dogleg curve.

To satisfy (6.4.12), η must be chosen so that the directional derivative along the line connecting C.P. and $\hat{N}$ is negative at every point on that line segment. Parametrize this line segment by

$$x_+(\lambda) = x_c + s^{C.P.} + \lambda(\eta s^N - s^{C.P.}) \qquad 0 \leq \lambda \leq 1.$$

The directional derivative of m_c along this line at $x_+(\lambda)$ is

$$\nabla m_c(x_+(\lambda))^T(\eta s^N - s^{C.P.}) = [\nabla f(x_c) + H_c(s^{C.P.} + \lambda(\eta s^N - s^{C.P.}))]^T(\eta s^N - s^{C.P.})$$

$$= (\nabla f(x_c) + H_c s^{C.P.})^T(\eta s^N - s^{C.P.})$$

$$+ \lambda(\eta s^N - s^{C.P.})^T H_c(\eta s^N - s^{C.P.}). \qquad (6.4.13)$$

Since H_c is positive definite, the right-hand side of (6.4.13) is a monotonically increasing function of λ. Therefore, we need only require that (6.4.13) be negative for $\lambda = 1$ to make it negative for $0 \leq \lambda \leq 1$. Some cancellation and substitution shows that this condition is equivalent to

$$0 > (1 - \eta)(\nabla f(x_c)^T(\eta s^N - s^{C.P.})) = (1 - \eta)(\gamma - \eta)(\nabla f(x_c)^T H_c^{-1} \nabla f(x_c)),$$

which is satisfied for any $\eta \in (\gamma, 1)$.

Thus $\hat{N}$ can be chosen as any point in the Newton direction $x_c + \eta s^N$ for which η is between 1 and γ given by (6.4.11). Powell's original choice was $\eta = 1$, giving the single dogleg curve. Computational testing, however, has shown that an earlier bias to the Newton direction seems to improve the performance of the algorithm. Therefore, Dennis and Mei (1979) suggest the choice $\eta = 0.8\gamma + 0.2$, leading to a double dogleg curve as in Figure 6.4.5.

The choices of C.P. and $\hat{N}$ completely specify the double dogleg curve. The selection of the x_+ on the curve such that $\| x_+ - x_c \|_2 = \delta_c$ is then a straightforward inexpensive algebraic problem, as illustrated in Example 6.4.4 below. Notice that the entire algorithm costs only $O(n^2)$ arithmetic operations after s_c^N has been calculated.

EXAMPLE 6.4.4 Let $f(x)$, x_c, H_c be given by Example 6.4.3, and let $\delta_c = 0.75$. Recall that

$$\nabla f(x_c) = (6, 2)^T, \qquad H_c = \begin{bmatrix} 14 & 0 \\ 0 & 2 \end{bmatrix}, \qquad s_c^N = (-\tfrac{3}{7}, -1)^T.$$

Since $\| s_c^N \|_2 = 1.088 > \delta_c$, the double dogleg algorithm first calculates the step to the Cauchy point. The reader can verify that it is given by

$$s^{C.P.} = -\left(\frac{40}{512}\right)\binom{6}{2} \cong \binom{-0.469}{-0.156}.$$

Since $\| s^{C.P.} \|_2 \cong 0.494 < \delta_c$, the algorithm next calculates the step to the point $\hat{N}$. The reader can verify that

$$\gamma = \frac{(40)^2}{(512)(\tfrac{32}{7})} \cong 0.684,$$

$$\eta = 0.8\gamma + 0.2 \cong 0.747,$$

$$s^{\hat{N}} \triangleq \eta s_c^N \cong \binom{-0.320}{-0.747}.$$

Since $\| s^{\hat{N}} \|_2 \cong 0.813 > \delta_c$, the double dogleg step must be along the line connecting C.P. and $\hat{N}$—that is, $s_c = s^{C.P.} + \lambda(s^{\hat{N}} - s^{C.P.})$ for the $\lambda \in (0, 1)$ for which $\| s_c \|_2 = \delta_c$. λ is calculated by solving for the positive root of the quadratic equation

$$\| s^{C.P.} + \lambda(s^{\hat{N}} - s^{C.P.}) \|_2^2 = \delta_c^2,$$

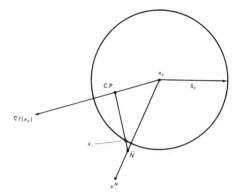

Figure 6.4.6   The double dogleg step of Example 6.4.4

which is $\lambda \cong 0.867$. Thus,

$$s_c = s^{\text{C.P.}} + 0.867(s^{\hat{N}} - s^{\text{C.P.}}) \cong \begin{pmatrix} -0.340 \\ -0.669 \end{pmatrix},$$

$$x_+ = x_c + s_c \cong \begin{pmatrix} 0.660 \\ 0.331 \end{pmatrix}.$$

The entire calculation is shown in Figure 6.4.6.

Our algorithm for selecting a point by the double dogleg strategy is given in Algorithm A6.4.4 in the appendix. Computational experience shows that the points selected by this strategy are, at worst, marginally inferior to those selected by the locally constrained optimal step of Algorithm A6.4.2. However, there is a trade-off: the complexity and running time of Algorithm A6.4.4 are considerably less than those of Algorithm A6.4.2. This makes the dogleg strategy attractive, especially for problems where the cost of function and derivative evaluation is not high.

Algorithm A6.4.3 contains the driver for a complete global step using the double dogleg algorithm to find candidate points x_+. Both algorithms contain the diagonal scaling of the variable space that is described in Section 7.1.

6.4.3   UPDATING THE TRUST REGION

To complete the global step given in Algorithm 6.4.2, one needs to decide whether the point x_+, found by using the techniques of Subsection 6.4.1 or 6.4.2, is a satisfactory next iterate. If x_+ is unacceptable, one reduces the size of the trust region and minimizes the same quadratic model on the smaller trust region. If x_+ is satisfactory, one must decide whether the trust region should be increased, decreased, or kept the same for the next step (of Algorithm 6.1.1). The basis for these decisions is discussed below.

The condition for accepting x_+ is the one developed in Section 6.3,

$$f(x_+) \le f(x_c) + \alpha g_c^T(x_+ - x_c), \tag{6.4.14}$$

where $g_c = \nabla f(x_c)$ or an approximation to it, and α is a constant in $(0, \frac{1}{2})$. In our algorithm we again choose $\alpha = 10^{-4}$, so that (6.4.14) is hardly more stringent than $f(x_+) < f(x_c)$. If x_+ does not satisfy (6.4.14), we reduce the trust region by a factor between $\frac{1}{10}$ and $\frac{1}{2}$ and return to the approximate solution of the locally constrained minimization problem by the locally constrained optimal step or double dogleg method. The reduction factor is determined by the same quadratic backtrack strategy used to decrease the line-search parameter in Algorithm A6.3.1. We model $f(x_c + \lambda(x_+ - x_c))$ by the quadratic $m_q(\lambda)$ that fits $f(x_c)$, $f(x_+)$, and the directional derivative $g_c^T(x_+ - x_c)$ of f at x_c in the direction $x_+ - x_c$. We then let the new trust radius δ_+ extend to the minimizer of this model, which occurs at

$$\lambda_* = \frac{-g_c^T(x_+ - x_c)}{2[f(x_+) - f(x_c) - g_c^T(x_+ - x_c)]}$$

(see Figure 6.4.7). Thus, $\delta_+ = \lambda_* \| x_+ - x_c \|_2$. If

$$\lambda_* \| x_+ - x_c \|_2 \notin [\tfrac{1}{10}\delta_c, \tfrac{1}{2}\delta_c],$$

we instead set δ_+ to the closer endpoint of this interval for the same reasons as in the line-search algorithm. Note that $\| x_+ - x_c \|_2 = \delta_c$ if x_+ is selected by the double dogleg strategy of Subsection 6.4.2, but that we only know that $\| x_+ - x_c \|_2 \in [\tfrac{3}{4}\delta_c, \tfrac{3}{2}\delta_c]$ if x_+ is selected by the locally constrained optimal strategy of Subsection 6.4.1.

Now suppose we have just found an x_+ that satisfies (6.4.14). If x_+ is a full Newton step from x_c, then we make the step, update δ, form the new model, and go to the next iteration. However, if $x_+ - x_c$ isn't the Newton step, we first consider whether we should try a larger step from x_c, using the current

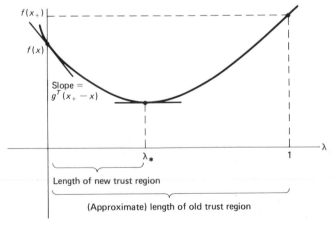

Figure 6.4.7   Reducing the trust region when x_+ is unacceptable.

model. The reason this may be worthwhile is that we may thereby avoid having to evaluate the gradient (and Hessian) at x_+, which is often the dominant cost in expensive problems. The reason a longer step may be possible is that the trust region may have become small during the course of the algorithm and may now need to be increased. This occurs when we leave a region where the step bound had to be small because the function was not well represented by any quadratic, and enter a region where the function is more nicely behaved.

To decide whether to attempt a larger step from x_c, we compare the actual reduction $\Delta f \triangleq f(x_+) - f(x_c)$ to the predicted reduction $\Delta f_{\text{pred}} \triangleq m_c(x_+) - f(x_c)$ and accept x_+ as the next iterate unless either the agreement is so good—i.e., $|\Delta f_{\text{pred}} - \Delta f| \le 0.1\,|\Delta f|$—that we suspect δ_c is an underestimate of the radius in which m_c adequately represents f, or the actual reduction in f is so large that the presence of negative curvature, and thus a continuing rapid decrease in $f(x)$, is implied—i.e., $f(x_+) \le f(x_c) + \nabla f(x_c)^T(x_+ - x_c)$. In both these cases we save x_+ and $f(x_+)$, but rather than move directly to x_+, we first double δ_c and compute a new x_+ using our current model. If (6.4.14) isn't satisfied for the new x_+, we drop back to the last good step we computed, but if it is, we consider doubling again. In practice we may save a significant number of gradient calls in this way.

An interesting situation in which the step bound must contract and then expand is when the algorithm passes close by a point that looks like a minimizer. The Newton steps shorten; the algorithm takes these Newton steps and behaves as though it were converging. Then the algorithm discovers a way out, the Newton steps lengthen, and the algorithm moves away. Such behavior is desirable, since a perturbed problem might indeed have a real minimizer at this point of distraction. We want to be able to increase δ_c rapidly in such a case. In one example, we saw six of these internal doublings of the step bound after a point of distraction.

Now suppose we are content with x_+ as our next iterate, and so we need to update δ_c to δ_+. We allow three alternatives: doubling, halving, or retaining the current step bound to get δ_+. The actual conditions are somewhat arbitrary; what is important is that if our current quadratic model is predicting the function well, we increase the trust region, but if it is predicting poorly, we decrease the trust region. If the quadratic model has predicted the actual function reduction sufficiently well—i.e., $\Delta f \le 0.75\,\Delta f_{\text{pred}}$—then we take $\delta_+ = 2\delta_c$. If the model has greatly overestimated the decrease in $f(x)$—i.e., $\Delta f > 0.1\,\Delta f_{\text{pred}}$—then we take $\delta_+ = \delta_c/2$. Otherwise $\delta_+ = \delta_c$.

EXAMPLE 6.4.5   Let $f(x)$, x_c, H_c, δ_c be given by Example 6.4.3, and suppose we have just taken the step determined in that example,

$$x_+ = x_c + s_c = \begin{pmatrix} 1 \\ 1 \end{pmatrix} - \begin{pmatrix} 0.334 \\ 0.335 \end{pmatrix} = \begin{pmatrix} 0.666 \\ 0.665 \end{pmatrix}.$$

Recall that

$$\nabla f(x_c) = (6,\ 2)^T, \qquad H_c = \nabla^2 f(x_c) = \begin{bmatrix} 14 & 0 \\ 0 & 2 \end{bmatrix}.$$

We want to decide whether x_+ is a satisfactory point, and update the trust region. First, we calculate

$$f(x_+) = 1.083,$$

$$f(x_c) + \alpha\ \nabla f(x_c)^T(x_+ - x_c) = 3 - 10^{-4}(2.673) = 2.9997.$$

Therefore, x_+ is acceptable. Next, we decide whether to try a larger step at the current iteration. We calculate

$$\Delta f = f(x_+) - f(x_c) = -1.917$$

and

$$\Delta f_{\text{pred}} = m_c(x_+) - f(x_c) = \nabla f(x_c)^T s_c + \tfrac{1}{2} s_c^T H_c s_c$$

$$= -2.673 + 0.892 = -1.781.$$

Since $|\Delta f - \Delta f_{\text{pred}}|/|\Delta f| = 0.071 < 0.1$, we double the trust region and go back to the locally constrained optimal step, Algorithm A6.4.1. The reader can confirm that with the new trust radius $\delta_c = 1$, Algorithm A6.4.1 will select the Newton step. It is an exercise to complete the updating of the trust region for this global iteration.

The algorithm for updating the trust region is Algorithm A6.4.5 in the appendix. It has several additional features.

1. It uses a minimum and maximum step length, discussed in Section 7.2. The trust radius is never allowed outside these bounds. The maximum step length is supplied by the user. The minimum step length is the quantity used to test for convergence in Algorithm 7.2.1. If x_+ is unsatisfactory, but the current trust region is already less than *minstep*, the global step is halted, because convergence would necessarily be detected at the end of the current global iteration. This situation may indicate convergence to a nonsolution, so a warning message should be printed.

2. When the approximate solution from Algorithm A6.4.2 or A6.4.4 to the constrained model problem is a Newton step that is shorter than the current trust radius, these algorithms immediately reduce the size of the trust region to the length of the Newton step. It is then still adjusted by Algorithm A6.4.5. This is an additional mechanism for regulating the trust region.

3. The algorithm is implemented using a diagonal scaling of the available space, discussed in Section 7.1.

Finally, we discuss how the initial estimate of the trust region radius, or step bound is obtained. Sometimes the user can supply a reasonable estimate based on his knowledge of the problem. If not, Powell (1970a) suggests using the length of the Cauchy step (see Subsection 6.4.2) as the initial trust region radius. Other strategies are possible. The updating strategy of Algorithm A6.4.5 does enable a trust region algorithm to recover in practice from a bad starting value of δ, but there is usually some cost in additional iterations. Therefore the initial trust region is reasonably important.

6.5 GLOBAL METHODS FOR SYSTEMS OF NONLINEAR EQUATIONS

We return now to the nonlinear equations problem:

$$\text{given} \quad F: \mathbb{R}^n \longrightarrow \mathbb{R}^n,$$

$$\text{find} \quad x_* \in \mathbb{R}^n \quad \text{such that} \quad F(x_*) = 0, \tag{6.5.1}$$

In this section we show how Newton's method for (6.5.1) can be combined with global methods for unconstrained optimization to produce global methods for (6.5.1).

The Newton step for (6.5.1) is

$$x_+ = x_c - J(x_c)^{-1}F(x_c), \tag{6.5.2}$$

where $J(x_c)$ is the Jacobian matrix of F at x_c. From Section 5.2 we know that (6.5.2) is locally q-quadratically convergent to x_*, but not necessarily globally convergent. Now assume x_c is not close to any solution x_* of (6.5.1). How would one decide then whether to accept x_+ as the next iterate? A reasonable answer is that $\| F(x_+)\|$ should be less than $\| F(x_c)\|$ for some norm $\| \cdot \|$, a convenient choice being the l_2 norm $\| F(x)\|_2^2 = F(x)^T F(x)$.

Requiring that our step result in a decrease of $\| F(x)\|_2$ is the same thing we would require if we were trying to find a minimum of the function $\| F(x)\|_2$. Thus, we have in effect turned our attention to the corresponding minimization problem:

$$\min_{x \in \mathbb{R}^n} f(x) = \tfrac{1}{2}F(x)^T F(x), \tag{6.5.3}$$

where the "$\tfrac{1}{2}$" is added for later algebraic convenience. Note that every solution to (6.5.1) is a solution to (6.5.3), but there may be local minimizers of (6.5.3) that are not solutions to (6.5.1) (see Figure 6.5.1). Therefore, although we could try to solve (6.5.1) simply by using a minimization routine on (6.5.3), it is better to use the structure of the orginal problem wherever possible, in particular to compute the Newton step (6.5.2). However, our global strategy for (6.5.1) will be based on a global strategy for the related minimization problem (6.5.3).

An important question to ask is, "What is a descent direction for prob-

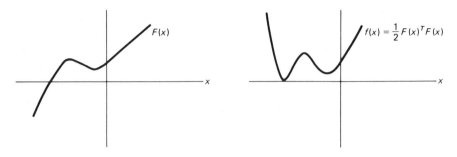

Figure 6.5.1   The nonlinear equations and corresponding minimization problem, in one dimension

lem (6.5.3)?" It is any direction p for which $\nabla f(x_c)^T p < 0$, where

$$\nabla f(x_c) = \frac{d}{dx} \sum_{i=1}^{n} \tfrac{1}{2}(f_i(x_c))^2 = \sum_{i=1}^{n} \nabla f_i(x_c) \cdot f(x_c) = J(x_c)^T F(x_c).$$

Therefore, the steepest-descent direction for (6.5.3) is along $-J(x_c)^T F(x_c)$. Furthermore, the Newton direction along $s^N = -J(x_c)^{-1} F(x_c)$ is a descent direction, since

$$\nabla f(x_c)^T s^N = -F(x_c)^T J(x_c) J(x_c)^{-1} F(x_c) = -F(x_c)^T F(x_c) < 0$$

as long as $F(x_c) \neq 0$. This may seem surprising, but it is geometrically reasonable. Since the Newton step yields a root of

$$M_c(x_c + s) = F(x_c) + J(x_c)s,$$

it also goes to a minimum of the quadratic function

$$\begin{aligned}\hat{m}_c(x_c + s) &\triangleq \tfrac{1}{2} M_c(x_c + s)^T M_c(x_c + s)\\ &= \tfrac{1}{2} F(x_c)^T F(x_c) + (J(x_c)^T F(x_c))^T s + \tfrac{1}{2} s^T (J(x_c)^T J(x_c))s,\end{aligned} \qquad (6.5.4)$$

because $\hat{m}_c(x_c + s) \geq 0$ for all s and $\hat{m}_c(x_c + s^N) = 0$. Therefore, s^N is a descent direction for $\hat{m}_c$, and since the gradients at x_c of $\hat{m}_c$ and f are the same, it is also a descent direction for f.

The above development motivates how we will create global methods for (6.5.1). They will be based on applying our algorithms of Section 6.3 or 6.4 to the quadratic model $\hat{m}_c(x)$ in (6.5.4). Since $\nabla^2 \hat{m}(x_c) = J(x_c)^T J(x_c)$, this model is positive definite as long as $J(x_c)$ is nonsingular, which is consistent with the fact that $x_c + s^N$ is the unique root of $M_c(x)$ and thus the unique minimizer of $\hat{m}_c(x)$ in this case. Thus, the model $\hat{m}_c(x)$ has the attractive properties that its minimizer is the Newton point for the original problem, and that all its descent directions are descent directions for $f(x)$ because $\nabla \hat{m}_c(x_c) = \nabla f(x_c)$. Therefore methods based on this model, by going downhill and trying to minimize $\hat{m}_c(x)$, will combine Newton's method for nonlinear equations with global methods for an associated minimization problem. Note that $\hat{m}_c(x)$ is not quite the same as the quadratic model of $f(x) = \frac{1}{2}F(x)^T F(x)$ around x_c,

$$m_c(x_c + s) = f(x_c) + \nabla f(x_c)^T s + \frac{1}{2} s^T \, \nabla^2 f(x_c) s,$$

because $\nabla^2 f(x_c) \neq J(x_c)^T J(x_c)$ (see Exercise 18 of Chapter 5).

The application of the global methods of Sections 6.3 and 6.4 to the nonlinear equations problem is now straightforward. As long as $J(x_c)$ is sufficiently well conditioned, then $J(x_c)^T J(x_c)$ is safely positive definite, and the algorithms apply without change if we define the objective function by $\frac{1}{2} \| F(x) \|_2^2$, the Newton direction by $-J(x)^{-1} F(x)$, and the positive definite quadratic model by (6.5.4). This means that in a line-search algorithm, we search in the Newton direction, looking for a sufficient decrease in $\| F(x) \|_2$. In trust region algorithms, we approximately minimize $\hat{m}_c(x_c + s)$ subject to $\| s \|_2 \leq \delta_c$. If $\delta_c \geq \| J(x_c)^{-1} F(x_c) \|_2$, then the step attempted is the Newton step; otherwise, for the locally constrained optimal step, it is

$$s = -(J(x_c)^T J(x_c) + \mu_c I)^{-1} J(x_c)^T F(x_c) \tag{6.5.5}$$

for μ_c such that $\| s \|_2 \cong \delta_c$. Using either global strategy, we expect to be taking Newton steps for $F(x) = 0$ eventually. Example 6.5.1 shows how our global strategy using a line search would work on the example of Section 5.4, started here from a different point, one at which the Newton step is unsatisfactory.

EXAMPLE 6.5.1   Let $F: \mathbb{R}^2 \longrightarrow \mathbb{R}^2$,

$$F(x) = \begin{bmatrix} x_1^2 + x_2^2 - 2 \\ e^{x_1 - 1} + x_2^3 - 2 \end{bmatrix},$$

which has the root $x_+ = (1, 1)^T$, and let $x_0 = (2, 0.5)^T$. Define $f(x) = \frac{1}{2} F(x)^T F(x)$. Then

$$J(x_0) = \begin{bmatrix} 4 & 1 \\ e & 0.75 \end{bmatrix}, \qquad F(x_0) \cong \begin{bmatrix} 2.25 \\ 0.843 \end{bmatrix},$$

and

$$s_0^N = -J(x_0)^{-1} F(x_0) \cong \begin{bmatrix} -3.00 \\ 9.74 \end{bmatrix}.$$

Our line-search algorithm A6.3.1 will calculate $x_+ = x_0 + \lambda_0 s_0^N$ starting with $\lambda_0 = 1$, decreasing λ_0 if necessary until $f(x_+) < f(x_0) + 10^{-4} \lambda_0 \, \nabla f(x_0)^T s_0^N$. For $\lambda_0 = 1$,

$$x_+ = x_0 + s_0^N \cong \begin{bmatrix} -1.00 \\ 10.24 \end{bmatrix}, \qquad F(x_+) \cong \begin{bmatrix} 104 \\ 1071 \end{bmatrix},$$

so that the Newton step clearly is unsatisfactory. Therefore we reduce λ_0 by a quadratic backtrack, calculating

$$\lambda_1 = \frac{-\nabla f(x_0)^T s_0^N}{2[f(x_+) - f(x_0) - \nabla f(x)^T s_0^N]}. \tag{6.5.6}$$

In this case, $f(x_+) \cong 5.79 \times 10^5$, $f(x_0) \cong 2.89$, $\nabla f(x)^T s_0^N = -F(x_0)^T F(x_0) \cong -5.77$, so that (6.5.6) gives $\lambda_1 \cong 4.99 \times 10^{-6}$. Since $\lambda_1 < 0.1$, Algorithm A6.3.1 sets $\lambda_1 = 0.1$,

$$x_+ = x_0 + 0.1 s_0^N \cong \begin{bmatrix} 1.70 \\ 1.47 \end{bmatrix}, \qquad F(x_+) \cong \begin{bmatrix} 3.06 \\ 3.21 \end{bmatrix}.$$

This is still unsatisfactory, and so Algorithm A6.3.2 does a cubic backtrack. The reader can verify that a backtrack yields $\lambda_2 \cong 0.0659$. Since $\lambda_2 > \frac{1}{2}\lambda_1$, the algorithm sets $\lambda_2 = \frac{1}{2}\lambda_1 = 0.05$,

$$x_+ = x_0 + 0.05 s_0^N \cong \begin{bmatrix} 1.85 \\ 0.987 \end{bmatrix}, \qquad F(x_+) \cong \begin{bmatrix} 2.40 \\ 1.30 \end{bmatrix}.$$

This point is still unsatisfactory, since $f(x_+) \cong 3.71 > f(x_0)$, so the algorithm does another cubic backtrack. It yields $\lambda_3 \cong 0.0116$, which is used since it is in the interval $[\lambda_2/10, \lambda_2/2] = [0.005, 0.025]$. Now

$$x_+ = x_0 + 0.0116 s_0^N \cong \begin{bmatrix} 1.965 \\ 0.613 \end{bmatrix}, \qquad F(x_+) \cong \begin{bmatrix} 2.238 \\ 0.856 \end{bmatrix}.$$

This point is satisfactory, since

$$f(x_+) \cong 2.87 < 2.89 \cong f(x_0) + 10^{-4}(0.0116)\,\nabla f(x_0)^T s_0^N,$$

so we set $x_1 = x_+$ and proceed to the next iteration.

It is interesting to follow this problem further. At the next iteration,

$$s_1^N = -J(x_1)^{-1} F(x_1) \cong \begin{bmatrix} -1.22 \\ 2.07 \end{bmatrix},$$

$$x_2^N = x_1 + s_1^N \cong \begin{bmatrix} 0.750 \\ 2.68 \end{bmatrix}, \qquad F(x_2^N) \cong \begin{bmatrix} 5.77 \\ 18.13 \end{bmatrix},$$

which is again unsatisfactory. However, the first backtrack step is successful: Algorithm A6.31 computes $\lambda_1 = 0.0156$, and, since this is less than 0.1, it sets $\lambda_1 = 0.1$,

$$x_+ = x_1 + 0.1 s_1^N \cong \begin{bmatrix} 1.84 \\ 0.820 \end{bmatrix}, \qquad F(x_+) \cong \begin{bmatrix} 2.07 \\ 0.876 \end{bmatrix},$$

and the step is accepted, since

$$f(x_+) \cong 2.53 < 2.87 \cong f(x_1) + 10^{-4}(0.1)\,\nabla f(x_1)^T s_1^N.$$

At the next iteration,

$$s_2^N = -J(x_2)^{-1}F(x_2) \cong \begin{bmatrix} -0.0756 \\ 0.0437 \end{bmatrix},$$

$$x_3^N = x_2 + s_2^N \cong \begin{bmatrix} 1.088 \\ 1.257 \end{bmatrix}, \qquad F(x_3^N) \cong \begin{bmatrix} 0.762 \\ 1.077 \end{bmatrix},$$

so the Newton step is very good. From here on, Newton's method converges q-quadratically to $x_* = (1, 1)^T$.

The only complication to this global strategy arises when J is nearly singular at the current point x_c. In this case we cannot accurately calculate the Newton direction $s^N = -J(x_c)^{-1}F(x_c)$, and the model Hessian $J(x_c)^T J(x_c)$ is nearly singular. To detect this situation, we perform the QR factorization of $J(x_c)$, and if R is nonsingular, estimate its condition number by Algorithm A3.3.1. If R is singular or its estimated condition number is greater than macheps$^{-1/2}$, we perturb the quadratic model to

$$\hat{m}_c(x_c + s) = \tfrac{1}{2}F(x_c)^T F(x_c) + (J(x_c)^T F(x_c))^T s + \tfrac{1}{2}s^T H_c s,$$

where

$$H_c = J(x_c)^T J(x_c) + (n \cdot \text{macheps})^{1/2} \| J(x_c)^T J(x_c) \|_1 \cdot I.$$

(The condition number of H_c is about macheps$^{-1/2}$; see Exercise 23 for a more precise justification.) From Section 6.4 we know that the Newton step to the minimizer of this model, $s^N = -H_c^{-1} J(x_c)^T F(x_c)$, solves

$$\min_{s \in \mathbb{R}^n} \hat{m}_c(x_c + s) = \| J(x_c)s + F(x_c) \|_2^2$$

subject to $\| s \|_2 \leq \delta$

for some $\delta > 0$. It is also a descent direction for $f(x) = \tfrac{1}{2}\| F(x) \|_2^2$. Therefore, we prefer this modified Newton step to some $s = -\hat{J}(x_c)^{-1}F(x_c)$ that would result from a perturbation $\hat{J}(x_c)$ of the QR factorization of $J(x_c)$. See Exercise 24 for further discussion of this step.

Driver D6.1.3 shows how we use the global algorithms for unconstrained minimization in solving systems of nonlinear equations. First at each iteration, Algorithm A6.5.1 calculates the $Q_c R_c$ factorization of $J(x_c)$ and estimates the condition number of R_c. If the condition number estimate is less than macheps$^{-1/2}$, it calculates the Newton step $-J(x_c)^{-1}F(x_c)$, otherwise it calculates the perturbed Newton step described above. Then the driver calls either the line-search, dogleg, or locally constrained optimal algorithm as selected by the user. All these global algorithms require $\nabla f(x_c) = J(x_c)^T F(x_c)$, which is also calculated by Algorithm D6.1.3. In addition, the trust region algorithms require the Cholesky factorization $L_c L_c^T$ of $H_c = J(x_c)^T J(x_c)$ (except for the case when H_c is perturbed as described above). Since we have $J(x_c) = Q_c R_c$, it is immediate that $L_c = R_c^T$. Finally, all our algorithms for nonlinear equations are implemented using a diagonal scaling matrix D_F on $F(x)$, discussed in Section 7.2. It causes

$$f(x) = \tfrac{1}{2}\| D_F F(x) \|_2^2, \qquad \nabla f(x) = J(x)^T D_F^2 F(x), \quad \text{and} \quad H = J(x)^T D_F^2 J(x).$$

There is one significant case when these global algorithms for nonlinear equations can fail. It is when a local minimizer of $f(x) = \tfrac{1}{2}\| F(x) \|_2^2$ is not a root of $F(x)$ (see Figure 6.5.1). A global minimization routine, started close to such a point, may converge to it. There is not much one can do in such a case, except report what has happened and advise the user to try to restart nearer to a root of $F(x)$ if possible. The stopping routine for nonlinear equations (Algorithm A7.2.3) detects this situation and produces such a message. Research is currently underway on methods that might be able to restart themselves from such local minimizers. See Allgower and Georg (1980) and Zirilli (1982).

6.6   EXERCISES

1. Let $f(x) = 3x_1^2 + 2x_1 x_2 + x_2^2$, $x_0 = (1, 1)^T$. What is the steepest-descent direction for f from x_0? Is $(1, -1)^T$ a descent direction?

2. Show that for a positive definite quadratic $f(x_k + s) = g^T s + \tfrac{1}{2} s^T \nabla^2 f(x_k) s$, the steepest-descent step from x_k is given by equation (6.2.5).

3. Given a $\hat{c} \in (0, 1)$, generalize Example 6.2.2 to an example where the sequence of points generated by the steepest-descent algorithm from a specific x_0 obeys $\| x_{k+1} - x_* \| = \hat{c} \| x_k - x_* \|$, $k = 0, 1, \ldots$.

4. Let $f(x) = x^2$. Show that the infinite sequence of points from Figure 6.3.2, $x_i = (-1)^i (1 + 2^{-i})$, $i = 0, 1, \ldots$, is not permitted by (6.3.3) for any $\alpha > 0$.

5. Let $f(x) = x^2$. Show that the infinite sequence of points $x_i = 1 + 2^{-i}$, $i = 0, 1, \ldots$, is permitted by (6.3.3) for any $\alpha \le \tfrac{1}{2}$ but is prohibited by (6.3.4) for any $\beta > 0$.

6. Show that if $f(x)$ is a positive definite quadratic, the Newton step from any $x_k \in \mathbb{R}^n$ satisfies condition (6.3.3) for $\alpha \le \tfrac{1}{2}$ and (6.3.4) for any $\beta > 0$.

7. Prove that if (6.3.4) is replaced by

$$f(x_{k+1}) \ge f(x_k) + \beta \, \nabla f(x_k)^T (x_{k+1} - x_k), \qquad \beta \in (\alpha, 1),$$

then Theorems 6.3.2 and 6.3.3 remain true; and that if in addition $\beta > \tfrac{1}{2}$, then Theorem 6.3.4 also remains true. This is Goldstein's original condition.

8. Let $f(x) = x_1^4 + x_2^2$, $x_c = (1, 1)^T$, and the search direction p_c be determined by the Newton step. What will x_+ be, using:
   (a) The Newton step?
   (b) Line-search Algorithm A6.3.1?
   (c) A "perfect line-search" algorithm that sets x_+ to the minimizer of $f(x)$ in the direction p_c from x_c? [Answer (c) approximately.]

9. Let $f(x) = \frac{1}{2}x_1^2 + x_2^2$, $x_c = (1, 1)^T$. What is the exact solution to (6.4.1) if $\delta = 2$? if $\delta = \frac{5}{6}$? [*Hint*: Try $\mu = 1$ for the second part.]

10. Prove that the locally constrained optimal curve $s(\mu)$ in Figure 6.4.2 is not necessarily planar by constructing a simple three-dimensional example in which x_c and some three points on the curve are not coplanar. [$f(x) = x_1^2 + 2x_2^2 + 3x_3^2$ will do.]

11. Let $H \in \mathbb{R}^{n \times n}$ be symmetric and positive definite, and let $v_1, \ldots, v_n$ be an orthonormal basis of eigenvectors for H with corresponding eigenvalues $\lambda_1, \ldots, \lambda_n$. Show that for

$$g = \sum_{j=1}^{n} \alpha_j v_j \quad \text{and} \quad \mu \geq 0, \quad (H + \mu I)^{-1}g = \sum_{j=1}^{n} \left(\frac{\alpha_j}{\lambda_j + \mu}\right)v_j.$$

How does this relate to the model $m_c(\mu)$ for $\|H + \mu I^{-1}g\| - \delta$ used in the locally constrained optimal algorithm?

12. Let H, g be given by Exercise 11, and define $s(\mu) = (H + \mu I)^{-1}g$ for $\mu \geq 0$ and $\eta(\mu) = \|s(\mu)\|_2$. Using the techniques of Exercise 11, show that

$$\frac{d}{d\mu}\eta(\mu) = -\frac{s(\mu)^T(H + \mu I)^{-1}s(\mu)}{\|s(\mu)\|_2}.$$

[You can also show that $(d^2/d\mu^2)\eta(\mu) > 0$ for all $\mu > 0$ using these techniques, but the proof is tedious.]

13. Let $f(x) = \frac{1}{2}x_1^2 + x_2^2$, $x_0 = (1, 1)^T$, $g = \nabla f(x_0)$, $H = \nabla^2 f(x_0)$. Calculate the Cauchy point of f from x_0 (see Subsection 6.4.2), and the point $\hat{N} = x_0 - (0.8\gamma + 0.2)H^{-1}g$ as used in our double dogleg algorithm. Then graph the double dogleg curve and show graphically the values of x_1 if $\delta = 1$; if $\delta = \frac{5}{4}$.

14. Let $H \in \mathbb{R}^{n \times n}$ be symmetric and positive definite, $g \in \mathbb{R}^n$. Prove that $(g^T g)^2 \leq (g^T H g)(g^T H^{-1}g)$. [*Hint*: Let $u = H^{1/2}g$, $v = H^{-1/2}g$, and apply the Cauchy-Schwarz inequality.]

15. Complete Example 6.4.5.

16. Let $F(x) = (x_1, 2x_2)^T$, $x_0 = (1, 1)^T$. Using the techniques of Section 6.5, what is the "steepest-descent" step from x_0 for $F = 0$? If all your steps were in steepest-descent directions, what rate of convergence to the root of F would you expect?

17. Let $F: \mathbb{R}^n \longrightarrow \mathbb{R}^n$ be continuously differentiable and $\hat{x} \in \mathbb{R}^n$. Suppose $\hat{x}$ is a local minimizer of $f(x) \triangleq \frac{1}{2}F(x)^T F(x)$ but $F(\hat{x}) \neq 0$. Is $J(\hat{x})$ singular? If $J(\hat{x})$ is singular, must $\hat{x}$ be a local minimizer of $f(x)$?

18. An alternative quadratic model for $F = 0$ to the one used in the global algorithms of Section 6.5 is the first three terms of the Taylor series of $\frac{1}{2}\|F(x_c + s)\|_2^2$. Show that this model is

$$m_c(x_c + s) = \frac{1}{2}F(x_c)^T F(x_c) + [J(x_c)^T F(x_c)]^T s$$

$$+ \frac{1}{2}s^T[J(x_c)^T J(x_c) + \sum_{i=1}^{n} f_i(x_c) \nabla^2 f_i(x_c)]s.$$

How does this compare with the model used in Section 6.5? Is the Newton step for minimizing $m_c(x)$ the same as the Newton step for $F(x) = 0$? Comment on the attractiveness of this model versus the one used in Section 6.5. (An analogous situation with different conclusions occurs in the solution of nonlinear least-squares problems—see Chapter 10.)

19. Why is the Newton step from x_0 in Example 6.5.1 bad? [*Hint*: What would happen if x_0 were changed to $(2, e/6)^T \cong (2, 0.45)^T$?]

20. What step would be taken from x_0 in Example 6.5.1 using the double dogleg strategy (Algorithm A6.4.4) with $\delta_0 = \frac{1}{2}$?

21. Find out what the conjugate gradient algorithm is for minimizing convex quadratic functions, and show that the dogleg with $\hat{N} = x_+^N$ is the conjugate gradient method applied to $m_c(x)$ on the subspace spanned by the steepest-descent and Newton directions.

22. One of the disadvantages often cited for the dogleg algorithm is that its steps are restricted to be in the two-dimensional subspace spanned by the steepest-descent and Newton directions at each step. Suggest ways to alleviate this criticism, based on Exercise 21. See Steihaug (1981).

23. Let $J \in \mathbb{R}^{n \times n}$ be singular. Show that

$$\frac{1}{n(\text{macheps})^{1/2}} \leq \kappa_2 (J^T J + (n \cdot \text{macheps})^{1/2} \| J^T J \|_1 I) - 1 \leq \frac{1}{(\text{macheps})^{1/2}}.$$

[*Hint*: Use (3.1.13) and Theorem 3.5.7.] Generalize this inequality to the case when $\kappa_2(J) \geq \text{macheps}^{-1/2}$.

24. Let $F \in \mathbb{R}^n$ be nonzero, $J \in \mathbb{R}^{n \times n}$ singular. Another step that has been suggested for the nonlinear equations problem is the s that solves

$$\min \{ \| s \|_2 : s \text{ solves } \min_{s \in \mathbb{R}^n} \| F + Js \|_2 \}. \tag{6.6.1}$$

It was shown in Section 3.6 that the solution to (6.6.1) is $s = -J^+ F$, where J^+ is the pseudoinverse of J. Show that this step is similar to $\hat{s} = -(J^T J + \alpha I)^{-1} J^T F$, where $\alpha > 0$ is small, by proving:

(a) $\lim_{\alpha \to 0 +} (J^T J + \alpha I)^{-1} J^T = J^+$.

(b) for any $\alpha > 0$ and $v \in \mathbb{R}^n$, both $(J^T J + \alpha I)^{-1} J^T v$ and $J^+ v$ are perpendicular to all vectors w in the null space of J.

25. The global convergence of the line search and trust region algorithms of Subsections 6.3.2, 6.4.1, and 6.4.2 does not follow directly from the theory in Subsection 6.3.1, because none of the algorithms implement condition (6.3.4). However, the bounds in the algorithms on the amount of each adjustment in the line search parameter λ or the trust region δ_c have the same theoretical effect. For a proof that all of our algorithms are globally convergent, see Shultz, Schnabel, and Byrd (1982).

7

Stopping, Scaling, and Testing

In this chapter we discuss three issues that are peripheral to the basic mathematical considerations in the solution of nonlinear equations and minimization problems, but essential to the computer solution of actual problems. The first is how to adjust for problems that are badly scaled in the sense that the dependent or independent variables are of widely differing magnitudes. The second is how to determine when to stop the iterative algorithms in finite-precision arithmetic. The third is how to debug, test, and compare nonlinear algorithms.

7.1 SCALING

An important consideration in solving many "real-world" problems is that some dependent or independent variables may vary greatly in magnitude. For example, we might have a minimization problem in which the first independent variable, x_1, is in the range $[10^2, 10^3]$ meters and the second, x_2, is in the range $[10^{-7}, 10^{-6}]$ seconds. These ranges are referred to as the *scales* of the respective variables. In this section we consider the effect of such widely disparate scales on our algorithms.

One place where scaling will effect our algorithms is in calculating terms such as $\| x_+ - x_c \|_2$, which we used in our algorithms in Chapter 6. In the above example, any such calculation will virtually ignore the second (time) variable. However, there is an obvious remedy: *rescale the independent variables*; that is, change their units. For example, if we change the units of x_1 to kilometers and x_2 to microseconds, then both variables will have range $[10^{-1}, 1]$ and the scaling problem in computing $\| x_+ - x_c \|_2$ will be eliminated. Notice that this corresponds to changing the independent variable to $\hat{x} = D_x x$, where D_x is the diagonal scaling matrix

$$D_x = \begin{bmatrix} 10^{-3} & 0 \\ 0 & 10^6 \end{bmatrix}. \tag{7.1.1}$$

This leads to an important question. Say we transform the units of our problem to $\hat{x} = D_x x$, or more generally, transform the variable space to $\hat{x} = Tx$, where $T \in \mathbb{R}^{n \times n}$ is nonsingular, calculate our global step in the new variable space, and then transform back. Will the resultant step be the same as if we had calculated it using the same globalizing strategy in the old variable space? The surprising answer is that the Newton step is unaffected by this transformation but the steepest-descent direction is changed, so that a line-search step in the Newton direction is unaffected by a change in units, but a trust region step may be changed.

To see this, consider the minimization problem and let us define $\hat{x} = Tx$, $\hat{f}(\hat{x}) = f(T^{-1}\hat{x})$. Then it is easily shown that

$$\nabla \hat{f}(\hat{x}) = T^{-T} \nabla f(x), \qquad \nabla^2 \hat{f}(\hat{x}) = T^{-T} \nabla^2 f(x) T^{-1},$$

so that the Newton step and steepest-descent direction in the new variable space are

$$\hat{s}^N = -(T^{-T} \nabla^2 f(x) T^{-1})^{-1}(T^{-T} \Delta f(x)) = -T \nabla^2 f(x)^{-1} \nabla f(x),$$

$$\hat{s}^{SD} = -T^{-T} \nabla f(x),$$

or, in the old variable space,

$$s^N = T^{-1} \hat{s}^N = -\nabla^2 f(x)^{-1} \nabla f(x),$$

$$s^{SD} = T^{-1} \hat{s}^{SD} = -T^{-1} T^{-T} \nabla f(x) = -(T^T T)^{-1} \nabla f(x).$$

These conclusions are really common sense. The Newton step goes to the lowest point of a quadratic model, which is unaffected by a change in units of x. (The Newton direction for systems of nonlinear equations is similarly unchanged by transforming the independent variable.) However, determining which direction is "steepest" depends on what is considered a unit step in each direction. The steepest-descent direction makes the most sense if a step of one unit in variable direction x_i has about the same relative length as a step of one unit in any other variable direction x_j.

For these reasons, we believe the preferred solution to scaling problems is for the user to choose the units of the variable space so that each component of x will have roughly the same magnitude. However, if this is troublesome, the *equivalent* effect can be achieved by a transformation in the algorithm of the variable space by a corresponding diagonal scaling matrix D_x. This is the scaling strategy on the independent variable space that is implemented in our algorithms. All the user has to do is set D_x to correspond to the desired change in units, and then the algorithms operate as if they were working in the transformed variable space. The algorithms are still written in the original variable space, so that an expression like $\| x_+ - x_c \|_2$ becomes $\| D_x(x_+ - x_c) \|_2$ and the steepest-descent and hook steps become

$$x_+ := x_c - \lambda D_x^{-2} \nabla f(x_c),$$

$$s(\mu) := -(H_c + \mu D_x^2)^{-1} \nabla f(x_c),$$

respectively (see Exercise 3). The Newton direction is unchanged, however, as we have seen.

The positive diagonal scaling matrix D_x is specified by the user on input by simply supplying n values typx_i, $i = 1, \dots, n$, giving "typical" magnitudes of each x_i. Then the algorithm sets $(D_x)_{ii} = (\text{typx}_i)^{-1}$, making the magnitude of each transformed variable $\hat{x}_i = (D_x)_{ii} x_i$ about 1. For instance, if the user inputs $\text{typx}_1 = 10^3$, $\text{typx}_2 = 10^{-6}$ in our example, then D_x will be (7.1.1). If no scaling of x_i is considered necessary, typx_i should be set to 1. Further instructions for choosing typx_i are given in Guideline 2 in the appendix. Naturally, our algorithms do not store the diagonal matrix D_x, but rather a vector S_x (S stands for scale), where $(S_x)_i = (D_x)_{ii} = (\text{typx}_i)^{-1}$.

The above scaling strategy is not always sufficient; for example, there are rare cases that need dynamic scaling because some x_i varies by many orders of magnitude. This corresponds to using D_x exactly as in all our algorithms, but recalculating it periodically. Since there is little experience along these lines, we have not included dynamic scaling in our algorithms, although we would need only to add a module to periodically recalculate D_x at the conclusion of an iteration of Algorithm D6.1.1 or D6.1.3.

An example illustrating the importance of considering the scale of the independent variables is given below.

EXAMPLE 7.1.1 A common test problem for minimization algorithms is the Rosenbrock banana function

$$f(x) = 100(x_1^2 - x_2)^2 + (1 - x_1)^2, \tag{7.1.2}$$

which has its minimum at $x_* = (1, 1)^T$. Two typical starting points are $x_0 = (-1.2, 1)^T$ and $x_0 = (6.39, -0.221)^T$. This problem is well scaled, but if $\alpha \neq 1$, then the scale can be made worse by substituting $\alpha \hat{x}_1$ for x_1, and $\hat{x}_2/\alpha$ for x_2 in (7.1.2), giving

$$\hat{f}(\hat{x}) = f\left(\alpha \hat{x}_1, \frac{\hat{x}_2}{\alpha}\right) = 100\left((\alpha \hat{x}_1)^2 - \frac{\hat{x}_2}{\alpha}\right)^2 - (1 - \alpha \hat{x}_1)^2$$

$$\hat{x}_* = \left(\frac{1}{\alpha}, \alpha\right)^T.$$

This corresponds to the transformation

$$\hat{x} = \begin{bmatrix} \dfrac{1}{\alpha} & 0 \\ 0 & \alpha \end{bmatrix} x.$$

If we run the minimization algorithms found in the appendix on $\hat{f}(\hat{x})$, starting from $\hat{x}_0 = (-1.2/\alpha, \alpha)^T$ and $\hat{x}_0 = (6.39/\alpha, \alpha(-0.221))^T$, use exact derivatives, the "hook" globalizing step, and the default tolerances, and neglect the scale by setting $\text{typx}_1 = \text{typx}_2 = 1$, then the number of iterations required for convergence with various values of α are as follows (the asterisk indicates failure to converge after 150 iterations):

α	Iterations from $x_0 = (-1.2/\alpha, \alpha)^T$	Iterations from $x_0 = (6.39/\alpha, \alpha(-0.221))^T$
0.01	150 + *	150 + *
0.1	94	47
1	24	29
10	52	48
100	150 + *	150 + *

However, if we set $\text{typx}_1 = 1/\alpha$, $\text{typx}_2 = \alpha$, then the output of the program is exactly the same as for $\alpha = 1$ in all cases, except that the x values are multiplied by

$$\begin{bmatrix} \dfrac{1}{\alpha} & 0 \\ 0 & \alpha \end{bmatrix}.$$

It is also necessary to consider the scale of the dependent variables. In minimization problems, the scale of the objective function f really only matters in the stopping conditions, discussed in Section 7.2. In all other calculations, such as the test $f(x_+) < f(x) + 10^{-4}\, \nabla f(x)^T(x_+ - x)$, a change in the units of f is of no consequence.

On the other hand, in solving systems of nonlinear equations, differing sizes among the component functions f_i can cause the same types of problems as differing sizes among the independent variables. Once again, the Newton step is independent of this scaling (see Exercise 4). However, the globalizing strategy for nonlinear equations requires a decrease in $\| F \|_2$, and it is clear that if the units of two component functions of $F(x)$ are widely different, then the smaller component function will be virtually ignored.

For this reason, our algorithms also use a positive diagonal scaling matrix D_F on the dependent variable $F(x)$, which works as D_x does on x. The diagonal matrix D_F is chosen so that all the components of $D_F F(x)$ will have about the same typical magnitude at points not too near the root. D_F is then used to scale F in all the modules for nonlinear equations. The affine model becomes $D_F M_c$, and the quadratic model function for the globalizing step becomes $\hat{m}_c = \frac{1}{2} \| D_F M_c \|_2^2$. All our interfaces and algorithms are implemented like this, and the user just needs to specify D_F initially. This is done by inputting values $\mathrm{typ} f_i$, $i = 1, \ldots, n$, giving typical magnitudes of each f_i at points not too near a root. The algorithm then sets $(D_F)_{ii} = \mathrm{typ} f_i^{-1}$. [Actually it stores $S_F \in \mathbb{R}^n$, where $(S_F)_i = (D_F)_{ii}$.] Further instructions on choosing $\mathrm{typ} f_i$ are given in Guideline 5 in the appendix.

7.2 STOPPING CRITERIA

In this section we discuss how to terminate our algorithms. The stopping criteria are the same common-sense conditions discussed in Section 2.5 for one-dimensional problems: "Have we solved the problem?" "Have we ground to a halt?" or "Have we run out of money, time, or patience?" The factors that need consideration are how to implement these tests in finite-precision arithmetic, and how to pay proper attention to the scales of the dependent and independent variables.

We first discuss stopping criteria for unconstrained minimization. The most important test is "Have we solved the problem?" In infinite precision, a necessary condition for x to be the exact minimizer of f is $\nabla f(x) = 0$, but in an iterative and finite-precision algorithm, we will need to modify this condition to $\nabla f(x) \cong 0$. Although $\nabla f(x) = 0$ can also occur at a maximum or saddle point, our globalizing strategy and our strategy of perturbing the model Hessian to be positive definite make convergence virtually impossible to maxima and saddle points. In our context, therefore, $\nabla f(x) = 0$ is considered a necessary and sufficient condition for x to be a local minimizer of f.

To test whether $\nabla f \cong 0$, a test such as

$$\| \nabla f(x_+) \| \le \varepsilon \tag{7.2.1}$$

is inadequate, because it is strongly dependent on the scaling of both f and x. For example, if $\varepsilon = 10^{-3}$ and f is always in $[10^{-7}, 10^{-5}]$, then it is likely that any value of x will satisfy (7.2.1); conversely if $f \in [10^5, 10^7]$, (7.2.1) may be overly stringent. Also, if x is inconsistently scaled—for example, $x_1 \in [10^6, 10^7]$ and $x_2 \in [10^{-1}, 1]$—then (7.2.1) is likely to treat the variables unequally. A common remedy is to use

$$| \nabla f(x_+)^T \, \nabla^2 f(x_+)^{-1} \, \nabla f(x_+) | \le \varepsilon. \tag{7.2.2}$$

Inequality (7.2.2) is invariant under any linear transformation of the independent variables and thus is independent of the scaling of x. However, it is still dependent on the scaling of f. A more direct modification of (7.2.1) is to define the relative gradient of f at x by

$$\text{relgrad } (x)_i = \frac{\text{relative rate of change in } f}{\text{relative rate of change in } x_i}$$

$$= \lim_{\delta \to 0} \frac{\dfrac{f(x + \delta e_i) - f(x)}{f(x)}}{\dfrac{\delta}{x_i}}$$

$$= \frac{\nabla f(x)_i \, x_i}{f(x)} \tag{7.2.3}$$

and test

$$\| \text{relgrad } (x_+) \|_\infty \leq \text{gradtol}. \tag{7.2.4}$$

Test (7.2.4) is independent of any change in the units of f or x. It has the drawback that the idea of relative change in x_i or f breaks down if x_i or $f(x)$ happen to be near zero. This problem is easily fixed by replacing x_i and f in (7.2.3) by max $\{|x_i|, \text{typ} x_i\}$ and max $\{|f(x)|, \text{typ} f\}$, respectively, where typ$f$ is the user's estimate of a typical magnitude of f. The resulting test,

$$\max_{1 \leq i \leq n} \left| \frac{\nabla f(x)_i \, \max \{|(x_+)_i|, \text{typ} x_i\}}{\max \{|f(x_+)|, \text{typ} f\}} \right| \leq \text{gradtol}, \tag{7.2.5}$$

is the one used in our algorithms.

It should be mentioned that the problem of measuring relative change when the argument z is near zero is commonly addressed by substituting $(|z| + 1)$ or max $\{|z|, 1\}$ for z. It is apparent from the above discussion that both these substitutions make the implicit assumption that z has scale around 1. They may also work satisfactorily if $|z|$ is much larger than 1, but they will be unsatisfactory if $|z|$ is always much smaller than 1. Therefore, if a value of typ$z$ is available, the substitution max $\{|z|, \text{typ} z\}$ is preferable.

The other stopping tests for minimization are simpler to explain. The test for whether the algorithm has ground to a halt, either because it has stalled or converged, is

$$\| \text{relative change in successive values of } x \|_\infty \leq \text{steptol}. \tag{7.2.6}$$

Following the above discussion, we measure the relative change in x_i by

$$\text{relx}_i = \frac{|(x_+)_i - (x_c)_i|}{\max \{|(x_+)_i|, \text{typ} x_i\}}. \tag{7.2.7}$$

Selection of steptol is discussed in Guideline 2; basically, if p significant digits of x_* are desired, steptol should be set to 10^{-p}.

As in most iterative procedures, we quantify available time, money, and patience by imposing an iteration limit. In real applications this limit is often governed by the cost of each iteration, which can be high if function evaluation is expensive. During debugging, it is a good idea to use a low iteration limit so that an erroneous program won't run too long. In a minimization algorithm one should also test for divergence of the iterates x_k, which can occur if f is unbounded below, or asymptotically approaches a finite lower bound from above. To test for divergence, we ask the user to supply a maximum step length, and if five consecutive steps are this long, the algorithm is terminated. (See Guideline 2.)

The stopping criteria for systems of nonlinear equations are similar. We first test whether x_+ approximately solves the problem—that is, whether $F(x_+) \cong 0$. The test $\| F(x_+) \| \le \varepsilon$ is again inappropriate, owing to problems with scaling, but since $(D_F)_{ii} = 1/\text{typ} f_i$ has been selected so that $(D_F)_{ii} F_i$ should have magnitude about 1 at points not near the root, the test

$$\| D_F F \|_\infty \le \text{fntol}$$

should be appropriate. Suggestions for fntol are given in Guideline 5; values around 10^{-5} are typical.

Next one tests whether the algorithm has converged or stalled at x_+, using the test (7.2.6–7.2.7). The tests for iteration limit and divergence are also the same as for minimization, though it is less likely for an algorithm for solving $F(x) = 0$ to diverge.

Finally, it is possible for our nonlinear equations algorithm to become stuck by finding a local minimum of the associated minimization function $f = \frac{1}{2} \| D_F F \|_2^2$ at which $F \ne 0$ (see Figure 6.5.1). Although convergence test (7.2.6–7.2.7) will stop the algorithm in this case, we prefer to test for it explicitly by checking whether the gradient of f at x_+ is nearly zero, using a relative measure of the gradient analogous to (7.2.5). If the algorithm has reached a local minimum of $\| D_F F \|_2^2$ at which $F \ne 0$, all that can be done is to restart the algorithm in a different place.

Algorithms A7.2.1 and A7.2.3 in the appendix contain the stopping criteria for unconstrained minimization and nonlinear equations, respectively. Algorithms A7.2.2 and A7.2.4 are used before the initial iteration to test whether the starting point x_0 is already a minimizer or a root, respectively. Guidelines 2 and 5 contain advice for selecting all the user-supplied parameters. In our software that implements these algorithms [Schnabel, Weiss, and Koontz (1982)], default values are available for all the stopping and scaling tolerances.

7.3   TESTING

Once a computer program for nonlinear equations or minimization has been written, it will presumably be tested to see whether it works correctly and how it compares with other software that solves the same problem. It is important to discuss two aspects of this testing process: (1) how should the software be tested and (2) what criteria should be used to evaluate its performance? It is perhaps surprising that there is no consensus on either of these important questions. In this section we indicate briefly some of the leading ideas.

The first job in testing is to see that the code is working correctly. By "correctly" we currently mean a general idea that the program is doing what it should, as opposed to the computer scientist's much more stringent definition of "correctness." This is certainly a nontrivial task for any program the size of those in this book. We strongly recommend a modular testing procedure, testing first each module as it is written, then the pieces the modules form, and finally the entire program. Taking the approach of testing the entire program at once can make finding errors extremely difficult. The difficulty with modular testing is that it may not be obvious how to construct input data to test some modules, such as the module for updating the trust region. Our advice is to start with data from the simplest problems, perhaps one or two dimensions with identity or diagonal Jacobians or Hessians, since it should be possible to hand-check the calculations. Then it is advisable to check the module on more complex problems. An advantage of this modular testing is that it usually adds to our understanding of the algorithms.

Once all the components are working correctly, one should test the program on a variety of nonlinear problems. This serves two purposes: to check that the entire program is working correctly, and then to observe its performance on some standard problems. The first problems to try are the simplest ones: linear systems in two or three dimensions for a program to solve systems of nonlinear equations, positive definite quadratics in two or three variables for minimization routines. Then one might try polynomials or systems of equations of slightly higher degree and small (two to five) dimension. When the program is working correctly on these, it is time to run it on some standard problems accepted in this field as providing good tests of software for nonlinear equations or minimization. Many of them are quite difficult. It is often useful to start these test problems from 10 or 100 times further out on the ray from the solution x_* to the standard starting point x_0, as well as from x_0; Moré, Garbow, and Hillstrom (1981) report that this often brings out in programs important differences not indicated from the standard starting points.

Although the literature on test problems is still developing, we provide some currently accepted problems in Appendix B. We give a nucleus of standard problems for nonlinear equations or minimization sufficient for class projects or preliminary research results and provide references to additional problems that would be used in a thorough research study. It should be noted that most of these problems are well scaled; this is indicative of the lack of attention that has been given to the scaling problem. The dimensions of the test problems in Appendix B are a reflection of the problems currently being solved. The supply of medium (10 to 100) dimensional problems is still inadequate, and the cost of testing on such problems is a significant factor.

The difficult question of how to evaluate and compare software for minimization or nonlinear equations is a side issue in this book. It is complicated by whether one is primarily interested in measuring the efficiency and reliability of the program in solving problems, or its overall quality as a piece of software. In the latter case, one is also interested in the interface between the software and its users (documentation, ease of use, response to erroneous input, robustness, quality of output), and between the software and the computing environment (portability). We will comment only on the first set of issues; for a discussion of all these issues, see, e.g., Fosdick (1979).

By reliability, we mean the ability of the program to solve successfully the problems it is intended for. This is determined first by its results on test problems, and ultimately by whether it solves the problems of the user community. For the user, efficiency refers to the computing bill incurred running the program on his or her problems. For minimization or nonlinear equations problems, this is sometimes measured by the running times of the program on test problems. Accurate timing data is difficult to obtain on shared computing systems, but a more obvious objection is the inherent assumption that the test problems are like those of the user. Another common measure of efficiency is the number of function and derivative evaluations the program requires to solve test problems. The justification for this measure is that it indicates the cost on those problems that are inherently expensive, namely those for which function and derivative evaluation is expensive. This measure is especially appropriate for evaluating secant methods (see Chapters 8 and 9), since they are used often on such problems. In minimization testing, the number of function and gradient evaluations used sometimes are combined into one statistic,

number of equivalent function evaluations

$$= \text{number of } f\text{-evaluations} + n \text{ (number of } \nabla f\text{-evaluations).}$$

This statistic indicates the number of function evaluations that would be used if the gradients were evaluated by finite differences. Since this is not always the case, it is preferable to report the function and gradient totals separately.

Some other possible measures of efficiency are number of iterations required, computational cost per iteration, and computer storage required. The number of iterations required is a simple measure, but is useful only if it is correlated to the running time of the problem, or the function and derivative evaluations required. The computational cost of an iteration, excluding function and derivative evaluations, is invariably determined by the linear algebra and is usually proportional to n^3, or n^2 for secant methods. When multiplied by the number of iterations required, it gives an indication of the running time for a problem where function and derivative evaluation is very inexpensive. Computer storage is usually not an issue for problems of the size discussed in this book; however, storage and computational cost per iteration become crucially important for large problems.

Using the above measures, one can compare two entirely different programs for minimization or nonlinear equations, but often one is interested only in comparing two or more versions of a particular segment of the algorithm—for example, the line search. In this case it may be desirable to test the alternative segments by substituting them into a modular program such as ours, so that the remainder of the program is identical throughout the tests. Such controlled testing reduces the reliance of the results on other aspects of the programs, but it is possible for the comparison to be prejudiced if the remainder of the program is more favorable to one alternative.

Finally, the reader should realize that we have discussed the evaluation of computer programs, not algorithms, in this section. The distinction is that a computer program may include many details that are crucial to its performance but are not part of the "basic algorithm." Examples are stopping criteria, linear algebra routines, and tolerances in line-search or trust region algorithms. The basic algorithm may be evaluated using measures we have already discussed: rate of local convergence, global convergence properties, performance on special classes of functions. When one tests a computer program, however, as discussed above, one must realize that a particular software implementation of a basic algorithm is being tested, and that two implementations of the same basic algorithm may perform quite differently.

7.4 EXERCISES

1. Consider the problem

$$\text{minimize}_{x \in \mathbb{R}^2} \quad (x_1 - 10^6)^4 + (x_1 - 10^6)^2 + (x_2 - 10^{-6})^4 + (x_2 - 10^{-6})^2.$$

   What problems might you encounter in applying an optimization algorithm without scaling to this problem? (Consider steepest-descent directions, trust regions, stopping criteria.) What value would you give to typx_1, typx_2 in our algorithms in order to alleviate these problems? What change might be even more helpful?

2. Let $f: \mathbb{R}^n \longrightarrow \mathbb{R}$, $T \in \mathbb{R}^{n \times n}$ nonsingular. For any $x \in \mathbb{R}^n$, define $\hat{x} = Tx$, $\hat{f}(\hat{x}) = f(T^{-1}\hat{x}) = f(x)$. Using the chain rule for multivariable calculus, show that

$$\nabla \hat{f}(\hat{x}) = T^{-T} \nabla f(x),$$

$$\nabla^2 \hat{f}(\hat{x}) = T^{-T} \nabla^2 f(x) T^{-1}.$$

3. Let $f \in R$, $g \in \mathbb{R}^n$, $H \in \mathbb{R}^{n \times n}$, H symmetric and positive definite, $D \in \mathbb{R}^{n \times n}$, D a positive diagonal matrix. Using Lemma 6.4.1, show that the solution to

$$\text{minimize}_{s \in \mathbb{R}^n} \quad f + g^T s + \tfrac{1}{2} s^T H s$$

$$\text{subject to} \quad \| Ds \|_2 \le \delta$$

   is given by

$$s(\mu) = -(H + \mu D^2)^{-1} g.$$

for some $\mu \geq 0$. [*Hint*: Make the transformation $\hat{s} = Ds$, use Lemma 6.4.1, and transform back.]

4. Let $F : \mathbb{R}^n \longrightarrow \mathbb{R}^n, T_1, T_2 \in \mathbb{R}^{n \times n}$ nonsingular. For any $x \in \mathbb{R}^n$ define $\hat{x} = T_1 x$, $\hat{F}(\hat{x}) = T_2 F(T_1^{-1}\hat{x}) = T_2 F(x)$. Show that the Jacobian matrix of $\hat{F}$ with respect to $\hat{x}$ is given by

$$\hat{J}(\hat{x}) = T_2 J(x) T_1^{-1}.$$

What is the Newton step in the $\hat{x}$ variable space? If this step is transformed back to the original variable space, how does it compare to the normal Newton step in the original variable space?

5. What are some situations in which the scaling strategy of Section 7.1 would be unsatisfactory? Suggest a dynamic scaling strategy that would be successful in these situations. Now give a situation in which your dynamic strategy would be unsuccessful.

6. Suppose our stopping test for minimization finds that $\nabla f(x_k) \approx 0$. How could you test whether x_k is a saddle point (or maximizer)? If x_k is a saddle point, how could you proceed in the minimization algorithm?

7. Write a program for unconstrained minimization or solving systems of nonlinear equations using the algorithms in Appendix A (and using exact derivatives). Choose one of globalizing strategies of Sections 6.3 and 6.4 to implement in your program. Debug and test your program as discussed in Section 7.3.

...And another program note

In the preceding chapters we have developed all the components of a system of complete quasi-Newton algorithms for solving systems of nonlinear equations and unconstrained minimization problems. There is one catch: we have assumed that we would compute the required derivative matrix, namely the Jacobian for nonlinear equations or the Hessian for unconstrained minimization, or approximate it accurately using finite differences. The problem with this assumption is that for many problems analytic derivatives are unavailable and function evaluation is expensive. Thus, the cost of finite-difference derivative approximations, n additional evaluations of $F(x)$ per iteration for a Jacobian or $(n^2 + 3n)/2$ additional evaluations of $f(x)$ for a Hessian, is high. In the next two chapters, therefore, we discuss a class of quasi-Newton methods that use cheaper ways of approximating the Jacobian or Hessian. We call these approximations *secant approximations*, because they specialize to the secant approximation to $f'(x)$ in the one-variable case, and we call the quasi-Newton methods that use them *secant methods*. We emphasize that only the method for approximating the derivative will be new; the remainder of the quasi-Newton algorithm will be virtually unchanged.

The development of secant methods has been an active research area since the mid 1960s. The result has been a class of methods very successful in practice and most interesting theoretically; we will try to transmit a feeling for both of these aspects. As in many active new fields, however, the development has been chaotic and sometimes confusing. Therefore, our exposition will be quite different from the way the methods were first derived, and we will even introduce some new names. The reason is to try to lessen the initial confusion the novice has traditionally had to endure to understand these methods and their interrelationships.

Two comprehensive references on this subject are Dennis and Moré (1977) and Dennis (1978). Another view of these approximations can be found in Fletcher (1980). Our naming convention for the methods is based on suggestions of Dennis and Tapia (1976).

8

Secant Methods for Systems of Nonlinear Equations

We start our discussion of secant methods with the nonlinear equations problem, because secant approximations to the Jacobian are simpler than secant approximations to the Hessian, which we discuss in Chapter 9. Recall that in Chapter 2 we saw that we could approximate $f'(x_+)$ at no additional cost in function evaluations by $a_+ = (f(x_+) - f(x_c))/(x_+ - x_c)$, and that the price we paid was a reduction in the local q-convergence rate from 2 to $(1 + \sqrt{5})/2$. The idea in multiple dimensions is similar: we approximate $J(x_+)$ using only function values that we have already calculated. In fact, multivariable generalizations of the secant method have been proposed which, although they require some extra storage for the derivative, do have r-order equal to the largest root of $r^{n+1} - r^n - 1 = 0$; but none of them seem robust enough for general use. Instead, in this chapter we will see the basic idea for a class of approximations that require no additional function evaluations or storage and that are very successful in practice. We will single out one that has a q-superlinear local convergence rate and r-order $2^{1/2n}$.

In Section 8.1 we introduce the most used secant approximation to the Jacobian, proposed by C. Broyden. The algorithm, analogous to Newton's method, but that substitutes this approximation for the analytic Jacobian, is called Broyden's method. In Section 8.2 we present the local convergence analysis of Broyden's method, and in Section 8.3 we discuss the implementation of a complete quasi-Newton method using this Jacobian approximation. We conclude the chapter with a brief discussion in Section 8.4 of other secant approximations to the Jacobian.

8.1   BROYDEN'S METHOD

In this section we present the most successful secant-method extension to solve systems of nonlinear equations. Recall that in one dimension, we considered the model

$$M_+(x) = f(x_+) + a_+(x - x_+),$$

which satisfies $M_+(x_+) = f(x_+)$ for any $a_+ \in \mathbb{R}$, and yields Newton's method if $a_+ = f'(x_+)$. If $f'(x_+)$ was unavailable, we instead asked the model to satisfy $M_+(x_c) = f(x_c)$—that is,

$$f(x_c) = f(x_+) + a_+(x_c - x_+)$$

—which gave the secant approximation

$$a_+ = \frac{f(x_+) - f(x_c)}{x_+ - x_c}.$$

The next iterate of the secant method was the x_{++} for which $M_+(x_{++}) = 0$—that is, $x_{++} = x_+ - f(x_+)/a_+$.

In multiple dimensions, the analogous affine model is

$$M_+(x) = F(x_+) + A_+(x - x_+), \tag{8.1.1}$$

which satisfies $M_+(x_+) = F(x_+)$ for any $A_+ \in \mathbb{R}^{n \times n}$. In Newton's method, $A_+ = J(x_+)$. If $J(x_+)$ is not available, the requirement that led to the one-dimensional secant method is $M_+(x_c) = F(x_c)$—that is,

$$F(x_c) = F(x_+) + A_+(x_c - x_+)$$

or

$$A_+(x_+ - x_c) = F(x_+) - F(x_c). \tag{8.1.2}$$

We will refer to (8.1.2) as the *secant equation*. Furthermore, we will use the notation $s_c = x_+ - x_c$ for the current step and $y_c = F(x_+) - F(x_c)$ for the yield of the current step, so that the secant equation is written

$$A_+ s_c = y_c. \tag{8.1.3}$$

The crux of the problem in extending the secant method to n dimensions is that (8.1.3) does not completely specify A_+ when $n > 1$. In fact, if $s_c \neq 0$, there is an $n(n - 1)$-dimensional affine subspace of matrices obeying (8.1.3). Constructing a successful secant approximation consists of selecting a good way to choose from among these possibilities. Logically, the choice should enhance the Jacobian approximation properties of A_+ or facilitate its use in a quasi-Newton algorithm.

Perhaps the most obvious strategy is to require the model (8.1.1) to interpolate $F(x)$ at other past points x_{-i}—that is,

$$F(x_{-i}) = F(x_+) + A_+(x_{-i} - x_+)$$

—which leads to the equations

$$A_+ s_{-i} = y_{-i}, \qquad i = 1, \ldots, m, \tag{8.1.4}$$

where

$$s_{-i} \triangleq x_{-i} - x_+, \quad y_{-i} \triangleq F(x_{-i}) - F(x_+).$$

If $m = n - 1$ and $s_c, s_{-1}, \ldots, s_{-(n-1)}$ are linearly independent, then the n^2 equations (8.1.3) and (8.1.4) uniquely determine the n^2 unknown elements of A_+. Unfortunately, this is precisely the strategy we were referring to in the introduction to this chapter, that has r-order equal to the largest root of $r^{n+1} - r^n - 1 = 0$; but is not successful in practice. One problem is that the directions $s_c, s_{-1}, \ldots, s_{-(n-1)}$ tend to be linearly dependent or close to it, making the computation of A_+ a poorly posed numerical problem. Furthermore, the strategy requires an additional n^2 storage.

The approach that leads to the successful secant approximation is quite different. We reason that aside from the secant equation we have no new information about either the Jacobian or the model, so we should preserve as much as possible of what we already have. Therefore, we will choose A_+ by trying to minimize the change in the affine model, subject to satisfying $A_+ s_c = y_c$. The difference between the new and old affine models at any $x \in \mathbb{R}^n$ is

$$\begin{aligned} M_+(x) - M_c(x) &= F(x_+) + A_+(x - x_+) - F(x_c) - A_c(x - x_c) \\ &= F(x_+) - F(x_c) - A_+(x_+ - x_c) + (A_+ - A_c)(x - x_c) \\ &= (A_+ - A_c)(x - x_c) \end{aligned}$$

with the last equality due to the secant equation (8.1.2). Now for any $x \in \mathbb{R}^n$, let us express

$$x - x_c = \alpha s_c + t,$$

where $t^T s = 0$. Then the term we wish to minimize becomes

$$M_+(x) - M_c(x) = \alpha(A_+ - A_c)s_c + (A_+ - A_c)t.$$

We have no control over the first term on the right side, since the secant equation implies $(A_+ - A_c)s_c = y_c - A_c s_c$. However, we can make the second term zero for all $x \in \mathbb{R}^n$ by choosing A_+ such that $(A_+ - A_c)t = 0$ for all t orthogonal to s_c. This requires that $A_+ - A_c$ be a rank-one matrix of the form $u s_c^T$, $u \in \mathbb{R}^n$. Now to fulfill the secant equation, which is equivalent to $(A_+ - A_c)s_c = y_c - A_c s_c$, u must be $(y_c - A_c s_c)/s_c^T s_c$. This gives

$$A_+ = A_c + \frac{(y_c - A_c s_c)s_c^T}{s_c^T s_c} \tag{8.1.5}$$

as the least change in the affine model consistent with $A_+ s_c = y_c$.

Equation (8.1.5) was proposed in 1965 by C. Broyden, and we will refer to it as *Broyden's update* or simply the *secant update*. The word *update* indicates that we are not approximating $J(x_+)$ from scratch; rather we are updating the approximation A_c to $J(x_c)$ into an approximation A_+ to $J(x_+)$. This updating characteristic is shared by all the successful multidimensional secant approximation techniques.

The preceding derivation is in keeping with the way Broyden derived the formula, but it can be made much more rigorous. In Lemma 8.1.1 we show that Broyden's update is the minimum change to A_c consistent with $A_+ s_c = y_c$, if the change $A_+ - A_c$ is measured in the Frobenius norm. We comment on the choice of norm after the proof. One new piece of notation will be useful: we denote

$$Q(y, s) = \{B \in \mathbb{R}^{n \times n} \,|\, Bs = y\}.$$

That is, $Q(y, s)$ is the set of matrices that act as quotients of y over s.

LEMMA 8.1.1   Let $A \in \mathbb{R}^{n \times n}$, $s, y \in \mathbb{R}^n$, $s \neq 0$. Then for any matrix norms $\|\cdot\|$, $\|\|\cdot\|\|$ such that

$$\| A \cdot B \| \leq \| A \| \cdot \|\| B \|\| \tag{8.1.6}$$

and

$$\left\|\left\| \frac{vv^T}{v^T v} \right\|\right\| = 1, \tag{8.1.7}$$

the solution to

$$\min_{B \in Q(y, s)} \| B - A \| \tag{8.1.8}$$

is

$$A_+ = A + \frac{(y - As)s^T}{s^T s}. \tag{8.1.9}$$

In particular, (8.1.9) solves (8.1.8) when $\|\cdot\|$ is the l_2 matrix norm, and (8.1.9) solves (8.1.8) uniquely when $\|\cdot\|$ is the Frobenius norm.

Proof.   Let $B \in Q(y, s)$; then

$$\| A_+ - A \| = \left\| \frac{(y - As)s^T}{s^T s} \right\| = \left\| \frac{(B - A)ss^T}{s^T s} \right\|$$

$$\leq \| B - A \| \left\|\left\| \frac{ss^T}{s^T s} \right\|\right\| \leq \| B - A \|.$$

If $\|\cdot\|$ and $\|\|\cdot\|\|$ are both taken to be the l_2 matrix norm, then (8.1.6) and (8.1.7) follow from (3.1.10) and (3.1.17), respectively. If $\|\cdot\|$ and $\|\|\cdot\|\|$ stand for the Frobenius and l_2 matrix norm, respectively, then (8.1.6) and (8.1.7) come from (3.1.15) and (3.1.17). To see that (8.1.9) is the unique solution to (8.1.8) in the Frobenius norm, we remind the reader that the Frobenius norm is strictly convex, since it is the l_2 vector norm of the matrix written as an n^2 vector. Since $Q(y, s)$ is a convex—in fact, affine—subset of $\mathbb{R}^{n \times n}$ or $\mathbb{R}^{n^2}$, the solution to (8.1.8) is unique in any strictly convex norm. $\square$

The Frobenius norm is a reasonable one to use in Lemma 8.1.1 because it measures the change in each component of the Jacobian approximation. An operator norm such as the l_2 norm is less appropriate in this case. In fact, it is an interesting exercise to show that (8.1.8) may have multiple solutions in the l_2 operator norm, some clearly less desirable than Broyden's update. (See Exercise 2.) This further indicates that an operator norm is inappropriate in (8.1.8).

Now that we have completed our affine model (8.1.1) by selecting A_+, the obvious way to use it is to select the next iterate to be the root of this model. This is just another way of saying that we replace $J(x_+)$ in Newton's method by A_+. The resultant algorithm is:

ALGORITHM 8.1.2   BROYDEN'S METHOD

   Given $F : \mathbb{R}^n \longrightarrow \mathbb{R}^n$, $x_0 \in \mathbb{R}^n$, $A_0 \in \mathbb{R}^{n \times n}$

   Do for $k = 0, 1, \ldots$:

   Solve $A_k s_k = -F(x_k)$ for s_k

   $x_{k+1} := x_k + s_k$

   $y_k := F(x_{k+1}) - F(x_k)$

   $$A_{k+1} := A_k + \frac{(y_k - A_k s_k)s_k^T}{s_k^T s_k}. \qquad (8.1.10)$$

We will also refer to this method as the *secant method*. At this point, the reader may have grave doubts whether it will work. In fact, it works quite well locally, as we suggest below by considering its behavior on the same problem that we solved by Newton's method in Section 5.1. Of course, like Newton's method, it may need to be supplemented by the techniques of Chapter 6 to converge from some starting points.

There is one ambiguity in Algorithm 8.1.1: how do we get the initial approximation A_0 to $J(x_0)$? In practice, we use finite differences this one time to get a good start. This also makes the minimum-change characteristic of Broyden's update more appealing. In Example 8.1.3, we assume for simplicity that $A_0 = J(x_0)$.

EXAMPLE 8.1.3   Let

$$F(x) = \begin{bmatrix} x_1 + x_2 - 3 \\ x_1^2 + x_2^2 - 9 \end{bmatrix},$$

which has roots $(0, 3)^T$ and $(3, 0)^T$. Let $x_0 = (1, 5)^T$, and apply Algorithm 8.1.2

with

$$A_0 = J(x_0) = \begin{bmatrix} 1 & 1 \\ 2 & 10 \end{bmatrix}.$$

Then

$$F(x_0) = \begin{bmatrix} 3 \\ 17 \end{bmatrix}, \qquad s_0 = -A_0^{-1}F(x_0) = \begin{bmatrix} -1.625 \\ -1.375 \end{bmatrix},$$

$$x_1 = x_0 + s_0 = \begin{bmatrix} -0.625 \\ 3.625 \end{bmatrix}, \qquad F(x_1) = \begin{bmatrix} 0 \\ 4.53125 \end{bmatrix}.$$

Therefore, (8.1.10) gives

$$A_1 = A_0 + \begin{bmatrix} 0 & 0 \\ -1.625 & -1.375 \end{bmatrix} = \begin{bmatrix} 1 & 1 \\ 0.375 & 8.625 \end{bmatrix}.$$

The reader can confirm that $A_1 s_0 = y_0$. Note that

$$J(x_1) = \begin{bmatrix} 1 & 1 \\ -1.25 & 7.25 \end{bmatrix},$$

so that A_1 is not very close to $J(x_1)$. At the next iteration,

$$s_1 = -A_1^{-1}F(x_1) \cong \begin{bmatrix} 0.549 \\ -0.549 \end{bmatrix}, \qquad x_2 = x_1 + s_1 \cong \begin{bmatrix} -0.076 \\ 3.076 \end{bmatrix},$$

$$F(x_2) \cong \begin{bmatrix} 0 \\ 0.466 \end{bmatrix}, \qquad A_2 \cong \begin{bmatrix} 1 & 1 \\ -0.799 & 8.201 \end{bmatrix}.$$

Again A_2 is not very close to

$$J(x_2) = \begin{bmatrix} 1 & 1 \\ -0.152 & 6.152 \end{bmatrix}.$$

The complete sequences of iterates produced by Broyden's method, and for comparison, Newton's method, are given below. For $k \geq 1$, $(x_k)_1 + (x_k)_2 = 3$ for both methods; so only $(x_k)_2$ is listed below.

Broyden's Method		Newton's Method
$(1, 5)^T$	x_0	$(1, 5)^T$
3.625	x_1	3.625
3.0757575757575	x_2	3.0919117647059
3.0127942681679	x_3	3.0026533419372
3.0003138243387	x_4	3.0000023425973
3.0000013325618	x_5	3.0000000000018
3.0000000001394	x_6	3.0
3.0	x_7	

Example 8.1.3 is characteristic of the local behavior of Broyden's method. If any components of $F(x)$ are linear, such as $f_1(x)$ above, then the corresponding rows of the Jacobian approximation will be correct for $k \geq 0$, and the corresponding components of $F(x_k)$ will be zero for $k \geq 1$ (Exercise 4). The rows of A_k corresponding to nonlinear components of $F(x)$ may not be very accurate, but the secant equation still gives enough good information that there is rapid convergence to the root. We show in Section 8.2 that the rate of convergence is q-superlinear, not q-quadratic.

8.2 LOCAL CONVERGENCE ANALYSIS OF BROYDEN'S METHOD

In this section we investigate the local convergence behavior of Broyden's method. We show that if x_0 is sufficiently close to a root x_*, where $J(x_*)$ is nonsingular, and if A_0 is sufficiently close to $J(x_0)$, then the sequence of iterates $\{x_k\}$ converges q-superlinearly to x_*. The proof is a special case of a more general proof technique that applies to the secant methods for minimization as well. We provide only the special case here, because it is simpler and easier to understand than the general technique and provides insight into why multi-dimensional secant methods work. The convergence results in Chapter 9 will then be stated without proof. The reader who is interested in a deeper treatment of the subject is urged to consult Broyden, Dennis, and Moré (1973) or Dennis and Walker (1981).

We motivate our approach by using virtually the same simple analysis that we used in analyzing the secant method in Section 2.6 and Newton's method in Section 5.2. If $F(x_*) = 0$, then from the iteration

$$x_{k+1} = x_k - A_k^{-1} F(x_k)$$

we have

$$x_{k+1} - x_* = x_k - x_* - A_k^{-1}(F(x_k) - F(x_*))$$

or

$$A_k(x_{k+1} - x_*) = A_k(x_k - x_*) - F(x_k) + F(x_*).$$

Defining $e_k \triangleq x_k - x_*$ and adding and subtracting $J(x_*)e_k$ to the right side of the above equation gives

$$A_k e_{k+1} = [-F(x_k) + F(x_*) + J(x_*)e_k] + (A_k - J(x_*))e_k. \qquad (8.2.1)$$

Under our standard assumptions,

$$\| -F(x_k) + F(x_*) + J(x_*)e_k \| = O\left(\|e_k\|^2\right)$$

so the key to the local convergence analysis of Broyden's method will be an analysis of the second term, $(A_k - J(x_*))e_k$. First, we will prove local q-linear convergence of $\{e_k\}$ to zero by showing that the sequence $\{\| A_k - J(x_*) \|\}$ stays bounded below some suitable constant. It may not be true that

$$\lim_{k \to \infty} \| A_k - J(x_*) \| = 0,$$

but we will prove local q-superlinear convergence by showing that

$$\lim_{k \to \infty} \frac{\| (A_k - J(x_*))e_k \|}{\| e_k \|} = 0.$$

This is really all we want out of the Jacobian approximation, and it implies that the secant step, $-A_k^{-1}F(x_k)$, converges to the Newton step, $-J(x_k)^{-1}F(x_k)$, in magnitude and direction.

Let us begin by asking how well we expect A_+ given by Broyden's update to approximate $J(x_*)$. If $F(x)$ is affine with Jacobian J, then J will always satisfy the secant equation—i.e., $J \in Q(y_c, s_c)$ (Exercise 5). Since A_+ is the nearest element in $Q(y_c, s_c)$ to A_c in the Frobenius norm, we have from the Pythagorian theorem that

$$\| A_+ - J \|_F^2 + \| A_+ - A_c \|_F^2 = \| A_c - J \|_F^2$$

—i.e., $\| A_+ - J \|_F \le \| A_c - J \|_F$ (see Figure 8.2.1). Hence Broyden's update cannot make the Frobenius norm of the Jacobian approximation error worse in the affine case. Unfortunately, this is not necessarily true for nonlinear functions. For example, one could have $A_c = J(x_*)$ but $A_c s_c \ne y_c$, which would guarantee $\| A_+ - J(x_*) \| > \| A_c - J(x_*) \|$. In the light of such an example, it is hard to imagine what useful result we can prove about how well A_k approximates $J(x_*)$. What is done is to show in Lemma 8.2.1 that if the approximation gets worse, then it deteriorates slowly enough for us to prove convergence of $\{x_k\}$ to x_*.

LEMMA 8.2.1   Let $D \subseteq \mathbb{R}^n$ be an open convex set containing x_c, x_+, with $x_c \ne x_*$. Let $F: \mathbb{R}^n \longrightarrow \mathbb{R}^n$, $J(x) \in \text{Lip}_\gamma(D)$, $A_c \in \mathbb{R}^{n \times n}$, A_+ defined by (8.1.5). Then for either the Frobenius or l_2 matrix norms,

$$\| A_+ - J(x_+) \| \le \| A_c - J(x_c) \| + \frac{3\gamma}{2} \| x_+ - x_c \|_2. \qquad (8.2.2)$$

Furthermore, if $x_* \in D$ and $J(x)$ obeys the weaker Lipschitz condition

$$\| J(x) - J(x_*) \| \le \gamma \| x - x_* \| \qquad \text{for all } x \in D,$$

then

$$\| A_+ - J(x_*) \| \le \| A_c - J(x_*) \|$$

$$+ \frac{\gamma}{2} (\| x_+ - x_* \|_2 + \| x_c - x_* \|_2). \qquad (8.2.3)$$

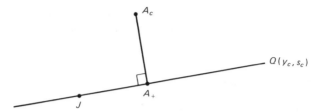

Figure 8.2.1 Broyden's method in the affine case

Proof. We prove (8.2.3), which we use subsequently. The proof of (8.2.2) is very similar.

Let $J_* \triangleq J(x_*)$. Subtracting J_* from both sides of (8.1.5),

$$A_+ - J_* = A_c - J_* + \frac{(y_c - A_c s_c)s_c^T}{s_c^T s_c}$$

$$= A_c - J_* + \frac{(J_* s_c - A_c s_c)s_c^T}{s_c^T s_c} + \frac{(y_c - J_* s_c)s_c^T}{s_c^T s_c}$$

$$= (A_c - J_*)\left[I - \frac{s_c s_c^T}{s_c^T s_c}\right] + \frac{(y_c - J_* s_c)s_c^T}{s_c^T s_c}. \qquad (8.2.4)$$

Now for either the Frobenius or l_2 matrix norm, we have from (3.1.15) or (3.1.10), and (3.1.17),

$$\| A_+ - J_* \| \leq \| A_c - J_* \| \left\| I - \frac{s_c s_c^T}{s_c^T s_c} \right\|_2 + \frac{\| y_c - J_* s_c \|_2}{\| s_c \|_2}.$$

Using

$$\left\| I - \frac{s_c s_c^T}{s_c^T s_c} \right\|_2 = 1$$

[because $I - (s_c s_c^T/s_c^T s_c)$ is a Euclidean projection matrix], and

$$\| y_c - J_* s_c \|_2 \leq \frac{\gamma}{2} (\| x_+ - x_* \|_2 + \| x_c - x_* \|_2)\| s_c \|_2$$

from Lemma 4.1.15, concludes the proof. □

Inequalities (8.2.2) and (8.2.3) are examples of a property called *bounded deterioration*. It means that if the Jacobian approximation gets worse, then it does so in a controlled way. Broyden, Dennis, and Moré (1973) have shown that any quasi-Newton algorithm whose Jacobian approximation rule obeys this property is locally *q-linearly* convergent to a root x_*, where $J(x_*)$ is nonsingular. In Theorem 8.2.2, we give the special case of their proof for Broyden's method. Later, we show Broyden's method to be locally *q-*

superlinearly convergent, by deriving a tighter bound on the norm of the term

$$(A_c - J_*)\left[I - \frac{s_c s_c^T}{s_c^T s_c}\right]$$

in (8.2.4).

For the remainder of this section we assume that $x_{k+1} \neq x_k$, $k = 0$, 1, ... Since we show below that our assumptions imply A_k nonsingular, $k = 0$, 1, ..., and since $x_{k+1} - x_k = -A_k^{-1}F(x_k)$, the assumption that $x_{k+1} \neq x_k$ is equivalent to assuming $F(x_k) \neq 0$, $k = 0, 1, \ldots$. Hence we are precluding the simple case when the algorithm finds the root exactly, in a finite number of steps.

THEOREM 8.2.2 Let all the hypotheses of Theorem 5.2.1 hold. There exist positive constants ε, δ such that if $\| x_0 - x_* \|_2 \leq \varepsilon$ and $\| A_0 - J(x_*) \|_2 \leq \delta$, then the sequence $\{x_k\}$ generated by Algorithm 8.1.2 is well defined and converges q-superlinearly to x_*. If $\{A_k\}$ is just assumed to satisfy (8.2.3), then $\{x_k\}$ converges at least q-linearly to x_*.

Proof. Let $\|\cdot\|$ designate the vector or matrix l_2 norm, $e_k \triangleq x_k - x_*$, $J_* \triangleq J(x_*)$, $\beta \geq \| J(x_*)^{-1} \|$, and choose ε and δ such that

$$6\beta\delta \leq 1, \tag{8.2.5}$$

$$3\gamma\varepsilon \leq 2\delta. \tag{8.2.6}$$

The local q-linear convergence proof consists of showing by induction that

$$\| A_k - J_* \| \leq (2 - 2^{-k})\delta, \tag{8.2.7}$$

$$\| e_{k+1} \| \leq \frac{\| e_k \|}{2}, \tag{8.2.8}$$

for $k = 0, 1, \ldots$. In brief, the first inequality is proven at each iteration using the bounded deterioration result (8.2.3), which gives

$$\| A_k - J_* \| \leq \| A_{k-1} - J_* \| + \frac{\gamma}{2}(\| e_k \| + \| e_{k-1} \|). \tag{8.2.9}$$

The reader can see that if

$$\sum_{j=0}^{k} \| e_j \|$$

is uniformly bounded above for all k, then the sequence $\{\| A_k - J_* \|\}$ will be bounded above, and using the two induction hypotheses and (8.2.6), we get (8.2.7). Then it is not hard to prove (8.2.8) by using (8.2.1), (8.2.7), and (8.2.5).

For $k = 0$, (8.2.7) is trivially true. The proof of (8.2.8) is identical to the proof at the induction step, so we omit it here.

Now assume that (8.2.7) and (8.2.8) hold for $k = 0, \ldots, i - 1$. For $k = i$, we have from (8.2.9), and the two induction hypotheses that

$$\| A_i - J_* \| \leq (2 - 2^{-(i-1)})\delta + \frac{3\gamma}{4} \| e_{i-1} \|. \tag{8.2.10}$$

From (8.2.8) and $\| e_0 \| \leq \varepsilon$ we get

$$\| e_{i-1} \| \leq 2^{-(i-1)} \| e_0 \| \leq 2^{-(i-1)}\varepsilon.$$

Substituting this into (8.2.10) and using (8.2.6) gives

$$\| A_i - J_* \| \leq (2 - 2^{-(i-1)})\delta + \frac{3\gamma}{4} \varepsilon \cdot 2^{-(i-1)}$$

$$\leq (2 - 2^{-(i-1)} + 2^{-i})\delta = (2 - 2^{-i})\delta,$$

which verifies (8.2.7).

To verify (8.2.8), we must first show that A_i is invertible so that the iteration is well defined. From $\| J(x_*)^{-1} \| \leq \beta$, (8.2.7) and (8.2.5),

$$\| J_*^{-1}(A_i - J_*) \| \leq \| J_*^{-1} \| \, \| A_i - J_* \| \leq \beta(2 - 2^{-i})\delta \leq 2\beta\delta \leq \tfrac{1}{3},$$

so we have from Theorem 3.1.4 that A_i is nonsingular and

$$\| A_i^{-1} \| \leq \frac{\| J_*^{-1} \|}{1 - \| J_*^{-1}(A_i - J_*) \|} \leq \frac{\beta}{1 - \tfrac{1}{3}} = \frac{3\beta}{2}. \tag{8.2.11}$$

Thus x_{i+1} is well defined and by (8.2.1),

$$\| e_{i+1} \| \leq \| A_i^{-1} \| \, [\| - F(x_i) + F(x_*) + J_* e_i \| + \| A_i - J_* \| \, \| e_i \|]. \tag{8.2.12}$$

By Lemma 4.1.12,

$$\| - F(x_i) + F(x_*) + J_* e_i \| \leq \frac{\gamma \| e_i \|^2}{2}.$$

Substituting this, (8.2.7), and (8.2.11) into (8.2.12) gives

$$\| e_{i+1} \| \leq \tfrac{3}{2}\beta \left[\frac{\gamma}{2} \| e_i \| + (2 - 2^{-i})\delta \right] \| e_i \|. \tag{8.2.13}$$

From (8.2.8), $\| e_0 \| \leq \varepsilon$, and (8.2.6), we have

$$\frac{\gamma \| e_i \|}{2} \leq 2^{-(i+1)}\gamma\varepsilon \leq \frac{2^{-i}\delta}{3},$$

which, substituted into (8.2.13), gives

$$\| e_{i+1} \| \leq \tfrac{3}{2}\beta[\tfrac{1}{3} \cdot 2^{-i} + 2 - 2^{-i}]\delta \| e_i \|$$

$$\leq 3\beta\delta \| e_i \|$$

$$\leq \frac{\| e_i \|}{2},$$

with the final inequality coming from (8.2.5). This proves (8.2.8) and completes the proof of q-linear convergence. We delay the proof of q-superlinear convergence until later in this section.   □

We have proven q-linear convergence of Broyden's method by showing that the bounded deterioration property (8.2.3) ensures that $\| A_k - J(x_*) \|$ stays sufficiently small. Notice that if all we knew about a sequence of Jacobian approximations $\{A_k\}$ was that they satisfy (8.2.3), then we could not expect to prove better than q-linear convergence; for example, the approximations $A_k = A_0 \neq J(x_*)$, $k = 0, 1, \ldots$, trivially satisfy (8.2.3), but from Exercise 11 of Chapter 5 the resultant method is at best q-linearly convergent. Thus the q-linear part of the proof of Theorem 8.2.2 is of theoretical interest partly because it shows us how badly we can approximate the Jacobian and get away with it. However, its real use is to ensure that $\{x_k\}$ converges to x_* with

$$\sum_{k=0}^{\infty} \| e_k \| < \infty,$$

which we will use in proving q-superlinear convergence.

We indicated at the beginning of this section that a sufficient condition for the q-superlinear convergence of a secant method is

$$\lim_{k \to \infty} \frac{\| (A_k - J(x_*))e_k \|}{\| e_k \|} = 0. \tag{8.2.14}$$

We will actually use a slight variation. In Lemma 8.2.3, we show that if $\{x_k\}$ converges q-superlinearly to x_*, then

$$\lim_{k \to \infty} \frac{\| s_k \|}{\| e_k \|} = 1,$$

where $s_k = x_{k+1} - x_k$ and $e_k = x_k - x_*$. This suggests that we can replace e_k by s_k in (8.2.14) and we might still have a sufficient condition for the q-superlinear convergence of a secant method; this is proven in Theorem 8.2.4. Using this condition, we prove the q-superlinear convergence of Broyden's method.

LEMMA 8.2.3   Let $x_k \in \mathbb{R}^n$, $k = 0, 1, \dots$. If $\{x_k\}$ converges q-superlinearly to $x_* \in \mathbb{R}^n$, then in any norm $\| \cdot \|$

$$\lim_{k \to \infty} \frac{\| x_{k+1} - x_k \|}{\| x_k - x_* \|} = 1.$$

Proof. Define $s_k = x_{k+1} - x_k$, $e_k = x_k - x_*$. The proof is really just the following picture:

Clearly, if

$$\lim_{k \to \infty} \frac{\| e_{k+1} \|}{\| e_k \|} = 0, \quad \text{then} \quad \lim_{k \to \infty} \frac{\| s_k \|}{\| e_k \|} = 1.$$

Mathematically,

$$\lim_{k \to \infty} \left| \frac{\| s_k \|}{\| e_k \|} - 1 \right| = \lim_{k \to \infty} \left| \frac{\| s_k \| - \| e_k \|}{\| e_k \|} \right|$$

$$\leq \lim_{k \to \infty} \left| \frac{\| s_k + e_k \|}{\| e_k \|} \right|$$

$$= \lim_{k \to \infty} \frac{\| e_{k+1} \|}{\| e_k \|} = 0,$$

where the final equality is the definition of q-superlinear convergence when $e_k \neq 0$ for all k. $\square$

Note that Lemma 8.2.3 is also of interest to the stopping criteria in our algorithms. It shows that whenever an algorithm achieves at least q-superlinear convergence, then any stopping test that uses s_k is essentially equivalent to the same test using e_k, which is the quantity we are really interested in.

Theorem 8.2.4 shows that (8.2.14), with e_k replaced by s_k, is a necessary and sufficient condition for q-superlinear convergence of a quasi-Newton method.

THEOREM 8.2.4 (Dennis-Moré, 1974) Let $D \subseteq \mathbb{R}^n$ be an open convex set, $F: \mathbb{R}^n \longrightarrow \mathbb{R}^n$, $J(x) \in \text{Lip}_\gamma(D)$, $x_* \in D$ and $J(x_*)$ nonsingular. Let $\{A_k\}$ be a sequence of nonsingular matrices in $\mathbb{R}^{n \times n}$, and suppose for some $x_0 \in D$ that the sequence of points generated by

$$x_{k+1} = x_k - A_k^{-1} F(x_k) \qquad (8.2.15)$$

remains in D, and satisfies $x_k \neq x_*$ for any k, and $\lim_{k \to \infty} x_k = x_*$. Then $\{x_k\}$ converges q-superlinearly to x_* in some norm $\| \cdot \|$, and $F(x_*) = 0$, if and only if

$$\lim_{k \to \infty} \frac{\| (A_k - J(x_*))s_k \|}{\| s_k \|} = 0, \qquad (8.2.16)$$

where $s_k \triangleq x_{k+1} - x_k$.

Proof. Define $J_* = J(x_*)$, $e_k = x_k - x_*$. First we assume that (8.2.16) holds, and show that $F(x_*) = 0$ and $\{x_k\}$ converges q-superlinearly to x_*. From (8.2.15)

$$0 = A_k s_k + F(x_k)$$
$$= (A_k - J_*)s_k + F(x_k) + J_* s_k$$

so that

$$-F(x_{k+1}) = (A_k - J_*)s_k + [-F(x_{k+1}) + F(x_k) + J_* s_k], \qquad (8.2.17)$$

$$\frac{\| F(x_{k+1}) \|}{\| s_k \|} \leq \frac{\| (A_k - J_*)s_k \|}{\| s_k \|} + \frac{\| -F(x_{k+1}) + F(x_k) + J_* s_k \|}{\| s_k \|}$$

$$\leq \frac{\| (A_k - J_*)s_k \|}{\| s_k \|} + \frac{\gamma}{2} (\| e_k \| + \| e_{k+1} \|) \qquad (8.2.18)$$

with the final inequality coming from Lemma 4.1.15. Using $\lim_{k \to \infty} \| e_k \| = 0$ and (8.2.16) in (8.2.18) gives

$$\lim_{k \to \infty} \frac{\| F(x_{k+1}) \|}{\| s_k \|} = 0. \qquad (8.2.19)$$

Since $\lim_{k \to \infty} \| s_k \| = 0$, this implies

$$F(x_*) = \lim_{k \to \infty} F(x_k) = 0.$$

From Lemma 4.1.16, there exist $\alpha > 0$, $k_0 \geq 0$, such that

$$\| F(x_{k+1}) \| = \| F(x_{k+1}) - F(x_*) \| \geq \alpha \| e_{k+1} \| \qquad (8.2.20)$$

for all $k \geq k_0$. Combining (8.2.19) and (8.2.20),

$$0 = \lim_{k \to \infty} \frac{\| F(x_{k+1}) \|}{\| s_k \|} \geq \lim_{k \to \infty} \alpha \frac{\| e_{k+1} \|}{\| s_k \|}$$

$$\geq \lim_{k \to \infty} \frac{\alpha \| e_{k+1} \|}{\| e_k \| + \| e_{k+1} \|} = \lim_{k \to \infty} \frac{\alpha r_k}{1 + r_k},$$

where $r_k \triangleq \| e_{k+1} \| / \| e_k \|$. This implies

$$\lim_{k \to \infty} r_k = 0,$$

which completes the proof of q-superlinear convergence.

This proof that q-superlinear convergence and $F(x_*) = 0$ imply (8.2.16) is almost the reverse of the above. From Lemma 4.1.16, there exist $\beta > 0$, $k_0 \geq 0$, such that

$$\| F(x_{k+1}) \| \leq \beta \| e_{k+1} \|$$

for all $k \geq k_0$. Thus q-superlinear convergence implies

$$0 = \lim_{k \to \infty} \frac{\| e_{k+1} \|}{\| e_k \|} \geq \lim_{k \to \infty} \frac{\| F(x_{k+1}) \|}{\beta \| e_k \|}$$

$$= \lim_{k \to \infty} \frac{1}{\beta} \frac{\| F(x_{k+1}) \|}{\| s_k \|} \cdot \frac{\| s_k \|}{\| e_k \|}. \tag{8.2.21}$$

Since $\lim_{k \to \infty} \| s_k \| / \| e_k \| = 1$ from Lemma 8.2.3, (8.2.21) implies that (8.2.19) holds. Finally, from (8.2.17) and Lemma 4.1.15,

$$\frac{\| (A_k - J_*)s_k \|}{\| s_k \|} \leq \frac{\| F(x_{k+1}) \|}{\| s_k \|} + \frac{\| -F(x_{k+1}) + F(x_k) + J_* s_k \|}{\| s_k \|}$$

$$\leq \frac{\| F(x_{k+1}) \|}{\| s_k \|} + \frac{\gamma}{2} (\| e_k \| + \| e_{k+1} \|),$$

which together with (8.2.19) and $\lim_{k \to \infty} \| e_k \| = 0$ proves (8.2.16). $\square$

Since $J(x)$ is Lipschitz continuous, it is easy to show that Lemma 8.2.4 remains true if (8.2.16) is replaced by

$$\lim_{k \to \infty} \frac{\| (A_k - J(x_k))s_k \|}{\| s_k \|} = 0. \tag{8.2.22}$$

This condition has an interesting interpretation. Since $s_k = -A_k^{-1}F(x_k)$, (8.2.22) is equivalent to

$$\lim_{k \to \infty} \frac{\| J(x_k)(s_k^N - s_k) \|}{\| s_k \|} = 0,$$

where $s_k^N = -J(x_k)^{-1}F(x_k)$, the Newton step from x_k. Thus the necessary and sufficient condition for the q-superlinear convergence of a secant method is that the secant steps converge, in magnitude and direction, to the Newton steps from the same points.

Now we complete the proof of Theorem 8.2.2. It is preceded by a technical lemma that is used in the proof.

LEMMA 8.2.5   Let $s \in \mathbb{R}^n$ be nonzero, $E \in \mathbb{R}^{n \times n}$, and let $\| \cdot \|$ denote the l_2 vector norm. Then

$$\left\| E\left(I - \frac{ss^T}{s^Ts}\right) \right\|_F = \left(\| E \|_F^2 - \left(\frac{\| Es \|}{\| s \|}\right)^2 \right)^{1/2} \qquad (8.2.23a)$$

$$\leq \| E \|_F - \frac{1}{2 \| E \|_F} \left(\frac{\| Es \|}{\| s \|} \right)^2. \qquad (8.2.23b)$$

Proof.   We remarked before that $I - (ss^T/s^Ts)$ is a Euclidean projector, and so is ss^T/s^Ts. Thus by the Pythagorian theorem,

$$\| E \|_F^2 = \left\| E \frac{ss^T}{s^Ts} \right\|_F^2 + \left\| E\left(I - \frac{ss^T}{s^Ts}\right) \right\|_F^2.$$

Since

$$\left\| E \frac{ss^T}{s^Ts} \right\|_F = \frac{\| Es \|}{\| s \|},$$

this proves (8.2.23a). Since for any $\alpha \geq |\beta| \geq 0$, $(\alpha^2 - \beta^2)^{1/2} \leq \alpha - \beta^2/2\alpha$, (8.2.23a) implies (8.2.23b). $\square$

Completion of the proof of Theorem 8.2.2 (q-superlinear convergence).   Define $E_k = A_k - J_*$, and let $\| \cdot \|$ denote the l_2 vector norm. From Theorem 8.2.4, a sufficient condition for $\{x_k\}$ to converge q-superlinearly to x_* is

$$\lim_{k \to \infty} \frac{\| E_k s_k \|}{\| s_k \|} = 0. \qquad (8.2.24)$$

From (8.2.4),

$$\| E_{k+1} \|_F \leq \left\| E_k \left(I - \frac{s_k s_k^T}{s_k^T s_k}\right) \right\|_F + \frac{\| (y_k - J_* s_k)s_k^T \|_F}{s_k^T s_k}. \qquad (8.2.25)$$

In the proof of Lemma 8.2.1, we showed that

$$\left\| \frac{(y_k - J_* s_k)s_k^T}{s_k^T s_k} \right\|_F \leq \frac{\gamma}{2} (\| e_k \| + \| e_{k+1} \|).$$

Using this, along with $\| e_{k+1} \| \leq \| e_k \|/2$ from (8.2.8) and Lemma 8.2.5 in (8.2.25), gives

$$\| E_{k+1} \|_F \leq \| E_k \|_F - \frac{\| E_k s_k \|^2}{2 \| E_k \|_F \| s_k \|^2} + \frac{3\gamma}{4} \| e_k \|,$$

or

$$\frac{\| E_k s_k \|^2}{\| s_k \|^2} \leq 2 \| E_k \|_F \left[\| E_k \|_F - \| E_{k+1} \|_F + \frac{3\gamma}{4} \| e_k \| \right]. \quad (8.2.26)$$

From the proof of Theorem 8.2.2, $\| E_k \|_F \leq 2\delta$ for all $k \geq 0$, and

$$\sum_{k=0}^{\infty} \| e_k \| \leq 2\varepsilon.$$

Thus from (8.2.26),

$$\frac{\| E_k s_k \|^2}{\| s_k \|^2} \leq 4\delta \left[\| E_k \|_F - \| E_{k+1} \|_F + \frac{3\gamma}{4} \| e_k \|_2 \right], \quad (8.2.27)$$

and summing the left and right sides of (8.2.27) for $k = 0, 1, \ldots, i,$

$$\sum_{k=0}^{i} \frac{\| E_k s_k \|^2}{\| s_k \|^2} \leq 4\delta \left[\| E_0 \|_F - \| E_{i+1} \| + \frac{3\gamma}{4} \sum_{k=0}^{i} \| e_k \|_2 \right]$$

$$\leq 4\delta \left[\| E_0 \|_F + \frac{3\gamma\varepsilon}{2} \right]$$

$$\leq 4\delta \left[\delta + \frac{3\gamma\varepsilon}{2} \right]. \quad (8.2.28)$$

Since (8.2.28) is true for any $i \geq 0$,

$$\sum_{i=0}^{\infty} \frac{\| E_k s_k \|^2}{\| s_k \|^2}$$

is finite. This implies (8.2.24) and completes the proof.   □

We present another example, which illustrates the convergence of Broyden's method on a completely nonlinear problem. We also use this example to begin looking at how close the final Jacobian approximation A_k is to the Jacobian $J(x_*)$ at the solution.

EXAMPLE 8.2.6   Let

$$F(x) = \begin{pmatrix} x_1^2 + x_2^2 - 2 \\ e^{x_1 - 1} + x_2^3 - 2 \end{pmatrix},$$

which has a root $x_* = (1, 1)^T$. The sequences of points generated by Broyden's method and Newton's method from $x_0 = (1.5, 2)^T$ with $A_0 = J(x_0)$ for Broyden's method, are shown below.

Broyden's Method			Newton's Method	
1.5	2.0	x_0	1.5	2.0
0.8060692	1.457948	x_1	0.8060692	1.457948
0.7410741	1.277067	x_2	0.8901193	1.145571
0.8022786	1.159900	x_3	0.9915891	1.021054
0.9294701	1.070406	x_4	0.9997085	1.000535
1.004003	1.009609	x_5	0.999999828	1.000000357
1.003084	0.9992213	x_6	0.99999999999992	1.0000000000002
1.000543	0.9996855	x_7	1.0	1.0
0.99999818	1.00000000389	x_8		
0.9999999885	0.999999999544	x_9		
0.99999999999474	0.99999999999998	x_{10}		
1.0	1.0	x_{11}		

The final approximation to the Jacobian generated by Broyden's method is

$$A_{10} \cong \begin{bmatrix} 1.999137 & 2.021829 \\ 0.9995643 & 3.011004 \end{bmatrix}, \quad \text{whereas} \quad J(x_*) = \begin{bmatrix} 2 & 2 \\ 1 & 3 \end{bmatrix}.$$

In the above example, A_{10} has a maximum relative error of 1.1% as an approximation to $J(x_*)$. This is typical of the final Jacobian approximation generated by Broyden's method. On the other hand, it is easy to show that in Example 8.1.3, $\{A_k\}$ does not converge to $J(x_*)$:

LEMMA 8.2.7   In Example 8.1.3,

$$\lim_{k \to \infty} A_k = \begin{bmatrix} 1 & 1 \\ 1.5 & 7.5 \end{bmatrix}, \quad \text{whereas} \quad J(x_*) = \begin{bmatrix} 1 & 1 \\ 0 & 6 \end{bmatrix}.$$

Proof. We showed in Example 8.1.3 that $(A_k)_{11} = (A_k)_{12} = 1$ for all $k \geq 0$, that

$$A_1 = \begin{bmatrix} 1 & 1 \\ 0.375 & 8.625 \end{bmatrix},$$

and that

$$F_1(x_k) = (x_k)_1 + (x_k)_2 - 3 = 0 \qquad (8.2.29)$$

for all $k \geq 1$. From (8.2.29), $(1, 1)^T s_k = 0$ for all $k \geq 1$. From the formula (8.1.10) for Broyden's update, this implies that $(A_{k+1} - A_k)(1, 1)^T = 0$ for all $k \geq 1$. Thus

$$(A_k)_{21} + (A_k)_{22} = (A_1)_{21} + (A_1)_{22} = 9 \qquad (8.2.30)$$

for all $k \geq 1$. Also, it is easily shown that the secant equation implies

$$\lim_{k \to \infty} (A_k)_{21} - (A_k)_{22} = -6 \qquad (8.2.31)$$

in this case. From (8.2.30) and (8.2.31),

$$\lim_{k \to \infty} (A_k)_{21} = 1.5 \quad \text{and} \quad \lim_{k \to \infty} (A_k)_{22} = 7.5. \qquad \square$$

The results of Lemma 8.2.7 are duplicated exactly on the computer; we got

$$A_7 \cong \begin{pmatrix} 1 & 1 \\ 1.5 - 10^{-6} & 7.5 + 10^{-6} \end{pmatrix}$$

in Example 8.1.1. The proof of Lemma 8.2.7 is easily generalized to show that

$$\lim_{k \to \infty} A_k \neq J(x_*)$$

for almost any partly linear system of equations that is solved using Broyden's method. Exercise 11 is an example of a completely nonlinear system of equations where $\{x_k\}$ converges to a root x_* but the final A_k is very different from $J(x_*)$.

In summary, when Broyden's method converges q-superlinearly to a root x_*, one cannot assume that the final Jacobian approximation A_k will approximately equal $J(x_*)$, although often it does.

If we could say how fast $\{\| E_k s_k \|/\| s_k \|\}$ goes to 0, then we could say more about the order of convergence of Broyden's method. The nearest to such a result is the proof given by Gay (1979). He proved that for any affine F, Broyden's method gives $x_{2n+1} = x_*$ which is equivalent to $\| E_{2n} s_{2n} \|/\| s_{2n} \| = 0$. Under the hypotheses of Theorem 8.2.2, this allowed him to prove $2n$-step q-quadratic convergence for Broyden's method on general nonlinear functions. As a consequence of Exercise 2.6, this implies r-order $2^{1/2n}$.

8.3 IMPLEMENTATION OF QUASI-NEWTON ALGORITHMS USING BROYDEN'S UPDATE

This section discusses two issues to consider in using a secant approximation, instead of an analytic or finite-difference Jacobian, in one of our quasi-Newton algorithms for solving systems of nonlinear equations: (1) the details involved in implementing the approximation, and (2) what changes, if any, need to be made to the rest of the algorithm.

The first problem in using secant updates to approximate the Jacobian is how to get the initial approximation A_0. We have already said that in practice we use a finite-difference approximation to $J(x_0)$, which we calculate using Algorithm A5.4.1. Moré's HYBRD implementation in MINPACK [Moré, Garbow, and Hillstrom (1980)] uses a modification of this algorithm that is more efficient when $J(x)$ has many zero elements; we defer its consideration to Chapter 11.

If Broyden's update is implemented directly as written, there is little to say about its implementation; it is simply

$$t = y - As,$$

$$\sigma = s^T s,$$

$$A_{ij} = A_{ij} + \frac{t_i s_j}{\sigma}, \qquad i = 1, \ldots, n, \quad j = 1, \ldots, n.$$

However, recall that the first thing our quasi-Newton algorithm will do with A_+ is determine its QR factorization in order to calculate the step $-A_+^{-1}F(x_+)$. Therefore, the reader may already have realized that since Broyden's update exactly fits the algebraic structure of equation (3.4.1), a more efficient updating strategy is to apply Algorithm A3.4.1 to update the factorization $Q_c R_c$ of A_c into the factorization $Q_+ R_+$ of A_+, in $O(n^2)$ operations. This saves the $O(n^3)$ cost of the QR factorization of A_k, at every iteration after the first. The reader can confirm that an entire iteration of a quasi-Newton algorithm, using the QR form of Broyden's update and a line-search or dogleg globalizing strategy, costs only $O(n^2)$ arithmetic operations.

We will refer to these two possible implementations of Broyden's update as the unfactored and factored forms of the update, respectively. Appendix A contains both forms, in Algorithms A8.3.1 and A8.3.2 respectively. The factored form leads to a more efficient algorithm and should be used in any production implementation. The unfactored form is simpler to code, and may be preferred for preliminary implementations.

Both forms of the update in the appendix contain two more features. First, the reader can confirm (Exercise 12) that the diagonal scaling of the independent variables discussed in Section 7.1 causes Broyden's update to become

$$A_+ = A_c + \frac{(y_c - A_c s_c)(D_x^2 s_c)^T}{s_c^T D_x^2 s_c}, \qquad (8.3.1)$$

and this is the update actually implemented. The reader can confirm that the update is independent of a diagonal scaling of the dependent variables.

Second, under some circumstances we decide not to change a row of A_c. Notice that if $(y_c - A_c s_c)_i = 0$, then (8.3.1) automatically causes row i of A_+ to equal row i of A_c. This makes sense, because Broyden's update makes the smallest change to row i of A_c consistent with $(A_+)_{i.}s_c = (y_c)_i$, and if

$(y_c - A_c s_c)_i = 0$, no change is necessary. Our algorithms also leave row i of A_c unchanged if $|(y_c - A_c s_c)_i|$ is smaller than the computational uncertainty in $(y_c)_i$, measured by $\eta(|f_i(x_c)| + |f_i(x_+)|)$, where η is the relative noise in function values that was discussed in Section 5.4. This condition is intended to prevent the update from introducing random noise rather than good derivative information into A_+.

There is another implementation of Broyden's method that also results in $O(n^2)$ arithmetic work per iteration. It uses the Sherman-Morrison-Woodbury formula given below.

LEMMA 8.3.1   (Sherman-Morrison-Woodbury) Let u, $v \in \mathbb{R}^n$,   and assume that $A \in \mathbb{R}^{n \times n}$ is nonsingular. Then $A + uv^T$ is nonsingular if and only if

$$1 + v^T A^{-1} u \triangleq \sigma \neq 0.$$

Furthermore,

$$(A + uv^T)^{-1} = A^{-1} - \frac{1}{\sigma} A^{-1} uv^T A^{-1}.$$

Proof.   Exercise 13.   □

Straightforward application of Lemma 8.3.1 shows that if one knows A_c^{-1}, then Broyden's update can be expressed as

$$A_+^{-1} = A_c^{-1} + \frac{(s_c - A_c^{-1} y_c) s_c^T A_c^{-1}}{s_c^T A_c^{-1} y_c}, \tag{8.3.2}$$

which requires $O(n^2)$ operations. The secant direction $-A_+^{-1} F(x_+)$ then can be calculated by a matrix-vector multiplication requiring an additional $O(n^2)$ operations. Therefore, until Gill and Murray (1972) introduced the notion of sequencing QR factorizations, algorithms using Broyden's update were implemented by initially calculating A_0^{-1} and then using (8.3.2). This implementation usually works fine in practice, but it has the disadvantage that it makes it hard to detect ill-conditioning in A_+. Since the factored update from A_c to A_+ doesn't have this problem, and since it requires essentially the same number of arithmetic operations as (8.3.2), it has replaced (8.3.2) as the preferred implementation in production codes.

The other question we answer in this section is, "What changes need to be made to the rest of our quasi-Newton algorithm for nonlinear equations when we use secant updates?" From the way we have presented the general quasi-Newton framework, we would like the answer to be "None," but there are two exceptions. The first is trivial: when the factored form of the update is used, the QR factorization of the model Jacobian at each iteration is omitted, and several implementation details are adjusted as discussed in Appendix A.

The other change is more significant: it is no longer guaranteed that the quasi-Newton direction, $s_c = -A_c^{-1} F(x_c)$, is a descent direction for $f(x) = \frac{1}{2} \| F(x) \|_2^2$. Recall from Section 6.5 that $\nabla f(x_c) = J(x_c)^T F(x_c)$, so that

$$s_c^T \nabla f(x_c) = -F(x_c)^T A_c^{-T} J(x_c)^T F(x_c),$$

which is nonpositive if $A_c = J(x_c)$, but not necessarily otherwise. Furthermore, there is no way we can check whether s_c is a descent direction for $f(x)$ in a secant algorithm, since we cannot calculate $\nabla f(x_c)$ without $J(x_c)$. This means that s_c may be an uphill direction for $f(x)$, and our global step may fail to produce any satisfactory point. Luckily, this doesn't happen often in practice, since A_c is an approximation to $J(x_c)$. If it does, there is an easy fix; we reset A_c, using a finite-difference approximation to $J(x_c)$, and continue the algorithm. This provision is included in the driver for our nonlinear equations algorithm; it is essentially what is done in Moré's MINPACK code as well.

8.4   OTHER SECANT UPDATES FOR NONLINEAR EQUATIONS

So far in this chapter we have discussed one secant update, Broyden's, for approximating the Jacobian. We saw that of all the members of $Q(y_c, s_c)$, the affine subspace of matrices satisfying the secant equation

$$A_+ s_c = y_c, \tag{8.4.1}$$

Broyden's update is the one that is closest to A_c in the Frobenius norm. In this section we mention some other proposed approximations from $Q(y_c, s_c)$.

Perhaps the best-known alternative was suggested by Broyden, in the paper where he proposed Broyden's update. The idea from the last section, of sequencing A_k^{-1} rather than A_k, suggested to Broyden the choice

$$A_+^{-1} = A_c^{-1} + \frac{(s_c - A_c^{-1} y_c) y_c^T}{y_c^T y_c}. \tag{8.4.2}$$

A straightforward application of Lemma 8.1.1 shows that (8.4.2) is the solution to

$$\min_{\substack{\text{nonsingular} \\ B \in Q(y_c, s_c)}} \| B^{-1} - A_c^{-1} \|_F. \tag{8.4.3}$$

From Lemma 8.3.1, (8.4.2) is equivalent to

$$A_+ = A_c + \frac{(y_c - A_c s_c) y_c^T A_c}{y_c^T A_c s_c} \tag{8.4.4}$$

—clearly a different update from Broyden's update.

Update (8.4.2) has the attractive property that it produces values of A_+^{-1}

directly, without problems of a zero denominator. Furthermore, it shares all the good theoretical properties of Broyden's update. That is, one can modify the proof of Lemma 8.2.1 to show that this method of approximating $J(x)^{-1}$ is of bounded deterioration as an approximation to $J(x_*)^{-1}$, and from there show that Theorem 8.2.2 holds using (8.4.2) in the place of Broyden's update. In practice, however, methods using (8.4.2) have been considerably less successful than the same methods using Broyden's update; in fact, (8.4.2) has become known as "Broyden's bad update."

Although we don't pretend to understand the lack of computational success with Broyden's bad update, it is interesting that Broyden's good update comes from minimizing the change in the affine model, subject to satisfying the secant equation, while Broyden's bad update is related to minimizing the change in the *solution* to the model. Perhaps the fact that we use information about the function, and not about its solution, in forming the secant model, is a reason why the former approach is more desirable. In any case, we will see an analogous situation in Section 9.2, where an update for unconstrained minimization derived from Broyden's good update outperforms a similar method derived from Broyden's bad update, although the two again have similar theoretical properties.

There are many other possible secant updates for nonlinear equations, some of which may hold promise for the future. For example, the update

$$A_{k+1} = A_k = \frac{(y_k - A_k s_k)v_k^T}{v_k^T s_k} \tag{8.4.5}$$

is well defined and satisfies (8.4.1) for any $v_k \in \mathbb{R}^n$ for which $v_k^T s_k \neq 0$. Broyden's good and bad updates are just the choices $v_k = s_k$ and $v_k = A_k^T y_k$, respectively. Barnes (1965) and Gay and Schnabel (1978) have proposed an update of the form (8.4.5), where v_k is the Euclidean projection of s_k orthogonal to $s_{k-1}, \ldots, s_{k-m}$, $0 \leq m < n$. This enables A_+ to satisfy $m + 1$ secant equations $A_{k+1}s_i = y_i$, $i = k - m, \ldots, k$, meaning that the affine model interpolates $F(x_i)$, $i = k - m, \ldots, k - 1$, as well as $F(x_k)$ and $F(x_{k+1})$. In implementations where $m < n$ is chosen so that $s_{k-m}, \ldots, s_k$ are very linearly independent, this update has slightly outperformed Broyden's update, but experience with it is still limited. Therefore, we still recommend Broyden's update as the secant approximation for solving systems of nonlinear equations.

8.5 EXERCISES

1. Suppose s_i, $y_i \in \mathbb{R}^n$, $i = 1, \ldots, n$, and that the vectors $s_1, \ldots, s_n$ are linearly independent. How would you calculate $A \in \mathbb{R}^{n \times n}$ to satisfy $As_i = y_i$, $i = 1, \ldots, n$? Why is this a bad way to form the model (8.1.1), if the directions $s_1, \ldots, s_n$ are close to being linearly dependent?

2. Show that if

$$A = \begin{bmatrix} 4 & 1 \\ -1 & 1 \end{bmatrix}, \qquad s = \begin{bmatrix} 1 \\ 0 \end{bmatrix}, \qquad \text{and} \qquad y = \begin{bmatrix} 5 \\ -2 \end{bmatrix},$$

then the solution to

$$\min_{M \in Q(y, s)} \| M - A \|_2$$

is

$$A_+ = A + \begin{bmatrix} 1 & \alpha \\ -1 & \alpha \end{bmatrix} \qquad \text{for any } \alpha \in [-1, 1].$$

What is the solution in the Frobenius norm? If $F(x)$ is linear in x_2, which solution is preferable?

3. (a) Carry out two iterations of Broyden's method on

$$F(x) = \begin{bmatrix} x_1 + x_2 - 3 \\ x_1^2 + x_2^2 - 9 \end{bmatrix},$$

   starting with $x_0 = (2, 7)^T$ and $A_0 = J(x_0)$.
   (b) Continue part (a) (on a computer) until $\| x_{k+1} - x_k \| \le (\text{macheps})^{1/2}$. What is the final value of A_k? How does it compare with $J(x_*)$?

4. Suppose Broyden's method is applied to a function $F: \mathbb{R}^n \longrightarrow \mathbb{R}^n$ for which f_1, $\ldots, f_m$ are linear, $m < n$. Show that if A_0 is calculated analytically or by finite differences, then for $k \ge 0$, $(A_k)_{i.} = J(x_k)_{i.}$, $i = 1, \ldots, m$, and that for $k \ge 1$, $f_i(x_k) = 0, i = 1, \ldots, m$.

5. Suppose $J \in \mathbb{R}^{n \times n}$, $b \in \mathbb{R}^n$, $F(x) = Jx + b$. Show that if x_c and x_+ are any two distinct points in $\mathbb{R}^n$ with $s_c = x_+ - x_c$, $y_c = F(x_+) - F(x_c)$, then $J \in Q(y_c, s_c)$.

6. Prove (8.2.2) using the techniques of the proof of Lemma 8.2.1.

7. Suppose that $s \in \mathbb{R}^n$, $s \ne 0$, and I is the $n \times n$ identity matrix. Show that

$$\left\| I - \frac{ss^T}{s^T s} \right\|_2 = 1.$$

8. Prove Theorem 8.2.4 with (8.2.22) in the place of (8.2.16).

9. Run Broyden's method to convergence on the function of Example 8.2.6, using the starting point $x_0 = (2, 3)^T$ (which was used with this function in Example 5.4.2). Why didn't we use $x_0 = (2, 3)^T$ in Example 8.2.6?

10. Generalize Lemma 8.2.7 as follows: let $1 \le m < n$,

$$F(x) = \begin{bmatrix} J_1 x + b_1 \\ F_2(x) \end{bmatrix},$$

where $J_1 \in \mathbb{R}^{m \times n}$, $b_1 \in \mathbb{R}^m$, and $F_2: \mathbb{R}^n \longrightarrow \mathbb{R}^{n-m}$ is nonlinear. Suppose Broyden's method is used to solve $F(x) = 0$, generating a sequence of Jacobian approximations $A_0, A_1, A_2, \ldots$. Let A_k be partitioned into

$$\begin{bmatrix} A_{k1} \\ A_{k2} \end{bmatrix}, \qquad A_{k1} \in \mathbb{R}^{m \times n}, \quad A_{k2} \in \mathbb{R}^{(n-m) \times n}.$$

Show that if A_0 is calculated analytically or by finite differences, then $A_{k1} = J_1$ for all $k \geq 0$, and $(A_{k2} - A_{12})J_1^T = 0$ for all $k \geq 1$. What does this imply about the convergence of the sequence $\{A_{k2}\}$ to the correct value $F'_2(x_*)$?

11. Computational examples of poor convergence of $\{A_k\}$ to $J(x_*)$ on completely nonlinear functions: Run Broyden's method to convergence on the function from Example 8.2.6, using the starting point $x_0 = (0.5, 0.5)^T$ and printing out the values of A_k. Compare the final value of A_k with $J(x_*)$. Do the same using the starting point $x_0 = (2, 3)^T$.

12. Suppose we transform the variables x and $F(x)$ by

$$\hat{x} = D_x \cdot x, \qquad \hat{F}(x) = D_F \cdot F(x),$$

where D_x and D_F are positive diagonal matrices, perform Broyden's update in the new variable and function space, and then transform back to the original variables. Show that this process is equivalent to using update (8.3.1) in the original variables x and $F(x)$. [*Hint*: The new Jacobian $\hat{J}(\hat{x})$ is related to the old Jacobian $J(x)$ by $\hat{J}(\hat{x}) = D_F J(x)D_x^{-1}$.]

13. Prove Lemma 8.3.1.

14. Using the algorithms from the appendix, program and run an algorithm for solving systems of nonlinear equations, using Broyden's method to approximate the Jacobian. When your code is working properly, try to find a problem where an "uphill" secant direction is generated [i.e., $-F(x_k)^T A_k^{-T} J(x_k)F(x_k) > 0$], so that it is necessary to reset A_k to $J(x_k)$.

15. Use Lemma 8.1.1 to show that (8.4.2) is the solution to (8.4.3). Then use Lemma 8.3.1 to show that (8.4.2) and (8.4.4) are equivalent, if A_c is nonsingular and $y_c^T A_c s_c \neq 0$.

16. (Hard) Come up with a better explanation of why Broyden's "good" method, (8.1.5), is more successful computationally than Broyden's bad method, (8.4.4).

Exercises 17 and 18 are taken from Gay and Schnabel (1978).

17. Let $A_k \in \mathbb{R}^{n \times n}$, s_i, $y_i \in \mathbb{R}^n$, $i = k - 1$, k, s_{k-1}, and s_k linearly independent, and assume that $A_k s_{k-1} = y_{k-1}$.
    (a) Derive a condition on v_k in (8.4.5) so that $A_{k+1}s_{k-1} = y_{k-1}$.
    (b) Show that the A_{k+1} given by (8.4.5), with

$$v_k = s_k - \frac{s_{k-1}(s_k^T s_{k-1})}{s_{k-1}^T s_{k-1}}$$

   is the solution to

$$\text{minimize } \| B - A_k \|_F$$
$$\scriptstyle B \in \mathbb{R}^{n \times n}$$

$$\text{subject to} \quad B \in Q(y_k, s_k) \cap Q(y_{k-1}, s_{k-1}).$$

18. Generalize Exercise 17 as follows: assume in addition that for some $m < n$, s_i, $y_i \in \mathbb{R}^n$, $i = k - m, \ldots, k$, with $s_{k-m}, \ldots, s_k$ linearly independent and $A_k s_i = y_i$, $i = k - m, \ldots, k - 1$.

(a)   Derive a condition on v_k in (8.4.5) so that $A_{k+1}s_i = y_i$, $i = k - m, \ldots, k - 1$.

(b)   Find a choice of v_k so that A_{k+1} given by (8.4.5) is the solution to

$$\underset{B \in \mathbb{R}^{n \times n}}{\text{minimize}} \; \| B - A_k \|_F$$

$$\text{subject to} \qquad B \in Q(y_i, s_i), \qquad i = k - m, \ldots, k.$$

9

Secant Methods
for Unconstrained Minimization

In this chapter we consider secant methods for the unconstrained minimization problem. The derivatives we have used in our algorithms for this
problem are the gradient, $\nabla f(x)$, and the Hessian, $\nabla^2 f(x)$. The gradient must be
known accurately in minimization algorithms, both for calculating descent
directions and for stopping tests, and the reader can see from Chapter 8 that
secant approximations do not provide this accuracy. Therefore, secant approximations to the gradient are not used in quasi-Newton algorithms. On the
other hand, the Hessian can be approximated by secant techniques in much
the same manner as the Jacobian was in Chapter 8, and this is the topic of the
present chapter. We will present the most successful secant updates to the
Hessian and the theory that accompanies them. These updates require no
additional function or gradient evaluations, and again lead to locally q-
superlinearly convergent algorithms.

Since the Hessian is the Jacobian of the nonlinear system of equations
$\nabla f(x) = 0$, it could be approximated using the techniques of Chapter 8. However, this would disregard two important properties of the Hessian: it is always
symmetric and often positive definite. The incorporation of these two properties into the secant approximation to the Hessian is the most important new
aspect of this chapter. In Section 9.1 we introduce a symmetric secant update,
and in Section 9.2 one that preserves positive definiteness as well. The latter
update, called the positive definite secant update (or the BFGS), is in practice
the most successful secant update for the Hessian. In Section 9.3 we present

another derivation of the positive definite secant update that leads to its local convergence proof. Implementation considerations are discussed in Section 9.4. We conclude the chapter with a global convergence result for the positive definite secant update in Section 9.5 and a brief discussion of other secant approximations to the Hessian in Section 9.6.

9.1   THE SYMMETRIC SECANT
UPDATE OF POWELL

Suppose that we are solving

$$\min_{x \in \mathbb{R}^n} f: \mathbb{R}^n \longrightarrow \mathbb{R},$$

that we have just taken a step from x_c to x_+, and that we wish to approximate $\nabla^2 f(x_+)$ by H_+ using secant techniques. Following the development in Section 8.1, we ask first what the secant equation should be. If we think of the Hessian as the derivative matrix of the system of nonlinear equations $\nabla f(x) = 0$, then the analogous secant equation to (8.1.3) is

$$H_+ s_c = y_c, \tag{9.1.1a}$$

where

$$s_c = x_+ - x_c, \qquad y_c = \nabla f(x_+) - \nabla f(x_c). \tag{9.1.1b}$$

We will refer to (9.1.1) as the secant equation for the unconstrained minimization problem. Notice that if H_+ satisfies (9.1.1), then the quadratic model

$$m_+(x) = f(x_+) + \nabla f(x_+)^T(x - x_+) + \tfrac{1}{2}(x - x_+)^T H_+(x - x_+)$$

satisfies $m_+(x_+) = f(x_+)$, $\nabla m_+(x_+) = \nabla f(x_+)$, and $\nabla m_+(x_c) = \nabla f(x_c)$. This is another way to motivate equation (9.1.1).

We could now use the secant update of Chapter 8 to approximate $\nabla^2 f(x_+)$ by

$$(H_+)_1 = H_c + \frac{(y_c - H_c s_c)s_c^T}{s_c^T s_c}, \tag{9.1.2}$$

where $H_c \in \mathbb{R}^{n \times n}$ is our approximation to $\nabla^2 f(x_c)$. The main problem with (9.1.2) is that even if H_c is symmetric, $(H_+)_1$ will not be unless $y_c - H_c s_c$ is a multiple of s_c. We have already seen several reasons why a Hessian approximation should be symmetric. (See, for example, Exercise 12 of Chapter 4.) Therefore we seek an H_+ that obeys $H_+ = H_+^T$ as well as (9.1.1).

Since Frobenius norm projection into $Q(y_c, s_c)$ proved to be a good way to update A_c in Chapter 8, a reasonable idea to construct a symmetric $(H_+)_2$ from $(H_+)_1$ is to make the Frobenius norm projection of $(H_+)_1$ into the subspace S of symmetric matrices. It is an easy exercise to show that this is accomplished by

$$(H_+)_2 = \tfrac{1}{2}((H_+)_1 + (H_+)_1^T) = H_c + \frac{(y_c - H_c s_c)s_c^T + s_c(y_c - H_c s_c)^T}{2s_c^T s_c}.$$

Now $(H_+)_2$ is symmetric (H_c is), but it does not in general obey the secant equation. However, one might consider continuing this process, generating $(H_+)_3, (H_+)_4, \ldots$ by

$$(H_+)_{2k+1} = (H_+)_{2k} + \frac{(y_c - (H_+)_{2k} s_c)s_c^T}{s_c^T s_c},$$

$$(H_+)_{2k+2} = \tfrac{1}{2}((H_+)_{2k+1} + (H_+)_{2k+1}^T).$$

Here each $(H_+)_{2k+1}$ is the closest matrix in $Q(y_c, s_c)$ to $(H_+)_{2k}$, and each $(H_+)_{2k+2}$ is the closest symmetric matrix to $(H_+)_{2k+1}$ (see Figure 9.1.1). This sequence of matrices was shown by M. J. D. Powell (1970b) to converge to

$$H_+ = H_c + \frac{(y_c - H_c s_c)s_c^T + s_c(y_c - H_c s_c)^T}{s_c^T s_c} - \frac{\langle y_c - H_c s_c, s_c \rangle s_c s_c^T}{(s_c^T s_c)^2}. \qquad (9.1.3)$$

We will call update (9.1.3) the *symmetric secant update*; it is also known as the Powell-symmetric-Broyden (PSB) update. It is an easy exercise to confirm that H_+ given by (9.1.3) obeys the secant equation (9.1.1), and that $H_+ - H_c$ is a symmetric matrix of rank two. In fact, Dennis and Moré (1977) prove the following theorem:

THEOREM 9.1.1   Let $H_c \in \mathbb{R}^{n \times n}$ be symmetric, $s_c, y_c \in \mathbb{R}^n$, $s_c \neq 0$. Then the unique solution to

$$\underset{H \in \mathbb{R}^{n \times n}}{\text{minimize}} \; \| H - H_c \|_F$$

$$\text{subject to} \qquad Hs_c = y_c, \quad (H - H_c) \text{ symmetric}$$

is given by (9.1.3).

The proof of Theorem 9.1.1 is similar to that of Theorem 8.1.1 and is indicated in Exercise 4. Theorem 9.1.1 also follows from a much more general result of Dennis and Schnabel (1979), given in Chapter 11, that guarantees the process of Figure 9.1.1 will converge to the closest matrix to H_c in the intersection of the two affine subspaces.

The local convergence properties of the quasi-Newton method using (9.1.3) are the same as for the secant method for systems of nonlinear equations, and so we just state the following theorem. The proof is a generalization of the techniques of Section 8.2; it is indicated in Exercise 5 and 6. Notice that

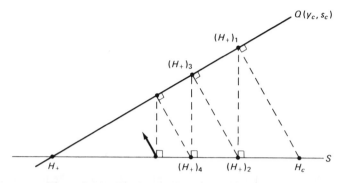

Figure 9.1.1   The Powell symmetrization process

the algorithm converges locally to any isolated zero of $\nabla f(x)$, since $\nabla^2 f(x_*)$ is required only to be symmetric and nonsingular, but not necessarily positive definite.

THEOREM 9.1.2   [Broyden, Dennis, and Moré (1973)]   Let $f: \mathbb{R}^n \longrightarrow \mathbb{R}$ be twice continuously differentiable in an open convex set $D \subset \mathbb{R}^n$, and let $\nabla^2 f \in \text{Lip}_\gamma(D)$. Assume there exists $x_* \in D$ such that $\nabla f(x_*) = 0$ and $\nabla^2 f(x_*)$ is nonsingular and that H_0 is symmetric. Then there exist positive constants ε, δ such that if $\|x_0 - x_*\|_2 < \varepsilon$ and $\|H_0 - \nabla^2 f(x_*)\|_2 \leq \delta$, the symmetric secant method

$$x_{k+1} = x_k - H_k^{-1} \nabla f(x_k), \tag{9.1.4a}$$

$$s_k = x_{k+1} - x_k, \qquad y_k = \nabla f(x_{k+1}) - \nabla f(x_k), \tag{9.1.4b}$$

$$H_{k+1} = H_k + \frac{(y_k - H_k s_k)s_k^T + s_k(y_k - H_k s_k)^T}{s_k^T s_k}$$

$$- \frac{\langle y_k - H_k s_k, s_k \rangle s_k s_k^T}{(s_k^T s_k)^2} \tag{9.1.4c}$$

is well defined, and $\{x_k\}$ remains in D and converges q-superlinearly to x_*.

At this point it might seem that we have accomplished everything that we want out of a secant approximation to the Hessian. However, the symmetric secant update has one important problem that hurts its performance in practice: H_+ given by (9.1.3) may not be positive definite, even if H_c is and there are positive definite matrices in $Q(y_c, s_c) \cap S$. Figure 9.1.2 illustrates the reason. Let PD be the set of symmetric and positive definite matrices, and $QS = Q(y_c, s_c) \cap S$, the affine space of symmetric matrices that obey the secant equation. Then we see that the closest matrix to H_c in QS may not be positive definite, even if $QS \cap PD$ is nonempty. Furthermore, since PD is an open convex cone, there may not even be a closest matrix to H_c in $QS \cap PD$.

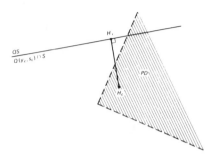

Figure 9.1.2   Why the symmetric secant update may not be positive definite

Since we have stressed the importance of having a positive definite model Hessian, we would like H_+ to be positive definite in the above situation. In Section 9.2 we see a useful way to accomplish this. The result is the secant update to the Hessian that is most successful in practice. It is also a symmetric, rank-two change to H_c.

To conclude this section, we mention a formula closely related to (9.1.3) that we will want to refer to in Section 9.3. J. Greenstadt (1970) suggested the update

$$
H_+^{-1} = H_c^{-1} + \frac{(s_c - H_c^{-1}y_c)y_c^T + y_c(s_c - H_c^{-1}y_c)^T}{y_c^T y_c} - \frac{\langle y_c, s_c - H_c^{-1}y_c\rangle y_c y_c^T}{(y_c^T y_c)^2}.
$$

(9.1.5)

The reader will easily guess that (9.1.5) is the result of applying the symmetrization procedure of Figure 9.1.1 to the bad Broyden update (8.4.2), and that H_+^{-1} is the nearest matrix in $Q(s_c, y_c) \cap S$ to H_c^{-1} in the Frobenius norm. Theorem 9.1.2 also holds for this update, but (9.1.5) shares with the symmetric secant update the deficiency of not necessarily giving a positive definite H_+ even when H_c is positive definite and $QS \cap PD$ is nonempty. However, we will see in Section 9.3 that (9.1.5) has an interesting relationship to the most successful secant update to the Hessian.

9.2 SYMMETRIC POSITIVE DEFINITE SECANT UPDATES

In solving the unconstrained minimization problem by a quasi-Newton algorithm, much has been made of the advantages in efficiency, numerical stability, and global convergence of having a symmetric and positive definite model

Hessian. In this section we show how to create symmetric and positive definite secant approximations to the Hessian.

We seek a solution to:

given s_c, $y_c \in \mathbb{R}^n$, $s_c \neq 0$,

find a symmetric and positive definite $H_+ \in \mathbb{R}^{n \times n}$ such that $H_+ s_c = y_c$.   (9.2.1)

To ensure that H_+ is symmetric and positive definite, we will use the analog of a common trick in mathematical programming. There, if a certain variable $h \in \mathbb{R}$ must be nonnegative, we replace it by j^2, where $j \in \mathbb{R}$ is a new unconstrained independent variable. In analogy, it is easy to prove that H_+ is symmetric and positive definite if and only if

$$H_+ = J_+ J_+^T \qquad \text{for some nonsingular } J_+ \in \mathbb{R}^{n \times n}.$$

Thus we will reexpress (9.2.1) as:

given s_c, $y_c \in \mathbb{R}^n$, $s_c \neq 0$,

find a nonsingular $J_+ \in \mathbb{R}^{n \times n}$ such that $J_+ J_+^T s_c = y_c$.   (9.2.2)

For any solution J_+ to (9.2.2), $H_+ = J_+ J_+^T$ will solve (9.2.1).

Even for $n = 1$, we see that there is no solution to (9.2.2) unless $y_c \neq 0$ has the same sign as s_c, so first we must determine the conditions under which (9.2.2) will have a solution. Lemma 9.2.1 contains the answer and also suggests a procedure for finding J_+.

LEMMA 9.2.1   Let s_c, $y_c \in \mathbb{R}^n$, $s_c \neq 0$. Then there exists a nonsingular $J_+ \in \mathbb{R}^{n \times n}$ such that $J_+ J_+^T s_c = y_c$, if and only if

$$s_c^T y_c > 0. \tag{9.2.3}$$

Proof.   Suppose there exists a nonsingular J_+ such that $J_+ J_+^T s_c = y_c$. Define $v_c = J_+^T s_c$. Then

$$0 < v_c^T v_c = s_c^T J_+ J_+^T s_c = s_c^T y_c.$$

This proves the *only if* portion of the lemma. The *if* part is proven by the derivation below, but first we digress to a discussion of the condition $s_c^T y_c > 0$.

Since $y_c = \nabla f(x_+) - \nabla f(x_c)$, condition (9.2.3) is equivalent to

$$s_c^T \nabla f(x_+) > s_c^T \nabla f(x_c), \tag{9.2.4}$$

which simply means that the directional derivative of $f(x)$ in the direction s_c is larger at x_+ than at x_c. This condition must be satisfied if the global algorithm implements our second step-acceptance condition from Section 6.3,

$$s_c^T \nabla f(x_+) > \beta s_c^T \nabla f(x_c), \qquad \beta \in (0, 1), \tag{9.2.5}$$

which we used to prevent excessively small steps. Even if the step-acceptance criterion does not explicitly enforce (9.2.5), the steps taken by our algorithms will usually satisfy (9.2.4), unless one is in a region with significant negative curvature.

Now we return to the first part of the proof of Lemma 9.2.1. It points out that for any J_+ that solves (9.2.2), there will exist a $v_c \in \mathbb{R}^n$ such that

$$J_+ v_c = y_c \quad \text{and} \quad J_+^T s_c = v_c.$$

This suggests the following procedure for solving (9.2.2):

1. Select $J_+ \in \mathbb{R}^{n \times n}$ such that $J_+ v_c = y_c$. J_+ will be a function of v_c and y_c.
2. Choose v_c so that $J_+^T s_c = v_c$.

At this point, let us remember where we are in our minimization algorithm when we want to find H_+. We have determined the step s_c using the Cholesky factorization $L_c L_c^T$ of the positive definite model Hessian H_c. It seems entirely in keeping with the spirit of Chapter 8 to select J_+ to be as close to L_c as possible. This suggests that we satisfy step 1 above by selecting

$$J_+ = L_c + \frac{(y_c - L_c v_c)v_c^T}{v_c^T v_c}, \tag{9.2.6}$$

which by Lemma 8.1.1 is the nearest matrix to L_c in $Q(y_c, v_c)$. Then step 2 requires that

$$v_c = J_+^T s_c = L_c^T s_c + v_c \frac{(y_c - L_c v_c)^T s_c}{v_c^T v_c}, \tag{9.2.7}$$

which can be satisfied only if

$$v_c = \alpha_c L_c^T s_c \tag{9.2.8}$$

for some $\alpha_c \in \mathbb{R}$. Substituting (9.2.8) into (9.2.7) and simplifying yields

$$\alpha_c^2 = \frac{y_c^T s_c}{s_c^T L_c L_c^T s_c} = \frac{y_c^T s_c}{s_c^T H_c s_c},$$

which has a real solution if and only if $y_c^T s_c > 0$, since $s_c^T H_c s_c > 0$. We will choose the positive square root,

$$v_c = + \left(\frac{y_c^T s_c}{s_c^T H_c s_c} \right)^{1/2} L_c^T s_c, \tag{9.2.9}$$

because this makes J_+ closer to L_c when $y_c \cong H_c s_c$. It is an easy exercise to show that J_+ given by (9.2.6, 9.2.9) is nonsingular, and thus solves

(9.2.2). This completes the proof of Lemma 9.2.1, since given any non-singular L_c, the if portion of Lemma 9.2.1 is satisfied by (9.2.6, 9.2.9).

$\square$

We have now derived a symmetric secant update that retains positive definiteness, whenever this is possible, by updating the Cholesky factor L_c of H_c to a J_+ for which $J_+ J_+^T = H_+$. However, a really useful form for J_+ would be as a lower triangular matrix L_+, since then we would have the Cholesky factorization of H_+. The reader may already have realized how to get L_+ from J_+ in $O(n^2)$ operations, by using the QR factorization updating techniques introduced in Section 3.4 and used in Section 8.3. If we use Algorithm A3.4.1 to get $J_+^T = Q_+ L_+^T$ in $O(n^2)$ operations, then $H_+ = J_+ J_+^T = L_+ Q_+^T Q_+ L_+^T = L_+ L_+^T$, and it is not even necessary to accumulate Q_+. Similarly, it is possible to get the $L_+ D_+ L_+^T$ factorization of H_+ directly from the $L_c D_c L_c^T$ factorization of H_c in $O(n^2)$ operations via the LDV factorization update.

We can also express (9.2.6, 9.2.9) as a direct update of H_c to H_+. If we form $H_+ = J_+ J_+^T$ using (9.2.6, 9.2.9) and recall $H_c = L_c L_c^T$, then we get

$$H_+ = H_c + \frac{y_c y_c^T}{y_c^T s_c} - \frac{H_c s_c s_c^T H_c}{s_c^T H_c s_c}. \tag{9.2.10}$$

This form of the update [or a form given in equation (9.3.5)] was discovered independently by Broyden, Fletcher, Goldfarb, and Shanno in 1970, and the update has become known as the *BFGS update*. We will refer to it simply as the *positive definite secant update*. The procedure for directly updating the Cholesky factor L_c to L_+ can be found in Goldfarb (1976). The reader can confirm that, like the symmetric secant update of Section 9.1, (9.2.10) obeys $H_+ s_c = y_c$, and $(H_+ - H_c)$ is a symmetric matrix of rank at most two.

Is this positive definite secant update at all useful? Yes—it is the best Hessian update currently known. This is illustrated in Example 9.2.2, where we employ quasi-Newton methods using the symmetric secant update and the positive definite secant update, respectively, to solve the same problem solved by Newton's method in Section 5.5. In analogy with Chapter 8, we set $H_0 = \nabla^2 f(x_0)$. However, we will see in Section 9.4 that a different strategy for choosing H_0 is recommended in practice.

EXAMPLE 9.2.2   Let $f(x_1, x_2) = (x_1 - 2)^4 + (x_1 - 2)^2 x_2^2 + (x_2 + 1)^2$, which has the minimizer $x_* = (2, -1)^T$. Then the sequence of points produced by the *positive definite secant method*,

$$x_{k+1} = x_k - H_k^{-1} \nabla f(x_k), \tag{9.2.11a}$$

$$s_k = x_{k+1} - x_k, \qquad y_k = \nabla f(x_{k+1}) - \nabla f(x_k), \tag{9.2.11b}$$

$$H_{k+1} = H_k + \frac{y_k y_k^T}{y_k^T s_k} - \frac{H_k s_k s_k^T H_k}{s_k^T H_k s_k}, \tag{9.2.11c}$$

starting from $x_0 = (1, 1)^T$ and $H_0 = \nabla^2 f(x_0)$, are (to eight figures on a CDC machine with 14 base-10 digits):

$$f(x_k)$$

$x_0 = (1 \qquad , \quad 1 \qquad)^T$	6.0
$x_1 = (1 \qquad , \; -0.5 \qquad)^T$	1.5
$x_2 = (1.45 \quad , \; -0.3875 \quad)^T$	5.12×10^{-1}
$x_3 = (1.5889290, -0.63729087)^T$	2.29×10^{-1}
$x_4 = (1.8254150, -0.97155747)^T$	3.05×10^{-2}
$x_5 = (1.9460278, -1.0705597 \;)^T$	8.33×10^{-3}
$x_6 = (1.9641387, -1.0450875 \;)^T$	3.44×10^{-3}
$x_7 = (1.9952140, -1.0017148 \;)^T$	2.59×10^{-5}
$x_8 = (2.0000653, -1.0004294 \;)^T$	1.89×10^{-7}
$x_9 = (1.9999853, -0.99995203)^T$	2.52×10^{-9}
$x_{10} = (2.0 \qquad , \; -1.0 \qquad)^T$	0

The sequence of points produced by the symmetric secant method (9.2.4), using the same x_0 and H_0, are

$$f(x_k)$$

$x_0 = (1 \qquad , \quad 1 \qquad)^T$	6.0
$x_1 = (1 \qquad , \; -0.5 \qquad)^T$	1.5
$x_2 = (1.3272727, -0.41818182)^T$	6.22×10^{-1}
$x_3 = (1.5094933, -0.60145121)^T$	3.04×10^{-1}
$x_4 = (1.7711111, -0.88948020)^T$	5.64×10^{-2}
$x_5 = (1.9585530, -1.0632382 \;)^T$	5.94×10^{-3}
$x_6 = (1.9618742, -1.0398148 \;)^T$	3.16×10^{-3}
$x_7 = (1.9839076, -0.99851156)^T$	2.60×10^{-4}
$x_8 = (1.9938584, -0.99730502)^T$	4.48×10^{-5}
$x_9 = (1.9999506, -0.99940456)^T$	3.57×10^{-7}
$x_{10} = (2.0000343, -0.99991078)^T$	9.13×10^{-9}
$x_{11} = (2.0000021, -0.99999965)^T$	1.72×10^{-11}
$x_{12} = (2.0 \qquad , \; -1.0 \qquad)^T$	0

Newton's method requires six or seven iterations to achieve the same accuracy.

The reasons why the positive definite secant update has been so successful are still not fully understood, although the next section offers some explanation. There we will also discuss another update that results from using a derivation analogous to the one in this section to find an $H_+^{-1} = J_+^{-T} J_+^{-1}$

such that $J_+^{-T} J_+^{-1} y_c = s_c$. The result is

$$J_+^{-T} = L_c^{-T} + \frac{(s_c - L_c^{-T} w_c) w_c^T}{w_c^T w_c},$$

$$w_c = \left(\frac{y_c^T s_c}{y_c^T H_c^{-1} y_c} \right)^{1/2} L_c^{-1} y_c, \tag{9.2.12}$$

where again $H_c = L_c L_c^T$. Using the Sherman-Morrison-Woodbury formula (Lemma 8.3.1), (9.2.12) can be shown to yield

$$H_+ = H_c + \frac{(y_c - H_c s_c) y_c^T + y_c (y_c - H_c s_c)^T}{y_c^T s_c} - \frac{\langle y_c - H_c s_c, s_c \rangle y_c y_c^T}{(y_c^T s_c)^2}.$$

$$\tag{9.2.13}$$

Update (9.2.13) is often called the *DFP update* for its discoverers Davidon (1959) and Fletcher and Powell (1963). It was the first secant update of any kind to be found, and the original derivation was by entirely different means than the above. From (9.2.12), it is easy to show that the DFP update also has the property that H_+ is positive definite if H_c is positive definite and $y_c^T s_c > 0$. Therefore we will sometimes refer to this update as the *inverse positive definite secant update*.

In the early 1970s a number of studies were made comparing quasi-Newton methods using the two secant updates introduced in this section [see, e.g., Dixon (1972) and Brodlie (1977)]. The consensus seems to be that the positive definite secant update (BFGS) performs better in conjunction with the algorithms of Chapter 6, and the DFP update sometimes produces numerically singular Hessian approximations. Nevertheless, the DFP is a seminal update that has been used in many successful computer codes, and it will also play an important role in our analysis in the next section.

9.3 LOCAL CONVERGENCE OF POSITIVE DEFINITE SECANT METHODS

In this section we give an alternate derivation of the positive definite secant update and the DFP update. The easy derivation of the last section has not yet led to a successful convergence analysis, but the derivation of this section has. We will see that if a natural rescaling of the variable space is applied to the minimization problem, then the two symmetric and positive definite updates of the last section result from the two symmetric updates of Section 9.1 applied to the rescaled problem. A very closely related rescaling leads to the convergence results.

We have seen in Chapter 7 how important it is to treat a well-scaled minimization problem. A quadratic minimization problem could be considered perfectly scaled if its Hessian were the identity, since this would result in circular contours, and the Newton and steepest-descent directions would be the same. This "perfect" scaling will be the key to our derivations.

We saw in Section 7.1 that if we rescale the variable space by

$$\hat{x} = Tx,$$

then in the new variable space, the quadratic model around $\hat{x}_+$ is

$$m_+(\hat{x}) = f(x_+) + [T^{-T} \nabla f(x)]^T(\hat{x} - \hat{x}_+) + \tfrac{1}{2}(\hat{x} - \hat{x}_+)^T[T^{-T}H_+ T^{-1}](\hat{x} - \hat{x}_+),$$

where H_+ is the model Hessian in the original variable space. That is, the gradient is transformed to $T^{-T} \nabla f(x)$, and the model Hessian to $T^{-T}H_+ T^{-1}$. Thus the rescaling that leads to an identity Hessian in the model around $\hat{x}_+$ satisfies $T_+^{-T}H_+ T_+^{-1} = I$, or $H_+ = T_+^T T_+$.

It would be nice to generate our secant update to H_c in the variable space transformed by this T_+, since we would be using the perfect rescaling for the new quadratic model. However, we don't know T_+ or H_+ a priori, so let us consider instead any arbitrary scale matrix $\bar{T}$ such that

$$\bar{T}^T \bar{T} = \bar{H} \in Q(y_c, s_c).$$

Now consider the symmetric secant update of Section 9.1 in this variable space. It is

$$\hat{H}_+ = \hat{H}_c + \frac{(\hat{y}_c - \hat{H}_c \hat{s}_c)\hat{s}_c^T + \hat{s}_c(\hat{y}_c - \hat{H}_c \hat{s}_c)^T}{\hat{s}_c^T \hat{s}_c}$$

$$- \frac{\langle \hat{y}_c - \hat{H}_c \hat{s}_c, \hat{s}_c \rangle \hat{s}_c \hat{s}_c^T}{(\hat{s}_c^T \hat{s}_c)^2}, \tag{9.3.1}$$

where

$$\hat{s}_c = \hat{x}_+ - \hat{x}_c = \bar{T}x_+ - \bar{T}x_c = \bar{T}s_c,$$

$$\hat{y}_c = \bar{T}^{-T} \nabla f(x_+) - \bar{T}^{-T} \nabla f(x_c) = \bar{T}^{-T}y_c,$$

$$\hat{H}_c = \bar{T}^{-T}H_c \bar{T}^{-1}, \qquad \hat{H}_+ = \bar{T}^{-T}H_+ \bar{T}^{-1}.$$

If we use the above relations to reexpress (9.3.1) in the original variable space, the reader can verify that we get

$$H_+ = H_c + \frac{(y_c - H_c s_c)s_c^T \bar{H} + \bar{H}s_c(y_c + H_c s_c)^T}{s_c^T \bar{H}s_c}$$

$$- \frac{\langle y_c - H_c s_c, s_c \rangle \bar{H}s_c s_c^T \bar{H}}{(s_c^T \bar{H}s_c)^2}. \tag{9.3.2}$$

Since $\bar{H}s_c = \bar{H}^T s_c = y_c$ for any symmetric $\bar{H} \in Q(y_c, s_c)$, (9.3.2) is independent of the particular $\bar{H}$ or $\bar{T}$ and becomes

$$H_+ = H_c + \frac{(y_c - H_c s_c)y_c^T + y_c(y_c - H_c s_c)^T}{y_c^T s_c}$$
$$- \frac{\langle y_c - H_c s_c, s_c \rangle y_c y_c^T}{(y_c^T s_c)^2}, \tag{9.3.3}$$

which is just the DFP update! Note that if we had known H_+ a priori, we could have used a T_+ such that $T_+^T T_+ = H_+$ in our derivation, and the result would have been the same.

Another way of stating the above is that if H_c is symmetric and $y_c^T s_c > 0$, then for any $\bar{T} \in \mathbb{R}^{n \times n}$ such that $\bar{T}^T \bar{T} \in Q(y_c, s_c)$, (9.3.3) is the unique solution to

$$\min_{H_+ \in \mathbb{R}^{n \times n}} \|\bar{T}^{-T}(H_+ - H_c)\bar{T}^{-1}\|_F \quad \text{subject to} \quad H_+ \in Q(y_c, s_c) \cap S. \tag{9.3.4}$$

That is, the DFP update is the projection of H_c onto $Q(y_c, s_c) \cap S$ in this weighted Frobenius norm.

If we apply the same rescaling, use Greenstadt's update (9.1.5) of $\hat{H}_c^{-1}$ to $\hat{H}_+^{-1}$, and then transform back, the result is

$$H_+^{-1} = H_c^{-1} + \frac{(s_c - H_c^{-1} y_c)s_c^T + s_c(s_c - H_c^{-1} y_c)^T}{y_c^T s_c}$$
$$- \frac{\langle s_c - H_c^{-1} y_c, y_c \rangle s_c s_c^T}{(y_c^T s_c)^2}. \tag{9.3.5}$$

It is straightforward but tedious to show that (9.3.5) is the expression that results from inverting both sides of equation (9.2.10), the H-form of the positive definite secant (BFGS) update. Thus, if H_c is symmetric and $y_c^T s_c > 0$, then for any $\bar{T} \in \mathbb{R}^{n \times n}$ such that $\bar{T}^T \bar{T} \in Q(y_c, s_c)$, the positive definite secant update is the unique solution to

$$\min_{H_+ \in \mathbb{R}^{n \times n}} \|\bar{T}(H_+^{-1} - H_c^{-1})\bar{T}^T\| \quad \text{subject to} \quad H_+ \in Q(y_c, s_c) \cap S. \tag{9.3.6}$$

The reader has probably noticed the intriguing duality that has run through the derivations of the positive definite and inverse positive definite secant updates in Sections 9.2 and 9.3. Since each of these updates can be derived using a scaling of the variable space that is different at every iteration, the methods that use them are sometimes called *variable metric methods*.

The local convergence proof for the positive definite secant method uses the rescaling idea in a central way. The proof shows that the sequence of Hessian approximations $\{H_k\}$ obeys a bounded deterioration property of the sort given in Chapter 8. To show this, the proof uses a weighted Frobenius norm like (9.3.6), but with $\bar{T}$ replaced by a T_* for which $T_*^T T_* = \nabla^2 f(x_*)$. This is just the rescaling that causes the Hessian matrix at the minimizer to be the identity. If the sequence of Hessian approximations converges to $\nabla^2 f(x_*)$, it is the limit of the rescalings used in the above derivations. The same techniques are used to prove the local convergence of the DFP method. In this case the norm from (9.3.4) is used, with $\bar{T}$ again replaced by T_*.

The proof of the following theorem is based on the techniques of Section 8.2; it can be found in Broyden, Dennis, and Moré (1973).

THEOREM 9.3.1 (Broyden-Dennis-Moré) Let the hypotheses of Theorem 9.1.2 hold, and assume in addition that $\nabla^2 f(x_*)$ is positive definite. Then there exist positive constants ε, δ such that if $\| x_0 - x_* \|_2 \leq \varepsilon$ and $\| H_0 - \nabla^2 f(x_*) \|_2 \leq \delta$, then the positive definite secant method (BFGS), (9.2.11), is well defined, $\{x_k\}$ remains in D and converges q-superlinearly to x_*. The same is true if (9.2.11c) is replaced by the inverse positive definite secant (DFP) update,

$$H_{k+1} = H_k + \frac{(y_k - H_k s_k)y_k^T + y_k(y_k - H_k s_k)^T}{y_k^T s_k} - \frac{\langle y_k - H_k s_k, s_k \rangle y_k y_k^T}{(y_k^T s_k)^2}.$$

This convergence result for the positive definite secant method is no stronger than Theorem 9.1.2, the convergence result for the symmetric secant method. However, the positive definite secant method often performs significantly better in practice, as we reiterate in Example 9.3.2 below. In this example, we also examine the convergence of the sequence of Hessian approximations $\{H_k\}$ to the Hessian at the minimizer, $\nabla^2 f(x_*)$. We see, at intermediate steps in the example, that the approximations generated by the positive definite secant update are more accurate than those generated by the symmetric secant update. This behavior, which is common in practice, partially explains the superiority of the positive definite secant method.

Another advantage of the positive definite secant method is related to the derivations that began this section. It is an easy exercise to show that the positive definite and inverse positive definite secant updates are invariant under linear transformations of the variable space, while the symmetric secant update is not (Exercise 13). This invariance property, which is shared by Newton's method but not by the method of steepest descent, allows the associated computer code to be nearly independent of the scaling of the minimization problem.

EXAMPLE 9.3.2   Let $f(x_1, x_2) = x_1^4 + (x_1 + x_2)^2 + (e^{x_2} - 1)^2$, which has its minimum at $x_* = (0, 0)^T$. In the cases $x_0 = (1, 1)^T$ and $x_0 = (-1, 3)^T$, the following methods require the indicated number of iterations to achieve $\|x_k - x_*\|_\infty \le 10^{-8}$:

	Iterations from	
Method	$x_0 = (1, 1)^T$	$x_0 = (-1, 3)^T$
Newton's method	6	11
Positive definite secant method, $H_0 = \nabla^2 f(x_0)$	13	20
Symmetric secant method, $H_0 = \nabla^2 f(x_0)$	14	41

When $x_0 = (-1, 3)^T$, the following values of H_k are produced by the two secant methods at intermediate and final values of x_k. The value of $\nabla^2 f(x)$ at the minimizer is

$$\begin{bmatrix} 2 & 2 \\ 2 & 4 \end{bmatrix}.$$

Values of H_k at the first iteration k for which $\|x_{k+1}\| \le \tau$:

τ	Positive Definite Secant Method	Symmetric Secant Method
10^{-1}	$H_{12} = \begin{bmatrix} 5.485 & 6.996 \\ 6.996 & 11.78 \end{bmatrix}$	$H_{24} = \begin{bmatrix} 16.86 & 11.35 \\ 11.35 & 9.645 \end{bmatrix}$
10^{-3}	$H_{16} = \begin{bmatrix} 2.365 & 2.264 \\ 2.264 & 4.160 \end{bmatrix}$	$H_{34} = \begin{bmatrix} 3.147 & 3.783 \\ 3.783 & 6.797 \end{bmatrix}$
10^{-8}	$H_{19} = \begin{bmatrix} 2.009 & 1.993 \\ 1.993 & 4.004 \end{bmatrix}$	$H_{40} = \begin{bmatrix} 2.003 & 2.021 \\ 2.021 & 4.143 \end{bmatrix}$

In Example 9.3.2, the value of the final positive definite secant approximation agrees with $\nabla^2 f(x_*)$ to within about 0.5%. This is typical of the behavior of secant updates to the Hessian: the final value of H_k often agrees with $\nabla^2 f(x_*)$ to a few significant digits. In practice, it seems to be less common for a final Hessian approximation to differ markedly from $\nabla^2 f(x_*)$, than for a final Jacobian approximation to differ markedly from $J(x_*)$. One reason is that, although we saw in Example 8.2.7 that the sequence of secant approximations to the Jacobian generally fails to converge to $J(x_*)$ when $F(x)$ is linear in some components, the Hessian approximations of this chapter do not share this problem in the analogous case when $\nabla f(x)$ has some linear components (see Exercise 14). The difference is due to the symmetry of the Hessian updates. For more information on this topic, see Schnabel (1982).

9.4 IMPLEMENTATION OF QUASI-NEWTON ALGORITHMS USING THE POSITIVE DEFINITE SECANT UPDATE

In this section we discuss the incorporation of secant updates into our quasi-Newton algorithms for unconstrained minimization. Since the most successful secant approximation for minimization has been the positive definite secant update (BFGS), it is the secant approximation provided in the algorithms in the appendix. We provide both the unfactored and factored forms of the update, in Algorithms A9.4.1 and A9.4.2, respectively. The unfactored algorithm simply implements the rank-two update formula (9.2.10). After the update one calls the model Hessian routine to obtain a Cholesky factorization of the Hessian approximation. The factored algorithm implements the update as it was derived in Section 9.2. It calculates the rank-one update J_+ to L_c [equations (9.2.6, 9.2.9)] and then calls Algorithm A3.4.1 to calculate the lower triangular $L_+ = J_+ Q_+$. Since $L_+ L_+^T$ is the Cholesky factorization of the new Hessian approximation, the model Hessian routine is omitted in the factored form of the update.

The main advantage of the factored update is that the total cost of updating the Hessian approximation and obtaining its Cholesky factorization is $O(n^2)$, whereas using the unfactored update it is $O(n^3)$. The disadvantage is that it is more work to code and debug, especially if one already has coded the model Hessian routine. Also, it is difficult to interface the factored update with the hook step, since this step needs H_c explicitly, and also because normally it may overwrite L_c with the Cholesky factor of $H_c + \alpha_c I$. Therefore, we recommend the factored update for production codes using a line search or dogleg, but not for the hook step. For class projects or preliminary research work, the unfactored form may be preferred in all cases. When using the factored update, one could estimate the condition number of L_c using Algorithm A3.3.1, and if the estimate is greater than (macheps)$^{-1/2}$, reform $H_c = L_c L_c^T$, use Algorithm A5.5.1 to augment H_c by a scaled multiple of the identity, and reset L_c to the Cholesky factor of this augmented Hessian. For simplicity, we have omitted this feature from the algorithms in the appendix.

The only additional feature of either update algorithm is that the update is skipped in two cases. First, it is omitted if $s_c^T y_c \leq 0$, since H_+ would not be positive definite. In fact, we skip the update if

$$s_c^T y_c \leq (\text{macheps})^{1/2} \| s_c \|_2 \| y_c \|_2. \tag{9.4.1}$$

In a line-search algorithm, this condition can virtually always be prevented by using line-search Algorithm A6.3.1 mod, which implements $s_c^T y_c \geq -0.1 \nabla f(x_c)^T s_c$; in a trust region algorithm, (9.4.1) may occur. Second, if y_c is sufficiently close to $H_c s_c$, there is no need to update. We skip the update if the magnitude of each component of $y_c - H_c s_c$ is less than the expected noise level in that component.

We must also discuss the choice of the initial Hessian approximation H_0. In contrast to the practice in solving systems of nonlinear equations, it is not usual to set H_0 to a finite-difference approximation to $\nabla^2 f(x_0)$. One reason is the cost of this approximation in function or gradient evaluations. Another is that secant methods for minimization are predicated on starting with and maintaining a positive-definite Hessian approximation, and so H_0 must be positive definite. However, there is no guarantee that $\nabla^2 f(x_0)$ will be positive definite. If not, it could be perturbed into a positive-definite matrix using the techniques of Section 5.5, but then it is questionable whether the cost of the finite-difference approximation was justified.

Instead, it is common to start secant algorithms for minimization with $H_0 = I$. This H_0 is certainly positive definite, and it results in the first step being in the steepest-descent direction. Following the discussion of scaling in Section 7.1, the reader can guess that we generalize this choice to $H_0 = D_x^2$, which causes the first step to be in the scaled steepest-descent direction.

The one problem with setting $H_0 = I$, or D_x^2, is that it doesn't consider the scale of $f(x)$. For this reason, $\| H_0 \|$ may differ from $\| \nabla^2 f(x_0) \|$ by many orders of magnitude, which can cause the algorithm to require an excessive number of iterations. Various strategies have been proposed to overcome this problem; Shanno and Phua (1978a) use a special case of a more general idea of Oren (1974), premultiplying H_0 by $\gamma_0 = (y_0^T s_0 / s_0^T H_0 s_0)$ immediately prior to the first update. This causes $\hat{H}_0 = \gamma_0 H_0$ to obey $s_0^T \hat{H}_0 s_0 = s_0^T y_0$, a weakened form of the secant equation. If $f(x)$ is quadratic, it guarantees that the range of eigenvalues of $\hat{H}_0$ and $\nabla^2 f(x_0)$ overlap, and in general it causes the magnitude of the diagonal elements of $\hat{H}_0$ to more closely resemble those of $\nabla^2 f(x_0)$. After this premultiplication, the normal updates are performed. Computational experience indicates that this modification may improve the performance of secant algorithms for minimization.

Because there are some small problems involved in using Shanno and Phua's modification in our modular system of algorithms, we use an alternate modification. We simply set

$$H_0 = \max \{| f(x_0)|, \, \text{typ} f\} \cdot D_x^2,$$

where $\text{typ} f$ is the user-supplied typical size of $f(x)$. This formula has the obvious drawback that it is sensitive to adding a constant to $f(x)$. However, in experiments by Frank and Schnabel (1982) it was the most successful among various strategies, including Shanno and Phua's. Our algorithms for calculating H_0 are A9.4.3 for the unfactored case and A9.4.4 for the factored case. Owing to the modular structure, the user can also set H_0 to a finite-difference approximation to $\nabla^2 f(x_0)$.

Example 9.4.1 uses the positive definite secant method to solve again the problem from Example 9.2.2, this time using $H_0 = I$ and $H_0 = | f(x_0)| I$. The performance using $H_0 = |f(x_0)|I$ is slightly better than with $H_0 = \nabla^2 f(x_0)$, whereas using $H_0 = I$ it is much worse. Example 9.4.2 shows the effect of using $H_0 = |f(x_0)|I$ instead of $H_0 = \nabla^2 f(x_0)$ on the problem from Example 9.3.2.

When the positive definite secant update is used in place of the analytic or finite-difference Hessian in our minimization algorithms, there are no other changes besides the ones we have mentioned in this section.

EXAMPLE 9.4.1 Let $f(x_1, x_2) = (x_1 - 2)^4 + (x_1 - 2)^2 x_2^2 + (x_2 + 1)^2$, which has the minimizer $x_* = (2, -1)^T$. The following number of iterations are required by the positive definite secant method (9.2.11) to achieve $\| x_k - x_* \| \le 10^{-8}$, using the specified values of H_0:

H_0	Iterations
$\nabla^2 f(x_0)$	10
$\|f(x_0)\|I$	8
I	24

EXAMPLE 9.4.2 Let $f(x_1, x_2) = x_1^4 + (x_1 + x_2)^2 + (e^{x_2} - 1)^2$, which has its minimum at $x_* = (0, 0)^T$. Starting from $x_0 = (1, 1)^T$ and $x_0 = (-1, 3)^T$, the indicated number of iterations are required by the positive definite secant method (9.2.11) to achieve $\| x_k - x_* \| \le 10^{-8}$, using the following values of H_0:

H_0	Iterations from $x_0 = (1, 1)^T$	Iterations from $x_0 = (-1, 3)^T$
$\nabla^2 f(x_0)$	13	20
$\|f(x_0)\|I$	11	18

If $H_0 = I$, the full secant step from x_0, $x_1 = x_0 - \nabla f(x_0)$, causes $f(x)$ to increase.

9.5 ANOTHER CONVERGENCE RESULT FOR THE POSITIVE DEFINITE SECANT METHOD

Powell (1976) has proven a strong convergence result for the positive definite secant method, in the case when it is applied to a strictly convex function and used in conjunction with a line search that obeys the conditions given in Subsection 6.3.1. This result is given below. Since a line search is included, this theorem is to some extent a global convergence result for the positive definite secant method. It has another attractive feature in that it makes no assumptions about the initial Hessian approximation H_0, except that it be positive definite.

We do not give the proof of Theorem 9.5.1 because it provides little insight, but the interested student is encouraged to study it further. To deepen the challenge, we mention that it is not known whether the results of this theorem hold for the DFP method.

THEOREM 9.5.1   [Powell (1976)]   Let the hypotheses of Theorem 9.3.1 hold and, in addition, let $\nabla^2 f(x)$ be positive definite for every x. Let $x_0 \in \mathbb{R}^n$ and H_0 be any positive definite symmetric matrix. Let the sequence $\{x_k\}$ be defined by

$$p_k = -H_k^{-1} \nabla f(x_k), \qquad x_{k+1} = x_k + \lambda_k p_k,$$

where λ_k is chosen to satisfy (6.3.3, 6.3.4), $\lambda_k = 1$ is used whenever it is a permissible value, and H_{k+1} is defined by (9.2.11b, c). Then the sequences $\{x_k\}$ and $\{H_k\}$ are well defined, and $\{x_k\}$ converges q-superlinearly to x_*.

9.6 OTHER SECANT UPDATES FOR UNCONSTRAINED MINIMIZATION

Many secant updates for minimization have been proposed besides those we have mentioned in this chapter. None of them seem to be superior to the positive definite secant update for general-purpose use. In this section we briefly mention some of these other updates, largely for reasons of historical interest.

So far, all our secant updates for minimization have been symmetric rank-two changes to H_c that obey the secant equation $H_+ s_c = y_c$. It is an easy exercise to show that there is one symmetric rank-one change to H_c that obeys the secant equation, the *symmetric rank-one* (SR1) update

$$H_+ = H_c + \frac{(y_c - H_c s_c)(y_c - H_c s_c)^T}{(y_c - H_c s_c)^T s_c}. \qquad (9.6.1)$$

The fatal flaw of this update is that it does not necessarily produce a positive definite H_+ when H_c is positive definite and $s_c^T y_c > 0$, and in fact the denominator is sometimes zero when the positive definite update is well behaved.

Broyden (1970) was the first to consider the continuous class of updates,

$$H_+(\phi) = H_c + \frac{(y_c - H_c s_c)y_c^T + y_c(y_c - H_c s_c)^T}{y_c^T s_c} - \frac{(y_c - H_c s_c)^T s_c\, y_c\, y_c^T}{(y_c^T s_c)^2}$$

$$- \phi(s_c^T H_c s_c)\left[\frac{y_c}{y_c^T s_c} - \frac{H_c s_c}{s_c^T H_c s_c}\right]\left[\frac{y_c}{y_c^T s_c} - \frac{H_c s_c}{s_c^T H_c s_c}\right]^T, \qquad (9.6.2)$$

that includes the DFP update when $\phi = 0$, the positive definite secant update (BFGS) when $\phi = 1$, and the SR1 update when $\phi = s_c^T H_c s_c/(s_c^T H_c s_c - s_c^T y_c)$. For any $\phi \in \mathbb{R}$, $H_+(\phi) - H_c$ is a symmetric rank-two matrix whose row and column spaces are both spanned by y_c and $H_c s_c$. There have been considerable theoretical investigations into this class of updates and some generalizations of it; they are discussed extensively in Fletcher (1980). However, these investigations have not produced an update that is superior to the BFGS in practice.

It is interesting that all the members of (9.6.2) for which $H_+(\phi) - H_c$ has one negative and one positive eigenvalue can equivalently be represented as

$$H_+ = H_c + \frac{(y_c - H_c s_c)v_c^T + v_c(y_c - H_c s_c)^T}{v_c^T s_c} - \frac{(y_c - H_c s_c)^T s_c v_c v_c^T}{(v_c^T s_c)^2}, \quad (9.6.3)$$

where v_c is a linear combination of y_c and $H_c s_c$ [see Schnabel (1977) or Exercise 22]. In particular, the positive definite secant update (BFGS) corresponds to the choice

$$v_c = \frac{1}{\alpha_c + 1} y_c + \frac{\alpha_c}{\alpha_c + 1} H_c s_c, \quad (9.6.4)$$

where $\alpha_c = (y_c^T s_c/s_c^T H_c s_c)^{1/2}$. Thus, the positive definite secant update can be derived from the weighted norm or scaling derivation of Section 9.3 by setting $T^T T$ to the appropriate convex combination of H_c and H_+.

Another class of secant updates for minimization that have received considerable attention are the "projected and optimally conditioned" updates of Davidon (1975) [see also Schnabel (1977)]. Interesting theoretical properties can also be proven about methods that use these updates, but in computational tests [see, e.g., Shanno and Phua (1978b)], they have not performed better than the BFGS.

9.7   EXERCISES

1. Give several reasons why a secant approximation to the Hessian should be symmetric.

2. Show that the problem

$$\underset{B \in \mathbb{R}^{n \times n}}{\text{minimize}} \, \| B - A \|_F \quad \text{subject to} \quad B \text{ symmetric}$$

   is solved by

$$B = \frac{A + A^T}{2}.$$

3. Let H_+ be the symmetric secant update (9.1.3). Confirm that $H_+ s_c = y_c$, and that $H_+ - H_c$ is a symmetric matrix of rank at most two.

4. Prove Theorem 9.1.1. [*Hint*: Use the techniques of the proof of Theorem 8.1.1 to show that if H_+ is given by (9.1.3) and $H \in \mathbb{R}^{n \times n}$ satisfies $H - H_c$ symmetric and $Hs_c = y_c$, then $(H - H_c)s_c = (H_+ - H_c)s_c$ and $\|(H - H_c)t\|_2 \geq \|(H_+ - H_c)t\|_2$ for any t satisfying $t^T s_c = 0$. Then use Exercise 11 of Chapter 3 to complete the proof.]

Exercises 5 and 6 indicate the proof of Theorem 9.1.2. They are taken from Broyden, Dennis, and Moré (1973).

5. Let H_+ be given by (9.1.3), $H_* = \nabla^2 f(x_*)$, $P_c = I - s_c s_c^T / s_c^T s_c$. Using the assumptions of Theorem 9.1.2 and the identity

$$H_+ - H_* = P_c (H_c - H_*) P_c + \frac{(y_c - H_* s_c) s_c^T}{s_c^T s_c} + \frac{s_c (y_c - H_* s_c)^T P_c}{s_c^T s_c}, \qquad (9.7.1)$$

show that H_+ obeys

$$\| H_+ - H_* \|_F \leq \| H_c - H_* \|_F + \frac{\gamma}{2} (\| x_+ - x_* \|_2 + \| x_c - x_* \|_2), \qquad (9.7.2)$$

the analogous equation to (8.2.3). Use (9.7.2) and the techniques from the linear convergence proof for the secant method (Theorem 8.2.2) to prove the local q-linear convergence of the symmetric secant method.

6. Use (9.7.1) to derive a stronger form of (9.7.2),

$$\| H_+ - H_* \|_F \leq \| E_c \|_F - \frac{1}{2 \| E_c \|_F} \left(\frac{\| E_c s_c \|_2}{\| s_c \|_2} \right)^2$$

$$+ \frac{\gamma}{2} (\| x_+ - x_* \|_2 + \| x_c - x_* \|_2), \qquad (9.7.3)$$

where $E_c = H_c - H_*$. Use (9.7.3) and the techniques from the superliner convergence proof for the secant method to prove the local q-superlinear convergence of the symmetric secant method.

7. Prove that $H \in \mathbb{R}^{n \times n}$ is positive definite and symmetric if and only if $H = JJ^T$ for some nonsingular $J \in \mathbb{R}^{n \times n}$.

8. Show that if J_+ is given by (9.2.6, 9.2.9) and $H_c = L_c L_c^T$, then $H_+ = J_+ J_+^T$ is given by (9.2.10), independent of the sign in (9.2.9).

9. Suppose $H_c s_c = y_c$. What is the value of J_+ given by (9.2.6, 9.2.9)? What if the sign on the right side of (9.2.9) is changed from plus to minus? How does this justify the choice of sign in (9.2.9)?

10. Let J_+ be given by (9.2.12), $H_+ = J_+ J_+^T$, $H_c = L_c L_c^T$. Show that (9.2.12) yields

$$H_+^{-1} = H_c^{-1} + \frac{s_c s_c^T}{s_c^T y_c} - \frac{H_c^{-1} y_c y_c H_c^{-1}}{y_c^T H_c^{-1} y_c}. \qquad (9.7.4)$$

Then show that (9.7.4) is the inverse of (9.2.13).

11. Show, under the assumptions following equation (9.3.1), that (9.3.1) and (9.3.2) are equivalent.

12. Work through the proof of Theorem 9.3.1 with the help of Broyden, Dennis, and Moré (1973).

13. (a)   Show that the positive definite secant update is invariant under linear transformations of the variable space (as discussed in Section 9.3) and that the symmetric secant update is not.

    (b)   Derive a nonsymmetric rank-one update that is invariant under linear transformations of the variable space.

14. (a)   Let

$$f(x_1, x_2) = x_1^2 + 2x_1x_2 + 2x_2^2 + x_2^4,$$

    which has the minimizer $x_* = (0, 0)^T$, and

$$\nabla^2 f(x_*) = \begin{bmatrix} 2 & 2 \\ 2 & 4 \end{bmatrix}.$$

    Show that if $x_0 = (1, -1)^T$ and $H_0 = \nabla^2 f(x_0)$, then the positive definite secant method generates a sequence of Hessian approximations $\{H_k\}$ obeying

$$\lim_{k \to \infty} H_k = \nabla^2 f(x_*),$$

    and the symmetric secant method generates the *identical* sequence of iterates $\{x_k\}$, but a sequence of Hessian approximations $\{\bar{H}_k\}$ obeying

$$\lim_{k \to \infty} \bar{H}_k = \begin{bmatrix} 5 & 5 \\ 5 & 7 \end{bmatrix}.$$

    [*Hint*: Generalize the techniques of the proof of Lemma 8.2.7]

    (b)   Generalize part (a) to any problem where $\nabla f(x)$ is linear in some of its components $\nabla f(x)_i$, and $\nabla f(x_0)$ is zero in all of those components.

    (c)   Determine experimentally how parts (a) and (b) are changed if $\nabla f(x_0)$ is not zero in the linear components of $\nabla f(x)$.

15. Let

$$f(x_1, x_2) = x_1^2 + x_2^2 + x_4^2,$$

    which has the minimizer $x_* = (0, 0)^T$, and

$$\nabla^2 f(x_*) = \begin{bmatrix} 2 & 0 \\ 0 & 2 \end{bmatrix}.$$

    Show that if $x_0 = (0, -1)^T$ and $H_0 = I$, then the positive definite secant method and the symmetric secant method produce identical sequences of points $\{x_k\}$ and of Hessian approximations $\{H_k\}$, and that

$$\lim_{k \to \infty} H_k = \begin{bmatrix} 1 & 0 \\ 0 & 2 \end{bmatrix}.$$

16. Program a minimization algorithm from Appendix A that uses the positive definite secant update. Run your algorithm on some of the test problems from Appendix B. Then replace the secant update in your algorithm by a finite-difference Hessian approximation (leaving the rest of the algorithm unchanged) and rerun your algorithm on the same problems. How do the numbers of iterations required by the two methods to solve the same problems compare? How do the numbers of

function and gradient evaluations compare? (Remember to count the function or gradient evaluations that you use in the finite-difference Hessian approximation.) How do the run times compare?

17. Program a minimization algorithm from Appendix A that uses the positive definite secant update and the modified line search. Run your algorithm on some test problems from Appendix B.

   (a) How often is the modification to the line search necessary; i.e., how often does the point x_+ returned by the regular line search fail to satisfy

$$\nabla f(x_+)^T(x_+ - x_c) \geq (0.9)\, \nabla f(x_c)^T(x_+ - x_c)?$$

   (b) How close is the final value of H_k to $\nabla^2 f(x_k)$ on your test problems? Try to find cases when there is a large discrepancy.

18. Let $f: \mathbb{R}^n \longrightarrow \mathbb{R}$ be a positive definite quadratic with Hessian $H \in \mathbb{R}^{n \times n}$, x_c and x_+ any two distinct points in $\mathbb{R}^n$, $s = x_+ - x_c$, $y = \nabla f(x_+) - \nabla f(x_c)$. Show that the range of eigenvalues of H includes the eigenvalue of $(s^T y / s^T s)I$.

19. What is the secant step $x_1 = x_0 - H_0^{-1} \nabla f(x_0)$ for the problems of Example 9.4.2 if $H_0 = |f(x_0)| I$? If $H_0 = I$? Comment on the comparison.

20. Show that H_+ given by (9.6.1) is the unique matrix for which $H_+ s_c = y_c$ and $(H_+ - H_c)$ is a symmetric matrix of rank one.

21. Confirm that the SR1 update (9.6.1) is the choice $\phi = s_c^T H_c s_c / (s_c^T H_c s_c - s_c^T y_c)$ in (9.6.2).

22. (a) Show that if $v_c = y_c + \sigma(y_c - H_c s_c)$, where σ is either solution to

$$\sigma(y_c - H_c s_c)^T s_c + y_c^T s_c = \pm(y_c^T s_c)\left[\frac{s_c^T H_c s_c}{s_c^T H_c s_c + \phi(y_c - H_c s_c)^T s_c}\right]^{1/2}$$

   then (9.6.3) is identical to (9.6.2).

   (b) Use (9.7.5) to show that the positive definite secant update [$\phi = 1$ in (9.6.2)] is equivalent to (9.6.3) with v_c given by (9.6.4)

A *final program note*

We have now finished the basic material of this book—methods for solving small-to-medium-size unconstrained optimization or nonlinear equations problems with no special structure. In the final two chapters we discuss extensions of these methods to problems with significant special structure. An important example is parameter estimation by nonlinear least squares, a special case of unconstrained minimization that is the topic of Chapter 10. In Chapter 11 we discuss the more general class of problems where part of the derivative matrix is readily available while another portion is not. Included in this category are problems where the derivative matrix is sparse. We discuss sparse problems briefly and then give a derivation and convergence analysis of least-change secant methods for any problem from this general class.

10

Nonlinear Least Squares

In this chapter we discuss the solution of the nonlinear least-squares problem that we introduced in Chapter 1. This problem is very closely related to the unconstrained minimization and nonlinear equations problems we have discussed already in this book, and our treatment will consist largely of an application of the techniques we have presented. Section 10.1 reintroduces the nonlinear least-squares problem and discusses the derivatives that will be important to us. In Sections 10.2 and 10.3 we explore two different approaches for nonlinear least-squares algorithms. The first is most closely related to solving systems of nonlinear equations, and it leads to the Gauss-Newton and Levenberg-Marquardt algorithms. The second, closely related to unconstrained minimization, leads to Newton's method for nonlinear least squares as well as the successful secant methods. Of course, the two types of methods are closely related to each other, and we explore this relationship. In Section 10.4 we mention briefly some other aspects of nonlinear least squares, including stopping criteria and the treatment of mixed linear-nonlinear least-squares problems.

10.1 THE NONLINEAR LEAST-SQUARES PROBLEM

The nonlinear least-squares problem is

$$\underset{x \in \mathbb{R}^n}{\text{minimize}} \, f(x) = \tfrac{1}{2} R(x)^T R(x) = \tfrac{1}{2} \sum_{i=1}^{m} r_i(x)^2 \qquad (10.1.1)$$

where $m > n$, the *residual function*

$$R: \mathbb{R}^n \longrightarrow \mathbb{R}^m$$

is nonlinear in x, and $r_i(x)$ denotes the ith component function of $R(x)$. If $R(x)$ is linear, (10.1.1) is a linear least-squares problem, whose solution was discussed in Section 3.6. The nonlinear least-squares problem arises most commonly from data-fitting applications, where one is attempting to fit the data (t_i, y_i), $i = 1, \ldots, m$, with a model $m(x, t)$ that is nonlinear in x. In this case $r_i(x) = m(x, t_i) - y_i$, and the nonlinear least-squares problem consists of choosing x so that the fit is as close as possible in the sense that the sum of the squares of the residuals—$r_i(x)$'s—is minimized. Commonly, m is much larger than n (e.g., $n = 5$, $m = 100$). The choice of the sum-of-squares measure for data fitting is justified by statistical considerations [see, e.g., Bard (1970)]; in statistical literature, however, the notation for the same problem is

$$\underset{\beta \in \mathbb{R}^p}{\text{minimize}} \ S(\beta) = \tfrac{1}{2} \sum_{i=1}^{n} (f(\beta, x_i) - y_i)^2,$$

where now β is the variable of p parameters, $f(\beta, x)$ is the model function, and there are n data points (x_i, y_i), $x_i \in \mathbb{R}^m$, $y_i \in \mathbb{R}$, $m \geq 1$. (Sometimes different symbols are used in place of S and β.) We will use the numerical analysis notation (10.1.1) to be consistent with the rest of this book.

The nonlinear least-squares problem is extremely closely related to the problems we have studied in this book. When $m = n$, it includes as a special case solving a system of nonlinear equations, and for any value of m, it is just a special case of unconstrained minimization. The reasons why we do not recommend that it be solved by general-purpose unconstrained minimization codes will become clear in this chapter; mainly, one wants to take the special structure of (10.1.1) into consideration. To start doing this, we investigate the derivatives of $R(x)$, and of $f(x) = \tfrac{1}{2}R(x)^T R(x)$. An example of a nonlinear least-squares problem and its derivatives is given at the end of this section.

The first-derivative matrix of $R(x)$ is simply the Jacobian matrix $J(x) \in \mathbb{R}^{m \times n}$, where $J(x)_{ij} = \partial r_i(x)/\partial x_j$. Thus an affine model of $R(x)$ around a point x_c is

$$M_c(x) = R(x_c) + J(x_c)(x - x_c),$$

the generalization to the case $m \neq n$ of the affine model for systems of nonlinear equations that we have used throughout this book. This model will be the basis for the methods discussed in Section 10.2, which are related to thinking of the nonlinear least-squares problem as an overdetermined system of equations.

We saw in Section 6.5 that the first derivative of $f(x) = \tfrac{1}{2}R(x)^T R(x)$ is

$$\nabla f(x) = \sum_{i=1}^{m} r_i(x) \cdot \nabla r_i(x) = J(x)^T R(x).$$

Similarly, the second derivative is

$$\nabla^2 f(x) = \sum_{i=1}^{m} (\nabla r_i(x) \cdot \nabla r_i(x)^T + r_i(x) \cdot \nabla^2 r_i(x))$$

$$= J(x)^T J(x) + S(x)$$

where

$$S(x) \triangleq \sum_{i=1}^{m} r_i(x) \cdot \nabla^2 r_i(x)$$

denotes the second-order information in $\nabla^2 f(x)$. Thus the quadratic model of $f(x)$ around x_c is

$$m_c(x) = f(x_c) + \nabla f(x_c)^T (x - x_c) + \tfrac{1}{2}(x - x_c)^T \nabla^2 f(x_c)(x - x_c)$$

$$= \tfrac{1}{2} R(x_c)^T R(x_c) + R(x_c)^T J(x_c)(x - x_c)$$

$$+ \tfrac{1}{2}(x - x_c)^T (J(x_c)^T J(x_c) + S(x_c))(x - x_c), \qquad (10.1.2)$$

the specialization of the Taylor series quadratic model for minimization to objective functions of form (10.1.1).

From (10.1.2), Newton's method applied to (10.1.1) is

$$x_+ = x_c - (J(x_c)^T J(x_c) + S(x_c))^{-1} J(x_c)^T R(x_c). \qquad (10.1.3)$$

Certainly (10.1.3) would be a fast local method for the nonlinear least-squares problem, since it is locally q-quadratically convergent under standard assumptions. The problem with the full Newton approach is that $S(x)$ is usually either unavailable or inconvenient to obtain, and it is too expensive to approximate by finite differences. Although frequently used in practice, a secant approximation to all of $\nabla^2 f(x)$ is undesirable, because the portion $J(x)^T J(x)$ of $\nabla^2 f(x)$ will already be readily available, since $J(x)$ must be calculated analytically or by finite differences to calculate $\nabla f(x)$. The methods of Section 10.3 will be related to unconstrained minimization methods, but we will use the structure of the quadratic model (10.1.2) in solving the nonlinear least-squares problem when $S(x)$ is not analytically available.

In discussing the various methods for nonlinear least squares, we will want to distinguish between *zero-residual*, *small-residual*, and *large-residual* problems. These terms refer to the value of $R(x)$ [or $f(x)$] at the minimizer x_* of (10.1.1). A problem for which $R(x_*) = 0$ is called a zero-residual problem; in data-fitting applications this means that the model $m(x_*, t)$ fits the data y_i exactly at each data point. The distinction between small- and large-residual problems will be clarified in Sections 10.2. It will turn out that the methods of Section 10.2 perform better on zero- or small-residual problems than on large-residual problems, while the methods of Section 10.3 are equally effective in all these cases.

EXAMPLE 10.1.1   Suppose we wish to fit the data (t_i, y_i), $i = 1, \ldots, 4$, with the model $m(x, t_i) = e^{t_i x_1} + e^{t_i x_2}$, using nonlinear least squares. Then $R : \mathbb{R}^2 \longrightarrow$

$\mathbb{R}^4$, $r_i(x) = e^{t_i x_1} + e^{t_i x_2} - y_i$, $i = 1, \ldots, 4$, and the nonlinear least-squares problem is to minimize $f(x) = \frac{1}{2}R(x)^T R(x)$. Thus we have $J(x) \in \mathbb{R}^{4 \times 2}$,

$$
J(x) = \begin{bmatrix}
t_1 e^{t_1 x_1} & t_1 e^{t_1 x_2} \\
t_2 e^{t_2 x_1} & t_2 e^{t_2 x_2} \\
t_3 e^{t_3 x_1} & t_3 e^{t_3 x_2} \\
t_4 e^{t_4 x_1} & t_4 e^{t_4 x_2}
\end{bmatrix}.
$$

Also $\nabla f(x) \in \mathbb{R}^2$,

$$
\nabla f(x)^T = R(x)^T J(x) = \left(\sum_{i=1}^{4} r_i(x) t_i e^{t_i x_1}, \quad \sum_{i=1}^{4} r_i(x) t_i e^{t_i x_2} \right),
$$

and $\nabla^2 f(x) \in \mathbb{R}^{2 \times 2}$. Using

$$
\nabla^2 r_i(x) = \begin{bmatrix}
t_i^2 e^{t_i x_1} & 0 \\
0 & t_i^2 e^{t_i x_2}
\end{bmatrix}
$$

and $S(x) = \sum_{i=1}^{4} r_i(x) \nabla^2 r_i(x)$, we have

$$
\nabla^2 f(x) = J(x)^T J(x) + S(x)
$$

$$
= \begin{bmatrix}
\sum_{i=1}^{4} t_i^2 e^{t_i x_1}(r_i(x) + e^{t_i x_1}) & \sum_{i=1}^{4} t_i^2 e^{t_i(x_1 + x_2)} \\
\sum_{i=1}^{4} t_i^2 e^{t_i(x_1 + x_2)} & \sum_{i=1}^{4} t_i^2 e^{t_i x_2}(r_i(x) + e^{t_i x_2})
\end{bmatrix}.
$$

10.2 GAUSS-NEWTON-TYPE METHODS

The first class of methods we consider for solving the nonlinear least-squares problem comes from using the affine model of $R(x)$ around x_c,

$$
M_c(x) = R(x_c) + J(x_c)(x - x_c), \tag{10.2.1}
$$

where $M_c : \mathbb{R}^n \longrightarrow \mathbb{R}^m$ and $m > n$.

We cannot in general expect to find an x_+ for which $M_c(x_+) = 0$, since this is an overdetermined system of linear equations. However, a logical way to use (10.2.1) to solve the nonlinear least-squares problem is to choose the next iterate x_+ as the solution to the linear least-squares problem

$$
\underset{x \in \mathbb{R}^n}{\text{minimize}} \; \frac{1}{2} \| M_c(x) \|_2^2 \triangleq \hat{m}_c(x). \tag{10.2.2}
$$

Let us assume for now that $J(x_c)$ has full column rank. Then we saw in Section 3.6 that the solution to (10.2.2) is

$$
x_+ = x_c - (J(x_c)^T J(x_c))^{-1} J(x_c)^T R(x_c). \tag{10.2.3}
$$

Of course, we would not usually compute x_+ by (10.2.3); instead we would solve (10.2.2) using the QR decomposition of $J(x_c)$.

The iterative method that consists of using (10.2.3) at each iteration is called the *Gauss-Newton method*. To get some insight into its behavior close to a solution x_* of (10.1.1), let us compare it to Newton's method for nonlinear least squares, equation (10.1.3). The two equations differ only in the term $S(x)$, included by Newton's method in the second-derivative matrix $J(x_c)^T J(x_c) + S(x_c)$ but omitted by the Gauss-Newton method. Equivalently, the only difference between the quadratic model $m_c(x)$, (10.1.2), from which we derived Newton's method, and the quadratic model $\hat{m}_c(x)$, (10.2.2), that gives the Gauss-Newton method, is that the portion $S(x_c)$ of $\nabla^2 f(x_c)$ is omitted from $\hat{m}_c(x)$. Since Newton's method is locally q-quadratically convergent under standard assumptions, the reader might guess that the success of the Gauss-Newton method will depend on whether the omitted term $S(x_c)$ is important— that is, whether it is a large part of $\nabla^2 f(x_c) = J(x_c)^T J(x_c) + S(x_c)$. This is essentially confirmed by Theorem 10.2.1 and Corollary 10.2.2. They show that if $S(x_*) = 0$, then the Gauss-Newton method is also q-quadratically convergent. This occurs when $R(x)$ is linear, or when we have a zero-residual problem. If $S(x_*)$ is small relative to $J(x_*)^T J(x_*)$, the Gauss-Newton method is locally q-linearly convergent. However, if $S(x_*)$ is too large, the Gauss-Newton method may not be locally convergent at all. We comment further on these results following Corollary 10.2.2.

THEOREM 10.2.1   Let $R: \mathbb{R}^n \longrightarrow \mathbb{R}^m$, and let $f(x) = \frac{1}{2} R(x)^T R(x)$ be twice continuously differentiable in an open convex set $D \subset \mathbb{R}^n$. Assume that $J(x) \in \text{Lip}_\gamma(D)$ with $\| J(x) \|_2 \leq \alpha$ for all $x \in D$, and that there exists $x_* \in D$ and $\lambda, \sigma \geq 0$, such that $J(x_*)^T R(x_*) = 0$, λ is the smallest eigenvalue of $J(x_*)^T J(x_*)$, and

$$\| (J(x) - J(x_*))^T R(x_*) \|_2 \leq \sigma \| x - x_* \|_2 \tag{10.2.4}$$

for all $x \in D$. If $\sigma < \lambda$, then for any $c \in (1, \lambda/\sigma)$, there exists $\varepsilon > 0$ such that for all $x_0 \in N(x_*, \varepsilon)$, the sequence generated by the Gauss-Newton method

$$x_{k+1} = x_k - (J(x_k)^T J(x_k))^{-1} J(x_k)^T R(x_k)$$

is well defined, converges to x_*, and obeys

$$\| x_{k+1} - x_* \|_2 \leq \frac{c\sigma}{\lambda} \| x_k - x_* \|_2 + \frac{c\alpha\gamma}{2\lambda} \| x_k - x_* \|_2^2 \tag{10.2.5}$$

and

$$\| x_{k+1} - x_* \|_2 \leq \frac{c\sigma + \lambda}{2\lambda} \| x_k - x_* \|_2 < \| x_k - x_* \|_2. \tag{10.2.6}$$

Proof.   The proof is by induction. We may assume that $\lambda > \sigma \geq 0$, since the conclusions of the theorem apply only in this case. Let c be a fixed constant in $(1, \lambda/\sigma)$, let us abbreviate $J(x_0)$, $R(x_0)$, and $R(x_*)$ by J_0, R_0, and R_*, respectively, and let $\|\cdot\|$ denote the vector or matrix l_2 norm. By a now familiar argument, there exists $\varepsilon_1 > 0$ such that $J_0^T J_0$ is non-singular and

$$\| (J_0^T J_0)^{-1} \| \leq \frac{c}{\lambda} \qquad \text{for } x_0 \in N(x_*, \varepsilon_1). \tag{10.2.7}$$

Let

$$\varepsilon = \min \left\{ \varepsilon_1, \frac{\lambda - c\sigma}{c\alpha\gamma} \right\}. \tag{10.2.8}$$

Then at the first step, x_1 is well defined and

$$\begin{aligned}
x_1 - x_* &= x_0 - x_* - (J_0^T J_0)^{-1} J_0^T R_0 \\
&= -(J_0^T J_0)^{-1} [J_0^T R_0 + J_0^T J_0(x_* - x_0)] \\
&= -(J_0^T J_0)^{-1} [J_0^T R_* - J_0^T(R_* - R_0 - J_0(x_* - x_0))].
\end{aligned} \tag{10.2.9}$$

By Lemma 4.1.12,

$$\| R_* - R_0 - J_0(x_* - x_0) \| \leq \frac{\gamma}{2} \| x_0 - x_* \|^2. \tag{10.2.10}$$

From (10.2.4), recalling that $J(x_*)^T R(x_*) = 0$,

$$\| J_0^T R_* \| \leq \sigma \| x - x_* \|. \tag{10.2.11}$$

Combining (10.2.9), (10.2.7), (10.2.10), and (10.2.11) and $\| J_0 \| \leq \alpha$ gives

$$\begin{aligned}
\| x_1 - x_* \| &\leq \|(J_0^T J_0)^{-1}\| [\| J_0^T R_* \| + \| J_0 \| \| R_* - R_0 - J_0(x_* - x_0)\|] \\
&\leq \frac{c}{\lambda} \left[\sigma \| x_0 - x_* \| + \frac{\alpha\gamma}{2} \| x_0 - x_* \|^2 \right],
\end{aligned}$$

which proves (10.2.5) in the case $k = 0$. From (10.2.8) and the above,

$$\begin{aligned}
\| x_1 - x_* \| &\leq \| x_0 - x_* \| \left[\frac{c\sigma}{\lambda} + \frac{c\alpha\gamma}{2\lambda} \| x_0 - x_* \| \right] \\
&\leq \| x_0 - x_* \| \left[\frac{c\sigma}{\lambda} + \frac{\lambda - c\sigma}{2\lambda} \right] \\
&= \frac{c\sigma + \lambda}{2\lambda} \| x_0 - x_* \| \\
&< \| x_0 - x_* \|,
\end{aligned}$$

which proves (10.2.6) in the case $k = 0$. The induction step now proceeds identically. $\square$

COROLLARY 10.2.2   Let the assumptions of Theorem 10.2.1 be satisfied. If $R(x_*) = 0$, then there exists $\varepsilon > 0$ such that for all $x_0 \in N(x_*, \varepsilon)$, the sequence $\{x_k\}$ generated by the Gauss-Newton method is well defined and converges q-quadratically to x_*.

Proof. If $R(x_*) = 0$, σ can be chosen equal to zero in (10.2.4). Therefore from (10.2.6), $\{x_k\}$ converges to x_*, and from (10.2.5) the rate of convergence is q-quadratic. $\square$

Theorem 10.2.1 shows why the Gauss-Newton method is less satisfactory than most of the local methods we have studied earlier in this book, because on many problems with $J(x_*)^T J(x_*)$ nonsingular it is not quickly locally convergent, and on some of these it is not locally convergent at all. The constant σ in (10.2.4) plays a crucial role in the convergence proof and deserves further examination.

It is convenient to view σ as denoting $\| S(x_*) \|_2$, since for x sufficiently close to x_*, it is an exercise to show that

$$(J(x) - J(x_*))^T R(x_*) \cong S(x_*)(x - x_*).$$

From this interpretation or from (10.2.4), σ is a combined *absolute* measure of the nonlinearity and residual size of the problem; if $R(x)$ is linear or $R(x_*) = 0$, it is immediate from (10.2.4) that $\sigma = 0$. The ratio σ/λ, which must be less than 1 to guarantee convergence, can be viewed as a combined *relative* measure of the nonlinearity and residual size of the problem. Thus Theorem 10.2.1 says that the speed of convergence of the Gauss-Newton method decreases as the relative nonlinearity or relative residual size of the problem increases; if either of these is too large, the method may not converge at all. Alternatively, it indicates that the larger $S(x_*)$ is in comparison to $J(x_*)^T J(x_*)$, the worse the Gauss-Newton method is likely to perform.

Even though the Gauss-Newton method has several problems, it is the basis of some important and successful practical methods for nonlinear least squares. In Table 10.2.3 we summarize the advantages of the Gauss-Newton method that we will want to retain and the disadvantages that we will want to overcome. Examples 10.2.4 and 10.2.5 show the behavior of the Gauss-Newton method on a simple, one-variable problem.

In Example 10.2.4 we examine the behavior of the Gauss-Newton method in fitting the model $e^{tx} = y$ to the (t, y) data $(1, 2)$, $(2, 4)$, $(3, y_3)$ for various values of y_3. When $y_3 = 8$, the model fits the data exactly with $x_* = \ln 2 \cong 0.69315$, and the Gauss-Newton method is quadratically convergent. As y_3 becomes smaller, the optimal value x_* becomes smaller, the residual at the solution $R(x_*)$ becomes larger, and the performance of the Gauss-Newton method deteriorates. In the cases $y_3 = 3$ and $y_3 = -1$, the Gauss-Newton

Table 10.2.3   *Advantages and disadvantages of the Gauss-Newton method*

Advantages

1. Locally q-quadratically convergent on zero-residual problems.
2. Quickly locally q-linearly convergent on problems that aren't too nonlinear and have reasonably small residuals.
3. Solves linear least-squares problems in one iteration.

Disadvantages

1. Slowly locally q-linearly convergent on problems that are sufficiently nonlinear or have reasonably large residuals.
2. Not locally convergent on problems that are very nonlinear or have very large residuals.
3. Not well defined if $J(x_k)$ doesn't have full column rank.
4. Not necessarily globally convergent.

method is still linearly convergent, although in the latter case the convergence is very slow. In the cases $y_3 = -4$ and $y_3 = -8$, the Gauss-Newton method does not converge locally.

For comparison, Example 10.2.4 also shows the behavior of Newton's method for nonlinear least squares, (10.1.3), on all these problems. While the performance of the Gauss-Newton method is strongly dependent on the residual size, Newton's method is not, and it does well on all these problems. For each problem we consider two starting points: $x_0 = 1$, and a value of x_0 within 0.1 of x_*. The behavior of the algorithms from the closer starting point gives the best indication of their local convergence properties.

EXAMPLE 10.2.4 Let $R: \mathbb{R}^1 \longrightarrow \mathbb{R}^4$, $r_i(x) = e^{t_i x} - y_i$, $i = 1, \ldots, 4$, $f(x) = \frac{1}{2} R(x)^T R(x)$, where $t_1 = 1$, $y_1 = 2$, $t_2 = 2$, $y_2 = 4$, $t_3 = 3$, and let y_3 and x_0 have the values shown on p. 226. Then the Gauss-Newton method, (10.2.3), and Newton's method, (10.1.3), require the following number of iterations to achieve $|\nabla f(x_k)| \le 10^{-10}$ on each of these problems, on a CDC machine with 14+ base-10 digits. The minimizer of $f(x)$, x_*, and the residual at the solution, $f(x_*)$, are also shown for each problem.

Example 10.2.5 examines more closely the cases in Example 10.2.4 when the Gauss-Newton method does not converge.

EXAMPLE 10.2.5 Let $f(x)$ be given by Example 10.2.4. If $y_3 = -4$, then $x_* = -0.3719287$. Let $x_0 = x_* + \delta$, then for $|\delta|$ sufficiently small, the Gauss-Newton method produces

$$x_1 \cong x_0 - (3.20)\delta,$$

y_3	x_0	Iterations required to achieve $\lvert \nabla f(x_k) \rvert \leq 10^{-10}$ by:		x_*	$f(x_*)$
		Gauss-Newton	Newton's method		
8	1	5	7	0.69315	0
	0.6	4	6		
3	1	12	9	0.44005	1.6390
	0.5	9	5		
-1	1	34	10	0.044744	6.9765
	0	32	4		
-4	1	*	12	-0.37193	16.435
	-0.3	*	4		
-8	1	*	12	-0.79148	41.145
	-0.7	*	4		

* Gauss-Newton method is not locally convergent (see Example 10.2.5).

so that $\lvert x_1 - x_* \rvert \cong (2.20) \lvert x_0 - x_* \rvert$. Thus x_1 is further than x_0 from x_*. For example, if $x_0 = -0.3719000$, then $x_1 = -0.3719919$. If $y_3 = -8$, $x_* = -0.7914863$, and now for $x_0 = x_* + \delta$ with $\lvert \delta \rvert$ sufficiently small, the Gauss-Newton method produces

$$x_1 \cong x_0 - (7.55)\delta,$$

so that $\lvert x_1 - x_* \rvert \cong (6.55) \lvert x_0 - x_* \rvert$. For example, if $x_0 = -0.7915000$, then $x_1 = -0.7913968$.

An interesting feature of Example 10.2.5 is that the Gauss-Newton method takes bad steps by taking steps that are too long, but in the correct direction. This may make the reader wonder whether the Gauss-Newton step is always in a descent direction, and indeed this is the case whenever the step is well defined. The proof is just a generalization of the proof in Section 6.5 that the Newton step for nonlinear equations is in a descent direction for the associated minimization problem. If $J(x_c)$ has full column rank, so that $J(x_c)^T J(x_c)$ is nonsingular and the Gauss-Newton step

$$s_G = -(J(x_c)^T J(x_c))^{-1} J(x_c)^T R(x_c)$$

is well defined, then $J(x_c)^T J(x_c)$ is positive definite and

$$\nabla f(x_c)^T s_G = [J(x_c)^T R(x_c)]^T [-(J(x_c)^T J(x_c))^{-1} J(x_c)^T R(x_c)]$$

$$= -[R(x_c)^T J(x_c)]^T (J(x_c)^T J(x_c))[J(x_c)^T R(x_c)] < 0,$$

which proves that the Gauss-Newton step is in a descent direction. This suggests two ways of improving the Gauss-Newton algorithm: using it with a line search or with a trust region strategy. These two approaches lead to two algorithms that are used in practice.

The algorithm that uses the Gauss-Newton method with a line search is simply

$$x_+ := x_c - \lambda_c (J(x_c)^T J(x_c))^{-1} J(x_c)^T R(x_c), \qquad (10.2.12)$$

where λ_c is chosen by the methods of Section 6.3. We will refer to (10.2.12) as the *damped Gauss-Newton* method; unfortunately, some references call it the Gauss-Newton method. Since the damped Gauss-Newton method always takes descent steps that satisfy the line-search criteria, it is locally convergent on almost all nonlinear least-squares problems, including large-residual or very nonlinear problems. In fact, by Theorem 6.3.3 it is usually globally convergent. However, it may still be very slowly convergent on the problems that the Gauss-Newton method had trouble with. For example, if (10.2.12) is used on the fourth problem from Example 10.2.4, using a simple halving line search (i.e., the first satisfactory value of λ_c in the sequence $1, \frac{1}{2}, \frac{1}{4}, \frac{1}{8}, \dots$ is used), then starting from $x_0 = -0.3$, the algorithm still requires 44 iterations, 45 evaluations of $J(x)$ and 89 evaluations of $R(x)$ to achieve $|\nabla f(x_c)| \leq 10^{-10}$, whereas Newton's method requires 4 iterations! Also, the damped Gauss-Newton algorithm still is not well defined if $J(x_c)$ doesn't have full column rank.

The other modification of the Gauss-Newton algorithm we suggested is to choose x_+ by the trust region approach:

$$\underset{x_+ \in \mathbb{R}^n}{\text{minimize}} \| R(x_c) + J(x_c)(x_+ - x_c) \|_2$$

$$\text{subject to} \qquad \| x_+ - x_c \|_2 \leq \delta_c. \qquad (10.2.13)$$

It is an easy corollary of Lemma 6.4.1 to show that the solution to (10.2.13) is

$$x_+ = x_c - (J(x_c)^T J(x_c) + \mu_c I)^{-1} J(x_c)^T R(x_c), \qquad (10.2.14)$$

where $\mu_c = 0$ if $\delta_c \geq \|(J(x_c)^T J(x_c))^{-1} J(x_c)^T R(x_c)\|_2$ and $\mu_c > 0$ otherwise. Formula (10.2.14) was first suggested by Levenberg (1944) and Marquardt (1963) and is known as the *Levenberg-Marquardt* method. Many versions of the Levenberg-Marquardt method have been coded using various strategies to choose μ_c; the implementation of (10.2.14) as a trust region algorithm, where μ_c and δ_c are chosen by the techniques of Sections 6.4.1 and 6.4.3, is due to Moré (1977). (Moré uses a scaled trust region as described in Section 7.1.) The step $x_+ - x_c$ given by (10.2.14) can be calculated more accurately by solving an equivalent linear least-squares problem (see Exercise 12).

The local convergence properties of the Levenberg-Marquardt method are similar to those of the Gauss-Newton method and are given in Theorem 10.2.6.

THEOREM 10.2.6   Let the conditions of Theorem 10.2.1 be satisfied, and let the sequence $\{\mu_k\}$ of nonnegative real numbers be bounded by $b > 0$. If $\sigma < \lambda$, then for any $c \in (1, (\lambda + b)/(\sigma + b))$, there exists $\varepsilon > 0$ such that for all $x_0 \in N(x_*, \varepsilon)$, the sequence generated by the Levenberg-Marquardt method

$$x_{k+1} = x_k - (J(x_k)^T J(x_k) + \mu_k I)^{-1} J(x_k)^T R(x_k)$$

is well defined and obeys

$$\| x_{k+1} - x_* \|_2 \leq \frac{c(\sigma + b)}{(\lambda + b)} \| x_k - x_* \|_2 + \frac{c\alpha\gamma}{2(\lambda + b)} \| x_k - x_* \|_2^2$$

and

$$\| x_{k+1} - x_* \|_2 \leq \frac{c(\sigma + b) + (\lambda + b)}{2(\lambda + b)} \| x_k - x_* \|_2 < \| x_k - x_* \|_2.$$

If $R(x_*) = 0$ and $\mu_k = O(\| J(x_k)^T R(x_k)\|_2)$, then $\{x_k\}$ converges q-quadratically to x_*.

Proof.   A straightforward extension of Theorem 10.2.1 and Corollary 10.2.2. See also Dennis (1977).   ☐

Theorem 10.2.5 suggests that a Levenberg-Marquardt algorithm may still be slowly locally convergent on large residual or very nonlinear problems, and sometimes this is the case. However, many implementations of this algorithm, particularly the one by Moré that is contained in MINPACK, have proven to be very successful in practice, and so this is one of the approaches that we recommend for the general solution of nonlinear least-squares problems. Several factors make Levenberg-Marquardt algorithms preferable to damped Gauss-Newton algorithms on many problems. One is that the Levenberg-Marquardt method is well defined even when $J(x_c)$ doesn't have full column rank. Another is that when the Gauss-Newton step is much too long, the Levenberg-Marquardt step is close to being in the steepest-descent direction $-J(x_c)^T R(x_c)$ and is often superior to the damped Gauss-Newton step. Several versions of the Levenberg-Marquardt algorithm have been proven globally convergent—for example, by Powell (1975), Osborne (1976), and Moré (1977).

In Section 10.3 we discuss another nonlinear least-squares approach that is somewhat more robust, but also more complex, than the Levenberg-Marquardt.

10.3   FULL NEWTON-TYPE METHODS

The other class of methods we consider for solving the nonlinear least-squares problem is based on the full Taylor-series quadratic model of $f(x)$ around x_c,

(10.1.2). Newton's method for nonlinear least squares consists of choosing x_+ to be the critical point of this model,

$$x_+ = x_c - (J(x_c)^T J(x_c) + S(x_c))^{-1} J(x_c)^T R(x_c).$$

By Theorem 5.2.1, it is locally q-quadratically convergent to the minimizer x_* of $f(x)$ as long as $\nabla^2 f(x) = J(x_c)^T J(x_c) + S(x_c)$ is Lipschitz continuous around x_c, and $\nabla^2 f(x_*)$ is positive definite. Thus the local convergence properties of Newton's method for nonlinear least squares are quite superior to those of the methods of the previous section, since Newton's method is quickly locally convergent on almost all problems, while the damped Gauss-Newton or Levenberg-Marquardt methods may be slowly locally convergent if $R(x)$ is very nonlinear or $R(x_*)$ is large.

The reason why Newton's method is rarely used for nonlinear least squares is that $J(x_c)^T J(x_c) + S(x_c)$ is rarely available analytically at a reasonable cost. If $J(x_c)$ is not available analytically, it must be approximated by finite differences in order that $\nabla f(x_c) = J(x_c)^T R(x_c)$ be known accurately. This costs n additional evaluations of $R(x)$ per iteration. However, the cost of approximating $\nabla^2 f(x_c)$ or $S(x_c)$ by finite differences, $(n^2 + 3n)/2$ additional evaluations of $R(x)$ per iteration for either one, is usually prohibitively expensive.

A straightforward application of the secant methods of Chapter 9 to nonlinear least squares is also not desirable, because these methods approximate all of $\nabla^2 f(x_c)$. However, since we have just said that nonlinear least-squares algorithms always should calculate $J(x_c)$ either analytically or by finite differences, the portion $J(x_c)^T J(x_c)$ of $\nabla^2 f(x_c)$ is always readily available, and we only need to approximate $S(x_c)$. Since in addition $J(x_c)^T J(x_c)$ often is the dominant portion of $\nabla^2 f(x_c)$, secant methods that disregard this information and approximate all of $\nabla^2 f(x_c)$ by a secant approximation have not been very efficient in practice.

Since

$$S(x_c) = \sum_{i=1}^{m} r_i(x_c) \nabla^2 r_i(x_c),$$

another alternative is to approximate each $n \times n$ matrix $\nabla^2 r_i(x_c)$ by the techniques of Chapter 9. This is unacceptable for a general-purpose algorithm because it would require the storage of m additional symmetric $n \times n$ matrices. However, it is related to the approach that is used.

The successful secant approach to nonlinear least squares approximates $\nabla^2 f(x_c)$ by $J(x_c)^T J(x_c) + A_c$, where A_c is a single secant approximation to $S(x_c)$. We will discuss the approach of Dennis, Gay, and Welsch (1981), but references to other approaches can be found in that paper. The determination of A_c is just an application of the techniques of Chapters 8 and 9. Suppose we have an approximation A_c to $S(x_c)$, have taken a step from x_c to x_+, and desire an approximation A_+ to $S(x_+)$. The first question is, "What secant equation should A_+ obey?" Recall that for minimization, H_+ approximated $\nabla^2 f(x_+)$, and the secant equation was $H_+(x_+ - x_c) = \nabla f(x_+) - \nabla f(x_c)$. Similarly, if we think temporarily of approximating

$$S(x_+) = \sum_{i=1}^{m} r_i(x_+) \nabla^2 r_i(x_+)$$

by

$$A_+ = \sum_{i=1}^{m} r_i(x_+)(H_i)_+, \tag{10.3.1}$$

where each $(H_i)_+$ approximates $\nabla^2 r_i(x_+)$, then each $(H_i)_+$ should obey

$$(H_i)_+(x_+ - x_c) = \nabla r_i(x_+) - \nabla r_i(x_c)$$

$$= (\text{row } i \text{ of } J(x_+))^T - (\text{row } i \text{ of } J(x_c))^T. \tag{10.3.2}$$

Combining (10.3.1) and (10.3.2) gives

$$A_+(x_+ - x_c) = \sum_{i=1}^{m} r_i(x_+)(H_i)_+(x_+ - x_c)$$

$$= \sum_{i=1}^{m} r_i(x_+)[(\text{row } i \text{ of } J(x_+))^T - (\text{row } i \text{ of } J(x_c))^T]$$

$$= J(x_+)^T R(x_+) - J(x_c)^T R(x_+) \triangleq y_c^{\#} \tag{10.3.3}$$

as the analogous secant equation for A_+. Note that while this equation was derived by considering individual approximations $(H_i)_+$, the result is a single secant equation for A_+, the only additional matrix we will store.

In the one-dimensional case, A_+ is completely determined by (10.3.3). Example 10.3.1 shows how the resulting secant method for nonlinear least squares performs on the same problems that we considered in Examples 10.2.4 and 10.2.5. Overall, we see that the secant method is only a little slower than Newton's method and that it can be superior to the Gauss-Newton method on the medium- and large-residual problems. In particular, it is quickly locally convergent on the examples for which the Gauss-Newton method did not converge at all.

EXAMPLE 10.3.1   Let $R(x)$, $f(x)$, t_1, y_1, t_2, y_2, t_3, and the choices of y_3 and x_0 be given by Example 10.2.4. Then the secant method for one-dimensional nonlinear least squares,

$$x_{k+1} = x_k - (J(x_k)^T J(x_k) + A_k)^{-1} J(x_k)^T R(x_k)$$

$$A_k = \begin{cases} 0, & k = 0, \\ \dfrac{[J(x_k) - J(x_{k-1})]^T R(x_k)}{x_k - x_{k-1}}, & k > 0, \end{cases} \tag{10.3.4}$$

requires the following number of iterations to achieve $|\nabla f(x_c)| \leq 10^{-10}$ on each of the following problems on a CDC machine with $14+$ base-10 digits. For comparison, the number of iterations required by Newton's method is repeated from Example 10.2.4, along with the solution, x_*, and the residual at the solution, $f(x_*)$.

		Iterations required to achieve $\|f(x_k)\| \leq 10^{-10}$ by:			
y_3	x_0	Secant method	Newton's method	x_*	$f(x_*)$
8	1	6	7	0.69315	0
	0.6	5	6		
3	1	8	9	0.44005	1.6390
	0.5	4	5		
-1	1	11	10	0.044744	6.9765
	0	5	4		
-4	1	13	12	-0.37193	16.435
	-0.3	6	4		
-8	1	15	12	-0.79148	41.145
	-0.7	8	4		

When $n > 1$, (10.3.3) doesn't completely specify A_+, and we complete the determination of A_+ using the techniques of Chapter 9. Since $S(x_+)$ is symmetric, A_+ should be too; however, in general $S(x)$ may not be positive definite even at the solution x_* (see Exercise 15), so we will not require A_+ to be positive definite. We have no other new information about A_+, so, as in Chapter 9, we will choose A_+ to be the matrix that is most similar to A_c among the set of permissible approximants, by selecting the A_+ that solves the least-change problem

$$\underset{A_+ \in \mathbb{R}^{n \times n}}{\text{minimize}} \ \| T^{-T}(A_+ - A_c)T^{-1} \|_F$$

$$\text{subject to} \quad A_+ - A_c \text{ symmetric,} \quad A_+ s_c = y_c^{\#}. \tag{10.3.5}$$

Here $s_c = x_+ - x_c$, $y_c^{\#}$ is defined by (10.3.3), and $T \in \mathbb{R}^{n \times n}$ is nonsingular. If we choose the same weighting matrix T as produced the DFP update, i.e., any T for which

$$T T^T s_c = y_c, \qquad y_c \triangleq \nabla f(x_+) - \nabla f(x_c) = J(x_+)^T R(x_+) - J(x_c)^T R(x_c),$$

then it is an easy extension of the techniques of Section 9.3 to show that

(10.3.5) is solved by

$$A_+ = A_c + \frac{(y_c^\# - A_c s_c)y_c^T + y_c(y_c^\# - A_c s_c)^T}{y_c^T s_c}$$

$$- \frac{\langle y_c^\# - A_c s_c, s_c \rangle y_c y_c^T}{(y_c^T s_c)^2}. \tag{10.3.6}$$

Update (10.3.6) was proposed by Dennis, Gay, and Welsch (1981) and is used to approximate $S(x)$ in their nonlinear least-squares code NL2SOL. As the reader might suspect, the resulting secant method for nonlinear least squares, which uses iteration formula (10.3.4) and updates A_k by (10.3.6), is locally q-superlinearly convergent under standard assumptions on $f(x)$ and A_0. This result is given by Dennis and Walker (1981) and is a special case of Theorem 11.4.2.

The NL2SOL code that implements the secant method we have just described has several additional interesting features. To make the algorithm globally convergent, a model-trust region strategy is used; at each iteration the problem

$$\underset{s \in \mathbb{R}^n}{\text{minimize}} \ \tfrac{1}{2} R(x_c)^T R(x_c) + s^T J(x_c)^T R(x_c) + \tfrac{1}{2} s^T [J(x_c)^T J(x_c) + A_c] s$$

$$\text{subject to} \quad \| s \|_2 \le \delta_c \tag{10.3.7}$$

is solved for s_c, so that

$$s_c = -(J(x_c)^T J(x_c) + A_c + \mu_c I)^{-1} J(x_c)^T R(x_c) \tag{10.3.8}$$

for some $\mu_c \ge 0$. (Again, a scaled trust region is actually used.) Second, the algorithm sometimes uses the Gauss-Newton model (10.2.2) that doesn't include A_c instead of the augmented model (10.3.7). Because of the above-mentioned difficulty in approximating $S(x)$ by finite differences and because the Gauss-Newton method tends to do well initially, NL2SOL uses the zero matrix for A_0, so that initially the two models are equivalent. Then at the point in each iteration where the step bound δ_c is updated to δ_+ by comparing the actual reduction $f(x_c) - f(x_+)$ to the reduction predicted by the quadratic model, NL2SOL calculates the reductions predicted by both quadratic models, regardless of which was used to make the step from x_c to x_+. The update of A_c to A_+ is always made, but the first trial step from x_+ is calculated using the model whose predicted reduction best matched the actual reduction from x_c to x_+. Typically, this adaptive modeling causes NL2SOL to use Gauss-Newton or Levenberg-Marquardt steps until A_c builds up useful second-order information, and then to switch to augmented steps defined by (10.3.7–10.3.8). For easy problems with small final residuals, convergence sometimes is obtained using only Gauss-Newton steps.

Finally, before each update (10.3.6) in NL2SOL, A_c is multiplied by the "sizing" factor

$$\gamma_c = \min \left\{ \frac{y_c^T s_c}{s_c^T A_c s_c}, 1 \right\}, \tag{10.3.9}$$

which is almost the same as the factor of Shanno and Phua mentioned in Section 9.4. The reason this sizing is necessary at each iteration is that the scalar components $r_i(x)$ of

$$S(x) = \sum_{i=1}^{m} r_i(x) \, \nabla^2 r_i(x)$$

sometimes change more quickly than the second-derivative components $\nabla^2 r_i(x)$, and the update (10.3.6) may not reflect these scalar changes quickly enough. This is particularly true if $R(x_*)$ is quite small. The sizing factor (10.3.9) tries to account for a reduction from $\| R(x_c) \|_2$ to $\| R(x_+) \|_2$, so that the approximation A_+ is more accurate in the small- or zero-residual case. For further details about the NL2SOL algorithm, see Dennis, Gay, and Welsch (1981).

In practice, the comparison between a good Levenberg-Marquardt code (e.g., Moré's in MINPACK) and a good secant code for nonlinear least squares (e.g., NL2SOL) seems to be close. On small-residual problems that are not very nonlinear, there usually isn't much difference between the two. On large-residual or very nonlinear problems, a secant code often requires fewer iterations and function evaluations. This is especially true if the solution x_* is required to high accuracy, since then the difference between slow linear convergence and q-superlinear convergence becomes important. On the other hand, a Levenberg-Marquardt code such as Moré's requires fewer lines of code and is less complex than a secant code with all the features described above. For these reasons, we recommend either for general-purpose use.

10.4 OTHER CONSIDERATIONS IN SOLVING NONLINEAR LEAST-SQUARES PROBLEMS

In this section we discuss briefly several other topics related to nonlinear least squares and refer the interested reader to more complete references. The first topic is stopping criteria for nonlinear least squares, which differ in several interesting ways from the criteria discussed in Section 7.2 for general unconstrained minimization problems. First, since the smallest possible value of $f(x) = \frac{1}{2} R(x)^T R(x)$ is zero, the test

$$f(x_+) \cong 0?$$

is an appropriate convergence criterion for zero-residual problems. This is implemented as $f(x_+) \leq \text{tol}_1$ where the tolerance tol_1 is appropriately chosen.

Second, the gradient convergence test for nonlinear least squares,

$$\nabla f(x_+) = J(x_+)^T R(x_+) \cong 0?, \tag{10.4.1}$$

has a special meaning due to the structure of $\nabla f(x)$, since it can be interpreted as asking whether $R(x_+)$ is nearly orthogonal to the linear subspace spanned by the columns of $J(x_+)$. It is an exercise to show that the cosine of the angle ϕ_+ between $R(x_+)$ and this subspace is

$$\cos \phi_+ = \frac{R_+^T J_+ (J_+^T J_+)^{-1} J_+^T R_+}{\| J_+ (J_+^T J_+)^{-1} J_+^T R_+ \|_2 \cdot \| R_+ \|_2}, \tag{10.4.2}$$

where $J_+ \triangleq J(x_+)$ and $R_+ \triangleq R(x_+)$. If $(J_+^T J_+)^{-1} J_+^T R_+$ is available, as it will be in a damped Gauss-Newton or Levenberg-Marquardt code, then the test $\cos \phi_+ \leq \text{tol}_2$ may be used instead of (10.4.1). Other stopping criteria, such as $(x_+ - x_c) \cong 0$, are the same as for unconstrained minimization. In addition, stopping criteria with statistical significance are sometimes used in data-fitting applications; a common one closely related to (10.4.2) is discussed in Exercise 18 and Pratt (1977). A discussion of stopping criteria for nonlinear least squares is given by Dennis, Gay, and Welsch (1981).

A second important topic is the solution of mixed linear-nonlinear least-squares problems, those problems where the residual function $R(x)$ is linear in some variables and nonlinear in others. A typical example is

$$f(x) = \tfrac{1}{2} \sum_{i=1}^{m} r_i(x)^2, \qquad r_i(x) = x_1 e^{t_i x_3} + x_2 e^{t_i x_4} - y_i, \quad i = 1, \dots, m.$$

Here x_1 and x_2 are the linear variables while x_3 and x_4 are the nonlinear variables. It would seem that we should be able to minimize $f(x)$ by solving a nonlinear least-squares problem in just the two nonlinear variables x_3 and x_4, since, given any values of x_3 and x_4, we can calculate the corresponding optimal values of x_1 and x_2 by solving a linear least-squares problem. This is indeed the case; the somewhat difficult analysis required to produce the algorithm was performed by Golub and Pereyra (1973) and is sketched in Exercises 19 and 20. Several computer codes are available that solve mixed linear-nonlinear least-squares problems in this manner, including the VARPRO algorithm of Kaufman (1975). Their advantages are that they usually solve these problems in less time and fewer function evaluations than standard nonlinear least-squares codes, and that no starting estimate of the linear variables is required.

Many interesting considerations related to the solution of data-fitting problems by nonlinear least squares are mainly outside the scope of this book. One is the estimation of the uncertainty in the answers. In linear least-squares problems $Ax \cong b$ that come from data fitting, the calculation of the solution is usually followed by the calculation of the variance-covariance matrix $\sigma^2 (A^T A)^{-1}$, where σ is an appropriate statistical constant. Under appropriate statistical assumptions, this matrix gives the variance and covariance of the answers x_i, given the expected uncertainty in the data measurements b.

In nonlinear least squares, several corresponding variance-covariance matrices are used, namely $\sigma^2(J_f^T J_f)^{-1}$, $\sigma^2(H_f^{-1})$, and $\sigma^2(H_f^{-1} J_f^T J_f H_f^{-1})$, where x_f is the final estimate of x_*, $J_f = J(x_f)$, $H_f = J(x_f)^T J(x_f) + S(x_f)$ or an approximation to it, and $\sigma = 2f(x)/(m - n)$ is the analogous constant to the linear case [see, e.g., Bard (1970)]. While a proper discussion of the variance-covariance matrix is beyond the scope of this book, it is important to warn users of nonlinear least-squares software that variance-covariance estimates for nonlinear least squares are not nearly as reliable as for linear least squares and should be used with caution. Considerable research has gone into improving this situation; see, e.g., Bates and Watts (1980).

Finally, as we mentioned in Section 1.2, many other measures of the size of the vector $R(x)$ besides least squares can be used for defining what is meant by a "close fit," and some of these are becoming increasingly important in practice. In fact, we can view nonlinear least squares as just the special case of

$$\underset{x \in \mathbb{R}^n}{\text{minimize}} \; f(x) = \rho(R(x)), \qquad R: \mathbb{R}^n \longrightarrow \mathbb{R}^m, \quad \rho: \mathbb{R}^m \longrightarrow \mathbb{R}, \qquad (10.4.3)$$

where $\rho(z) = \frac{1}{2} z^T z$. Other obvious possibilities are $\rho(z) = \| z \|_1$ and $\rho(z) = \| z \|_\infty$, referred to as l_1 and minimax data fitting, respectively. [See, e.g., Murray and Overton (1980, 1981) and Bartels and Conn (1982).] When bad data points ("outliers") are present, one sometimes becomes interested in other measures $\rho(z)$ that don't weigh very large components of z as heavily. Two examples are the Huber (1973) loss function

$$\rho(z) = \sum_{i=1}^m \rho_1(z_i), \qquad \rho_1: \mathbb{R} \longrightarrow \mathbb{R},$$

$$\rho_1(z_i) = \begin{cases} \dfrac{z_i^2}{2}, & |z_i| \leq k, \\[2mm] k|z_i| - \dfrac{k^2}{2}, & |z_i| > k, \end{cases} \qquad (10.4.4)$$

which is linear in z_i for $|z_i| > k$, and the Beaton-Tukey (1974) loss function

$$\rho(z_i) = \sum_{i=1}^m \rho_2(z_i), \qquad \rho_2: \mathbb{R} \longrightarrow \mathbb{R},$$

$$\rho_2(z_i) = \begin{cases} \left(\dfrac{k^2}{6}\right)\left(1 - \left(1 - \left(\dfrac{z_i}{k}\right)^2\right)^3\right), & |z_i| \leq k, \\[3mm] \left(\dfrac{k^2}{6}\right), & |z_i| > k, \end{cases} \qquad (10.4.5)$$

which is constant for $|z_i| > k$; in each case, the constant k must be chosen appropriately. These are examples of "robust" data-fitting measures and a good reference is Huber (1981). In Chapter 11 we will refer briefly to some issues involved in the minimization of a twice continuously differentiable function of form (10.4.3).

10.5 EXERCISES

1. Let $R(x)$: $\mathbb{R}^4 \longrightarrow \mathbb{R}^{20}$, $r_i(x) = x_1 + x_2 e^{-(t_i + x_3)^2/x_4} - y_i$, $i = 1, \ldots, 20$, $f(x) = \frac{1}{2} R(x)^T R(x)$. (This problem was discussed in Section 1.2.) What is $J(x)$? $\nabla f(x)$? $S(x)$? $\nabla^2 f(x)$?

2. Prove that Theorem 10.2.1 remains true if the constant $(c\sigma + \lambda)/(2\lambda)$ in (10.2.6) is replaced by any other constant between $c\sigma/\lambda$ and 1.

3. Show that under the assumptions of Theorem 10.2.1 and for x sufficiently close to x_*,

$$[J(x) - J(x_*)]^T R(x_*) = S(x_*)(x - x_*) + O(\| x - x_* \|_2^2).$$

   [*Hint*: Write

$$J(x)^T R(x_*) = \sum_{i=1}^{m} r_i(x_*) \nabla r_i(x),$$

   and use a Taylor series expansion of $\nabla r_i(x)$ around x_*.]

4. Calculate, by hand or calculator, the first step of the Gauss-Newton method for the problem of Example 10.2.4 when $y_3 = 8$ and $x_0 = 1$. Also calculate the first step of Newton's method, and the first two steps of the secant method given in Example 10.3.1.

5. Repeat Exercise 4 in the case $y_3 = -8$, $x_0 = -0.7$.

6. Let $R_c = R(x_c)$, $J_c = J(x_c)$, $S_c = S(x_c)$, $x_+^G = x_c - (J_c^T J_c)^{-1} J_c^T R_c$, $x_+^N = x_c - (J_c^T J_c + S_c)^{-1} J_c^T R_c$, $s_G = x_+^G - x_c$, $s_N = x_+^N - x_c$. Show that $s_G - s_N = (J_c^T J_c)^{-1} S_c s_N$. Use this to show that if $\nabla^2 f(x) = J(x)^T J(x) + S(x)$ is Lipschitz continuous around the minimizer x_* of $f(x) = \frac{1}{2} R(x)^T R(x)$, $\nabla^2 f(x_*)$ is nonsingular, and x_c is sufficiently close to x_*, then

$$\| x_+^G - x_* \|_2 \le \|(J_c^T J_c)^{-1}\|_2 \| S_c \|_2 \| x_c - x_* \|_2 + O(\| x_c - x_* \|_2^2).$$

   In the case $n = 1$, you should be able to show that

$$[(x_+^G - x_*) - (S_c(J_c^T J_c)^{-1})(x_c - x_*)] = O(| x_c - x_* |^2). \tag{10.5.1}$$

7. Calculate $J(x_*)^T J(x_*)$ and $S(x_*)$ for the two problems of Example 10.2.5. Then use (10.5.1) to derive the two constants of repulsion, 2.20 and 6.55, that are given in that example.

8. Let $R(x)$: $\mathbb{R}^2 \longrightarrow \mathbb{R}^4$, $r_i(x) = e^{x_1 + t_i x_2} - y_i$, $i = 1, \ldots, 4$, $f(x) = \frac{1}{2} R(x)^T R(x)$. Suppose $t_1 = -2$, $t_2 = -1$, $t_3 = 0$, $t_4 = 1$, $y_1 = 0.5$, $y_2 = 1$, $y_3 = 2$, $y_4 = 4$. [$f(x) = 0$ at $x_* = (\ln 2, \ln 2)$.] Calculate one iteration of the Gauss-Newton method and one iteration of Newton's method starting from $x_0 = (1, 1)^T$. Do the same if the values of y_1 and y_4 are changed to 5 and -4, respectively.

9. Program a damped Gauss-Newton algorithm using the line-search algorithm A6.3.1, and run it on the problems from Example 9.3.2. How do your results compare with those of Newton's method and the secant method on these problems?

10. Use Lemma 6.4.1 to show that the solution to (10.2.13) is (10.2.14).

11. Show that for any $\mu \geq 0$, the Levenberg-Marquardt step $x_+ - x_c$ given by (10.2.14) is a descent direction for the nonlinear least-squares function.

12. Given $R \in \mathbb{R}^m$, $J \in \mathbb{R}^{m \times n}$, show that $s = -(J^T J + \mu I)^{-1} J^T R$ is the solution to the linear least-squares problem

$$\underset{s \in \mathbb{R}^n}{\text{minimize}} \; \| As + b \|_2,$$

$$A \in \mathbb{R}^{(m+n) \times n}, \qquad A = \boxed{\begin{array}{c} J \\ \hline \mu^{1/2} I \end{array}},$$

$$b \in \mathbb{R}^{m+n}, \qquad b = \boxed{\begin{array}{c} R \\ \hline 0 \end{array}}.$$

13. Prove Theorem 10.2.6, using the techniques of the proofs of Theorem 10.2.1 and Corollary 10.2.2.

14. Solve the problems of Example 10.3.1 by the secant method for unconstrained minimization,

$$x_{k+1} = x_k - \frac{f'(x_k)}{a_k}, \qquad a_k = \frac{f'(x_k) - f'(x_{k-1})}{x_k - x_{k-1}}, \qquad \text{where } a_0 = f''(x_0).$$

Compare your results to those given in Example 10.3.1 for the secant method for nonlinear least squares.

15. Let $f(x)$ be given by Example 10.2.4, with $y_3 = 10$. Calculate x_* (by calculator or computer) and show that $S(x_*) < 0$.

16. Use the techniques of Section 9.3 to show that if $TT^T s_c = y_c$, then the solution to (10.3.5) is (10.3.6).

17. Let $R \in \mathbb{R}^m$, $J \in \mathbb{R}^{m \times n}$ have full column rank. Show that the Euclidean projection of R onto the linear subspace spanned by the columns of J is $J(J^T J)^{-1} J^T R$. [*Hint*: How is this problem related to the linear least-squares problem, minimize $\| Jx - R \|_2$?] Then show that the cosine of the angle between R and this linear subspace is given by (10.4.2).

Exercise 18 indicates a statistically based stopping test for nonlinear least squares.

18. (a) Let $\alpha \in \mathbb{R}$, $s \in \mathbb{R}^n$, $M \in \mathbb{R}^{n \times n}$. Show that the maximum value of

$$\phi(v) \triangleq \frac{\alpha(v^T s)}{(v^T M v)^{1/2}} \tag{10.5.2}$$

is $\alpha(s^T M^{-1} s)^{1/2}$. Assume M positive definite and $\alpha > 0$.
(b) If M is the variance-covariance matrix for nonlinear least squares, then $\phi(v)$ given by (10.5.2) is proportional to the component of s in the direction v divided by the uncertainty of the nonlinear least-squares solution in the direc-

tion v, and so the stopping test

$$\max \{\phi(v) \,|\, v \in \mathbb{R}^n\} \leq \text{tol} \tag{10.5.3}$$

is a statistically meaningful way of measuring whether the step s is small. Show that if $s = -(J^T J)^{-1} J^T R$ and $M = (J^T J)^{-1}$, then (10.5.3) is equivalent to

$$\alpha \| J(J^T J)^{-1} J^T R \|_2 \leq \text{tol}. \tag{10.5.4}$$

Show that a similar expression is true if $s\cdot = -H^{-1} J^T R$ and $M = H^{-1}$, where $H \in \mathbb{R}^{n \times n}$ is $\nabla^2 f(x)$ or an approximation to it. Relate (10.5.4) to the cosine stopping test, $\cos \phi_+ \leq \text{tol}$, where $\cos \phi_+$ is given by (10.4.2).

Exercises 19 and 20 indicate the mixed linear-nonlinear least-squares algorithm of Golub and Pereyra (1973).

19. Let $R: \mathbb{R}^n \longrightarrow \mathbb{R}^m$, $m > n$. Furthermore, let $x = (u, v)^T$, $u \in \mathbb{R}^{n-p}$, $v \in \mathbb{R}^p$, $0 < p < n$, and let $R(x) = N(v) \cdot u - b$, where $N: \mathbb{R}^p \longrightarrow \mathbb{R}^{m \times (n-p)}$, $b \in \mathbb{R}^m$. For example, if $R: \mathbb{R}^4 \longrightarrow \mathbb{R}^{20}$, $r_i(x) = x_1 e^{t_i x_3} + x_2 e^{t_i x_4} - y_i$, then $u = (x_1, x_2)$, $v = (x_3, x_4)$, row i of $N(v) = (e^{t_i x_3}, e^{t_i x_4})$, $b_i = y_i$. Let the minimizer of $f(x) = \frac{1}{2} R(x)^T R(x)$ be $x_* = (u_*, v_*)$. Prove that $u_* = N(v_*)^+ b$. Furthermore, prove that the problem "minimize$_{x \in \mathbb{R}^n} f(x)$" is equivalent to

$$\text{minimize} \; \tfrac{1}{2} \| N(v)N(v)^+ b - b \|_2^2. \tag{10.5.5}$$
$$v \in \mathbb{R}^p$$

20. With the help of Golub and Pereyra (1973), develop a damped Gauss-Newton or Levenberg-Marquardt algorithm for solving (10.5.5). In particular, describe how the required derivatives are obtained analytically (this is the hardest part), and how the QR decomposition is used.

21. Modify the algorithms in Appendix A for solving a system of nonlinear equations, using the locally constrained optimal ("hook") global step and an analytic or finite-difference Jacobian, to create a Levenberg-Marquardt algorithm for nonlinear least squares. You will not have to make many modifications to the existing algorithms. Algorithms A5.4.1 (finite-difference Jacobian), A6.5.1 (formation of the model Jacobian), and A3.2.1–A3.2.2 (QR decomposition and solve) will differ only in that J has m rather than n rows, and the QR algorithm now calculates a linear least-squares solution [see Stewart (1973) for the minor changes required]. The hook-step algorithms are unchanged [however, see Moré (1977) for a more efficient linear algebra implementation]. The stopping criteria must be changed as discussed in Section 10.4; see also Moré (1977) or Dennis, Gay, and Welsch (1981). You can test your code on many of the problems in Appendix B, since many of them are easily recast as nonlinear least-squares problems.

11

Methods for Problems
with Special Structure

This chapter is almost like an appendix to the main material of the book in that it is concerned with ways to apply the techniques already given to problems where the Jacobian or Hessian matrix has an important type of structure that is encountered frequently in practice.

The incorporation of special structure into our quasi-Newton methods is not new to us. For example, in Chapter 9 we incorporated first hereditary symmetry, and then hereditary positive definiteness, into the secant update from Chapter 8. In Chapter 10 we modified our unconstrained minimization methods to use the special structure of the nonlinear least-squares problem.

Our concern in this chapter is for the case when part of the derivative matrix is easily and accurately obtained, while another part is not. The most obvious and probably most important example is when we know that many entries of the derivative matrix are zero. In this case the Jacobian or Hessian is said to be sparse. Sections 11.1 and 11.2 discuss the finite-difference Newton method and secant methods for sparse problems, respectively.

The generalization of Section 11.2 is the case when the Jacobian or Hessian matrix can be represented as

$$J(x) = C(x) + A(x),$$

where $C(x)$ is obtained accurately by some method, and we wish to use secant approximations for $A(x)$, a matrix with some special structure. We have already seen an example of this type of structure in the secant method for nonlinear least squares in Section 10.3. In Section 11.3 we give several additional examples of this type of structure and present a general theory that shows how to construct secant approximations for any problem that fits this very general form. In Section 11.4 we present a general convergence theory for any secant method that is derived by the techniques of Section 11.3. These results actually include as special cases all the derivations and convergence results of Chapters 8 and 9.

11.1   THE SPARSE FINITE-DIFFERENCE NEWTON METHOD

In this section we consider the problem of solving a system of nonlinear equations $F(x) = 0$, where we also know that each single equation depends on relatively few of the unknowns. In other words, the ith row of the derivative matrix has very few nonzero elements, and their positions are independent of the particular value of x. We could consider just ignoring these zeros and trying to solve the problem with the methods of Chapters 5 through 9. For many problems, however, taking advantage of the zero or sparsity structure of $J(x)$ is crucial in order that we can even consider solving the problem. In the first place, n may be so large that storage of $J(x)$ is out of the question unless we take advantage of not having to store the zeros. Furthermore, the savings in arithmetic operations achieved by taking advantage of the zeros in solving

$$J(x_c)s^N = -F(x_c) \qquad (11.1.1)$$

for the Newton step can be very significant.

If the nonzero elements of $J(x)$ are readily computed analytically, then there is not much to say. An efficient storage scheme for $J(x)$ should be used, along with an efficient sparse linear solver for (11.1.1). Several sparse linear equations packages are available by now, for example in the Harwell library or the Yale Sparse Matrix Package available through IMSL. The global strategies of Chapter 6 can be adapted to the sparse problem without much difficulty, except that the multiple linear systems required by the locally constrained optimal step make it less attractive.

The remainder of this section is devoted to efficient techniques for approximating the nonzero elements of $J(x)$ by finite differences when they are not available analytically. It is based mainly on Curtis, Powell, and Reid (1974).

The discussion of finite differences in Chapter 4, including the choice of the step sizes h_i, is still valid here. The only change is that we will try to use fewer than n additional evaluations of $F(x)$ for an approximation of $J(x)$, by obtaining more than one column of the finite-difference Jacobian from a single additional value of $F(x)$. To see how this can be done, consider the following example.

Let e_j denote the jth unit vector, and

$$F(x_1, x_2) = \begin{pmatrix} 2x_1 + x_1^2 \\ x_2 - 3 \end{pmatrix}, \qquad x_c = \begin{pmatrix} 1 \\ 10 \end{pmatrix}$$

for which

$$F(x_c) = \begin{pmatrix} 3 \\ 7 \end{pmatrix}, \qquad J(x_c) = \begin{pmatrix} 4 & 0 \\ 0 & 1 \end{pmatrix}.$$

We see immediately that the (1, 2) and (2, 1) elements of the Jacobian are zero, and so the two Jacobian columns have the key property that there is at most one nonzero in each row. On a hypothetical machine with macheps $= 10^{-6}$, the method of Section 5.4 would choose $h = (10^{-3}, 10^{-2})^T$, evaluate

$$v_1 = F(x_c + h_1 e_1) = F(1.001, 10) = \begin{pmatrix} 3.004001 \\ 7 \end{pmatrix},$$

$$v_2 = F(x_c + h_2 e_2) = F(1, 10.01) = \begin{pmatrix} 3 \\ 7.01 \end{pmatrix},$$

and then approximate the first and second columns of $J(x_c)$ by

$$\frac{v_1 - F(x_c)}{h_1} = \begin{pmatrix} 4.001 \\ 0 \end{pmatrix} \quad \text{and} \quad \frac{v_2 - F(x_c)}{h_2} = \begin{pmatrix} 0 \\ 1 \end{pmatrix}.$$

The important point is that since we were interested only in A_{11} and A_{22}, only the first element $f_1(x_c + h_1 e_1)$ of v_1 and the second element $f_2(x_c + h_2 e_2)$ of v_2 were required. Furthermore, since f_1 doesn't depend on x_2 and f_2 doesn't depend on x_1, these two values could have been obtained from the single function evaluation

$$w = F(x_c + h_1 e_1 + h_2 e_2) = F(1.001, 10.01)^T = \begin{pmatrix} 3.004001 \\ 7.01 \end{pmatrix}.$$

Then $A_{11} = (w_1 - f_1(x_c))/h_1$, $A_{22} = (w_2 - f_2(x_c))/h_2$, and we have approximated $J(x_c)$ with just one evaluation of $F(x)$ in addition to $F(x_c)$.

The general case works the same way. First we find a set of column indices $\Gamma \subset \{1, \dots, n\}$ with the property that among this subset of the columns of the Jacobian, there is at most one nonzero element in each row. Then we evaluate

$$w = F\left(x_c + \sum_{j \in \Gamma} h_j e_j\right),$$

and for each nonzero row i of a column $j \in \Gamma$, approximate $J(x_c)_{ij}$ by

$$A_{ij} = \frac{w_i - f_i(x_c)}{h_j}.$$

The total number of additional evaluations of $F(x)$ required to approximate

$J(x_c)$ is just the number of such index sets Γ required to span all the columns. For example, if $J(x)$ is a tridiagonal matrix

$$J(x)_{ij} \neq 0, \qquad |i - j| \leq 1,$$

$$J(x)_{ij} = 0, \qquad \text{otherwise,}$$

then it is an easy exercise to show that the three index sets $\{j \mid j = k \bmod 3\}$, $k = 0,\ 1,\ 2$, will suffice for any value of n. Thus only three, rather than n additional evaluations of $F(x)$ are required at each iteration to approximate $J(x_c)$.

In general, it is easy to group the columns of a banded matrix into the smallest number of groups of such index sets, and to generalize the above tridiagonal example (Exercise 1). Unfortunately, there does not seem to be an efficient way to get the smallest number of groups for a general sparse Jacobian. Nevertheless, simple heuristic algorithms have proven to be very effective. Recent work by Coleman and Moré (1983) has improved upon the heuristics of Curtis, Powell, and Reid and shown the general problem to be NP-complete. Note that if the evaluation of $F(x)$ can be decomposed naturally into the evaluation of its n component functions, then the above techniques are unnecessary, since we can simply approximate each nonzero $J(x_c)_{ij}$ by $(f_i(x_c + h_j e_j) - f_i(x_c))/h_j$. However, in practice, if $J(x_c)$ isn't analytically available. then $F(x_c)$ is not usually cheaply decomposable in this manner.

These techniques extend to the unconstrained minimization problem, where the derivative matrix is the Hessian $\nabla^2 f(x_c)$. Again, a quasi-Newton method for large sparse minimization problems should store $\nabla^2 f(x_c)$ efficiently and use a symmetric (and positive definite, if applicable) sparse linear solver. If $\nabla^2 f(x_c)$ is approximated by finite differences using function values, then one simply approximates the nonzero elements using the methods of Section 5.6. Sometimes in practice, however, $\nabla f(x)$ is analytically available at little additional cost when $f(x)$ is calculated. In this case, $\nabla^2 f(x_c)$ should be approximated from additional values of $\nabla f(x)$, using a procedure like the above to make the process efficient. One can simply use the techniques of Curtis, Powell, and Reid and then average the resultant matrix with its transpose to get the sparse symmetric Hessian approximation. Recent research by Powell and Toint (1979) and by Coleman and Moré (1982) has centered on exploiting symmetry to obtain this approximation more efficiently (see Exercise 2).

11.2 SPARSE SECANT METHODS

In the previous section we saw how to use sparsity to reduce the number of function evaluations required to approximate $J(x)$ by finite differences. If $F(x)$ is expensive to evaluate, however, this may still be too costly, and we may want to approximate $J(x)$ by secant techniques. We will want these secant approximations to be as sparse as the actual Jacobian to achieve the savings in storage and arithmetic operations mentioned in Section 11.1. In this section we discuss a sparse secant update that is just the specialization of the secant update of Chapter 8 to the sparse case. We also discuss briefly sparse symmetric secant updates for minimization problems.

Recall that the secant update for nonlinear equations (Broyden's update),

$$A_+ = A_c + \frac{(y_c - A_c s_c)s_c^T}{s_c^T s_c}, \tag{11.2.1}$$

is the solution to the least-change problem

$$\min_{B \in \mathbb{R}^{n \times n}} \| B - A_c \|_F \quad \text{subject to} \quad Bs_c = y_c \tag{11.2.2}$$

where A_c, $A_+ \in \mathbb{R}^{n \times n}$ approximate $J(x_c)$ and $J(x_+)$, respectively, $s_c = x_+ - x_c$, and $y_c = F(x_+) - F(x_c)$ (for convenience, we will drop the subscript c from A, s, and y for the remainder of this chapter). In the sparse case, let $Z \in \mathbb{R}^{n \times n}$ be the 0-1 matrix denoting the pattern of zeros in $J(x)$—i.e.,

$$Z_{ij} = \begin{cases} 0, & J(x)_{ij} = 0 \quad \text{for all } x \in \mathbb{R}^n \\ 1, & \text{otherwise,} \end{cases}$$

and let SP(Z) denote the set of $n \times n$ matrices with this zero pattern—i.e.,

$$\text{SP}(Z) = \{ M \in \mathbb{R}^{n \times n} : M_{ij} = 0 \text{ if } Z_{ij} = 0, 1 \le i, j \le n \}. \tag{11.2.3}$$

If we assume that $A \in \text{SP}(Z)$, then the natural extension of (11.2.2) to the sparse case is

$$\min_{B \in \mathbb{R}^{n \times n}} \| B - A \|_F \quad \text{subject to } Bs = y, \quad B \in \text{SP}(Z). \tag{11.2.4}$$

Of course, there may be no element of SP(Z) for which $Bs = y$, as in the case when a whole row of B is required to be zero but the corresponding component of y is not zero. When the sparsity required is consistent with $Bs = y$, then the solution to (11.2.4) is easy to find, and is given in Theorem 11.2.1. It also can be obtained using the more general Theorem 11.3.1.

We require the following additional notation: we define the matrix projection operator $P_Z : \mathbb{R}^{n \times n} \longrightarrow \mathbb{R}^{n \times n}$ by

$$(P_Z(M))_{ij} = \begin{cases} 0, & Z_{ij} = 0, \\ M_{ij}, & Z_{ij} = 1. \end{cases} \tag{11.2.5}$$

That is, P_Z zeros out the elements of M corresponding to the zero positions of the sparsity pattern Z, while otherwise leaving M unchanged. It is an exercise to show that $P_Z(M)$ is the closest matrix in SP(Z) to M in the Frobenius norm. Similarly for $v \in \mathbb{R}^n$, we define $v_i \in \mathbb{R}^n$ by

$$(v_i)_j = \begin{cases} 0, & Z_{ij} = 0, \\ v_j, & Z_{ij} = 1. \end{cases} \tag{11.2.6}$$

That is, v_i is the result of imposing on v the sparsity pattern of the ith row of Z.

THEOREM 11.2.1   Let $Z \in \mathbb{R}^{n \times n}$ be a 0-1 matrix, and let SP(Z) be defined by (11.2.3) and P_Z by (11.2.5). Let $A \in$ SP(Z), s, $y \in \mathbb{R}^n$, and define s_i, $i = 1, \ldots, n$ by (11.2.6). If $s_i = 0$ only when $y_i = 0$, then the solution to (11.2.4) is the *sparse secant update*

$$A_+ = A + P_Z(D^+(y - As)s^T), \tag{11.2.7a}$$

where D^+ is defined by (3.6.6) for $D \in \mathbb{R}^{n \times n}$, a diagonal matrix with

$$D_{ii} = s_i^T s_i, \qquad i = 1, \ldots, n. \tag{11.2.7b}$$

Proof.   Let $A_{i.}$, $B_{i.} \in \mathbb{R}^n$ denote the ith rows of A and B, respectively. Then since

$$\| B - A \|_F^2 = \sum_{i=1}^{n} \| B_{i.} - A_{i.} \|_2^2,$$

(11.2.4) can be solved by choosing each $B_{i.}$ to solve

$$\min_{B_{i.} \in \mathbb{R}^n} \| B_{i.} - A_{i.} \|_2 \quad \text{subject to} \quad (B_{i.} - A_{i.})^T s_i = (y - As)_i,$$

$$(B_{i.} - A_{i.}) \in \text{SP}(Z)_{i.} \tag{11.2.8}$$

where

$$\text{SP}(Z)_{i.} \triangleq \{v \in \mathbb{R}^n : v_j = 0 \text{ if } Z_{ij} = 0\}.$$

It is an easy exercise to show that the solution to (11.2.8) is

$$B_{i.} = A_{i.} + (s_i^T s_i)^+ (y - As)_i s_i^T. \tag{11.2.9}$$

Since for any u, $v \in \mathbb{R}^n$, row i of $P_Z(uv^T)$ is $(u_i)v_i$, then $B_{i.}$ given by (11.2.9) is just the ith row of A_+ given by (11.2.7). $\square$

EXAMPLE 11.2.2   Let

$$Z = \begin{bmatrix} 1 & 0 & 1 \\ 1 & 1 & 0 \\ 0 & 1 & 1 \end{bmatrix},$$

$s = (1, -1, 2)^T$, $y = (2, 1, 3)^T$, $A = I$. Then $s_1 = (1, 0, 2)^T$, $s_2 = (1, -1, 0)^T$, $s_3 = (0, -1, 2)^T$, and (11.2.7) gives

$$A_+ = A + \begin{bmatrix} 0.2 & 0 & 0.4 \\ 1 & -1 & 0 \\ 0 & -0.2 & 0.4 \end{bmatrix}.$$

By comparison, Broyden's update (11.2.1) would be

$$A_+ = A + \begin{bmatrix} \frac{1}{6} & -\frac{1}{6} & \frac{1}{3} \\ \frac{1}{3} & -\frac{1}{3} & \frac{2}{3} \\ \frac{1}{6} & -\frac{1}{6} & \frac{1}{3} \end{bmatrix}.$$

Update (11.2.7) was suggested independently by Schubert (1970) and Broyden (1971), and Theorem 11.2.1 was proved independently by Reid (1973) and Marwil (1979). The update seems effective in quasi-Newton algorithms for sparse nonlinear equations, although there is little published computational experience. If some $s_i = 0$, then (11.2.4) has a solution only if the corresponding $y_i = 0$; however, when $J(x) \in \text{SP}(Z)$, $s = x_+ - x_c$, and $y = F(x_+) - F(x_c)$, this is always the case (Exercise 5). The more general case is addressed by Theorem 11.3.1. Theorem 11.4.1 will show that a quasi-Newton method using update (11.2.7) on problems with $J(x) \in \text{SP}(Z)$ is locally q-superlinearly convergent under standard assumptions.

Unfortunately, the adaptation of sparse updating to unconstrained minimization where the approximations must also be symmetric and likely positive definite has not been as satisfying. Marwil (1978) and Toint (1977, 1978) were the first to generalize the symmetric secant update to the sparse case, solving

$$\min_{B \in \mathbb{R}^{n \times n}} \| B - A \|_F \quad \text{subject to} \quad Bs = y, \quad (B - A) \in \text{SP}(Z),$$

$$(B - A) \text{ symmetric}.$$

We give this update in Section 11.3, and the local q-superlinear convergence of the quasi-Newton method using it is proven in Section 11.4. However, this update has several major disadvantages that we mention below, one being that the approximations generated are not necessarily positive definite. In fact, the effort to generate positive definite sparse symmetric updates has as yet been unsuccessful. Shanno (1980) and Toint (1981) have produced a sparse update that reduces to the BFGS in the nonsparse case, but it doesn't necessarily preserve positive definiteness in sparse situations.

The fact that none of these sparse symmetric updates preserves positive definiteness already eliminates a major advantage of secant updates for minimization. In addition, the calculation of any of the known sparse symmetric updates requires the solution of an additional sparse symmetric system of linear equations at each iteration. Finally, the limited computational experience with methods using these updates has not been good. For these reasons, secant methods may not be a promising technique for large sparse minimization problems.

Instead, it appears likely that large minimization problems will be solved using either the finite-difference methods of Section 11.1 or *conjugate gradient* methods. Conjugate gradient methods are somewhat different from quasi-Newton methods in their basic structure, and we do not cover them in this book. For small problems they are usually less efficient than the methods of this book, but for large problems they are a leading contender. Like secant methods, they use only function and gradient information in minimizing a function, but a major difference is that they do not store any approximation to the second-derivative matrix. For further information, see Buckley (1978), Shanno (1978), Fletcher (1980), Hestenes (1980) or Gill, Murray, and Wright (1981). For a straight linear algebra treatment, see Golub and Van Loan (1983). Truncated Newton methods [see, e.g., Dembo, Eisenstat, and Steihaug (1982)] combine some features of conjugate gradient methods with the trust region methods of Chapter 6. Another recent approach is given in the papers by Griewank and Toint (1982a, b, c).

We conclude our discussion of secant methods for sparse problems by mentioning an advantage of secant methods for nonsparse problems that does not carry over to the sparse case. We saw in Chapters 8 and 9 that we could use secant updates to reduce the arithmetic cost per iteration from $O(n^3)$ to $O(n^2)$ operations, either by updating an approximation to the inverse directly or by sequencing a QR or LL^T factorization. In the sparse case, the first option is unattractive because the inverse of a sparse matrix is rarely sparse, while the second is unavailable because the update usually constitutes a rank-n change to A_c. On the other hand, losing this advantage may not be important in the sparse case, because the factorization and solution of a sparse linear system often requires only $O(n)$ operations.

11.3 DERIVING LEAST-CHANGE SECANT UPDATES

For the remainder of this chapter we consider problems $F: \mathbb{R}^n \longrightarrow \mathbb{R}^n$ where the derivative matrix $J(x) \triangleq F'(x)$ has the form

$$J(x) = C(x) + A(x) \tag{11.3.1}$$

and $C(x)$ is readily available while $A(x)$ needs to be approximated using secant updates. [This includes optimization problems, where $F(x) = \nabla f(x)$ and $J(x) = \nabla^2 f(x)$.] In this section we give an extended abstract of the results of Dennis and Schnabel (1979) for deriving least-change updates in this general framework. We will leave out the lengthy proof of the main theorem and include only the simplest applications, since the specialist reader can consult the original reference.

We have already seen several examples of problem structure (11.3.1). In the sparse problems considered in Sections 11.1 and 11.2, where the full generality of this section is not needed, $C(x) = 0$ and $A(x)$ contains the nonzero elements of $J(x)$ and must be approximated by matrices having the same sparsity pattern. Two slight variations are when some of the equations of $F(x)$ are linear, or when $F(x)$ is linear in some of its variables. An example of the first case is

$$F(x_1, x_2) = \begin{pmatrix} 2x_1 + 3x_2 - 5 \\ f_2(x_1, x_2) = \text{black box } (x_1, x_2) \end{pmatrix},$$

$$J(x) = \begin{pmatrix} 2 & 3 \\ \dfrac{\partial f_2(x)}{\partial x_1} & \dfrac{\partial f_2(x)}{\partial x_2} \end{pmatrix}$$

$$= \begin{pmatrix} 2 & 3 \\ 0 & 0 \end{pmatrix} + \begin{pmatrix} 0 & 0 \\ \dfrac{\partial f_2(x)}{\partial x_1} & \dfrac{\partial f_2(x)}{\partial x_2} \end{pmatrix}. \tag{11.3.2}$$

An example of the second case is

$$F(x_1, x_2, x_3) = \begin{pmatrix} 3x_1 + \text{mess}_1(x_2, x_3) \\ -x_1 + \text{mess}_2(x_2, x_3) \\ +5x_1 + \text{mess}_3(x_2, x_3) \end{pmatrix}$$

$$J(x_1, x_2, x_3) = \begin{pmatrix} 3 & \dfrac{\partial m_1(x_2, x_3)}{\partial x_2} & \dfrac{\partial m_1(x_2, x_3)}{\partial x_3} \\ -1 & \dfrac{\partial m_2(x_2, x_3)}{\partial x_2} & \dfrac{\partial m_2(x_2, x_3)}{\partial x_3} \\ 5 & \dfrac{\partial m_3(x_2, x_3)}{\partial x_2} & \dfrac{\partial m_3(x_2, x_3)}{\partial x_3} \end{pmatrix}$$

$$= \begin{pmatrix} 3 & 0 & 0 \\ -1 & 0 & 0 \\ 5 & 0 & 0 \end{pmatrix} + \begin{pmatrix} 0 & \dfrac{\partial m_1(x_2, x_3)}{\partial x_2} & \dfrac{\partial m_1(x_2, x_3)}{\partial x_3} \\ 0 & \dfrac{\partial m_2(x_2, x_3)}{\partial x_2} & \dfrac{\partial m_2(x_2, x_3)}{\partial x_3} \\ 0 & \dfrac{\partial m_3(x_2, x_3)}{\partial x_2} & \dfrac{\partial m_3(x_2, x_3)}{\partial x_3} \end{pmatrix}$$

$$= C(x) + A(x). \tag{11.3.3}$$

In each case, $C(x)$ is a known constant matrix, while $A(x)$ may need to be approximated by secant methods.

Another example is the nonlinear least-squares problem we saw in Chapter 10, where $C(x) = J(x)^T J(x)$ and

$$A(x) = \sum_{i=1}^{m} r_i(x) \, \nabla^2 r_i(x).$$

Remember that $J(x)$ is always calculated analytically or by finite differences in order to compute $\nabla f(x) = J(x)^T R(x)$, and so $C(x)$ is readily available, but $A(x)$

is usually approximated by zero or by secant techniques. Even more interesting is the general data-fitting function

$$f(x) = \rho(R(x)), \qquad \rho: \mathbb{R}^m \longrightarrow \mathbb{R}^1, \quad R: \mathbb{R}^n \longrightarrow \mathbb{R}^m, \qquad (11.3.4)$$

that we mentioned in Section 10.4. Here

$$\nabla f(x) = J(x)^T \nabla \rho(R(x))$$

$$\nabla^2 f(x) = J(x)^T \nabla^2 \rho(R(x)) J(x) + \sum_{i=1}^{m} \frac{\partial \rho(R(x))}{\partial r_i} \nabla^2 r_i(x)$$

$$= C(x) + A(x),$$

where the $m \times n$ matrix $J(x) \triangleq R'(x)$. In practice, $\nabla \rho$ and $\nabla^2 \rho$ are usually trivial to obtain and $J(x)$ is evaluated analytically or by finite differences to calculate $\nabla f(x)$, but obtaining the roughly $mn^2/2$ different second partial derivatives of the residuals is out of the question. Thus we readily have $C(x)$ but need to approximate $A(x)$.

In all these examples, the secant approximation to $A(x)$ should incorporate the special structure of the particular matrix being approximated. The idea we will use in finding an appropriate approximation to $A(x_+)$ at some iterate x_+ is just a logical extension of what we have done so far. We will decide on a secant condition, say $A_+ \in Q(y^\#, s) \triangleq \{M \in \mathbb{R}^{n \times n}: Ms = y^\#\}$. In addition, we will choose an affine subspace $\mathscr{A}$ defined by properties such as symmetry or sparsity that all the approximants A_k should have. Then we will ask A_+ to be the nearest matrix in $\mathscr{A} \cap Q(y^\#, s)$ to $A \in \mathscr{A}$, in an appropriate norm. That is, A_+ will be the "least-change update" to A with the desired properties.

Let us consider the choices of $Q(y^\#, s)$, $\mathscr{A}$, and the least-change metric for some of our examples. We again use $s \triangleq x_+ - x_c$ and $y \triangleq F(x_+) - F(x_c)$, and we remind the reader that the secant equation from Chapter 8, $A_+ s = y$, came from the fact that $y = J(x_+)s$ if $F(x)$ is linear and $y \cong J(x_+)s$ otherwise. Here we will want $y^\# \cong A(x_+)s$.

For the first example, function (11.3.2), we see that $\mathscr{A} = \{A \in \mathbb{R}^{2 \times 2}: A_{1.} = (0, 0)\}$ and the Frobenius norm seems reasonable as a measure of change. The secant condition is also very simple, since

$$A(x_+)s = \begin{pmatrix} 0 \\ \nabla f_2(x_+)^T s \end{pmatrix} \cong \begin{pmatrix} 0 \\ f_2(x_+) - f_2(x_c) \end{pmatrix},$$

so we take $y^\# = (0, f_2(x_+) - f_2(x_c))^T = (0, y_2)^T$.

In the second example, function (11.3.3), $\mathscr{A} = \{A \in \mathbb{R}^{3 \times 3}: A_{.1} = (0, 0, 0)^T\}$, and again the Frobenius norm seems reasonable, but the secant condition is a bit more interesting. If we can evaluate the messy portion Mess(x_2, x_3) of $F(x_1, x_2, x_3)$ directly, then there is no difficulty, since clearly $y^\# = $ Mess$((x_+)_2, (x_+)_3) - $ Mess$((x_c)_2, (x_c)_3)$ is the logical choice. If we cannot separate the function F in this way, we use instead a general technique that *always* works and that reduces to the same choice of $y^\#$ in this case as well as in the first example.

If the structure of $A(x)$ doesn't suggest to us an inexpensive choice of $y^\#$, then, since we are computing $C(x_+)$, we can always default to the secant equation

$$A_+ s = y^\# = y - C(x_+)s, \tag{11.3.5}$$

which is equivalent to asking our complete Jacobian approximation $J_+ = C(x_+) + A_+$ to satisfy the standard secant equation

$$J_+ s = (C(x_+) + A_+)s = y.$$

The reader can verify that (11.3.5) is equivalent to our choice of $y^\#$ in the above two examples.

Our final example, function (11.3.4), is much more interesting because the default $y^\#$ is not the same as the $y^\#$ we obtain by looking directly at $A(x_+)s$. From (11.3.5), the default $y^\#$ is

$$y^\# = J(x_+)^T \nabla\rho(R(x_+)) - J(x_c)^T \nabla\rho(R(x_c)) - J(x_+)^T \nabla^2\rho(R(x_+))J(x_+)s. \tag{11.3.6}$$

On the other hand, a straightforward generalization of the derivation in equations (10.3.1–10.3.3) leads us to suggest

$$
\begin{aligned}
A(x_+)s &= \sum_{i=1}^{m} \frac{\partial\rho(R(x_+))}{\partial r_i} \nabla^2 r_i(x_+)s \\
&\cong \sum_{i=1}^{m} \frac{\partial\rho(R(x_+))}{\partial r_i} [\nabla r_i(x_+) - \nabla r_i(x_c)] \\
&= J(x_+)^T \nabla\rho(R(x_+)) - J(x_c)^T \nabla\rho(R(x_+)) = y^\#,
\end{aligned} \tag{11.3.7}
$$

which in general is different from (11.3.6). It is satisfying that the latter choice of $y^\#$ is a little better in practice.

To complete specifying the secant approximation conditions for function (11.3.4), we choose $\mathscr{A} = \{A \in \mathbb{R}^{n \times n}: A = A^T\}$, since $A(x)$ is always symmetric. We could certainly choose the least-change metric to be the Frobenius norm, which would reduce our update to the symmetric secant update (9.1.3) if $C(x_+) = 0$. However, since we are solving a minimization problem, the discussion of scaling in Section 9.3 is applicable here and leads us to use a Frobenius norm after scaling by $\bar{J}$, where $\bar{J}\bar{J}^T s = y$ if $y^T s > 0$. This reduces our update to a DFP update if $C(x_+) = 0$.

Given a choice of the secant condition $A_+ \in Q(y^\#, s)$, the subspace of approximants $\mathscr{A}$, and the weighted Frobenius norm $\| \cdot \|$, Theorem 11.3.1 produces the update satisfying

$$\min_{B \in \mathbb{R}^{n \times n}} \| B - A \| \quad \text{subject to} \quad B \in Q(y^\#, s) \cap \mathscr{A},$$

if such a B exists. In fact, Theorem 11.3.1 considers the more general case when $Q(y^\#, s) \cap \mathscr{A}$ may be empty. In this case, A_+ is chosen to be the matrix in $\mathscr{A}$ that is nearest to A among the set of nearest matrices of $\mathscr{A}$ to $Q(y^\#, s)$. For simplicity, we state the theorem for the unweighted Frobenius norm, but it remains true for any weighted Frobenius norm. Dennis and Schnabel's proof uses an iterated projection technique that is the straightforward generalization of the Powell symmetrization procedure of Section 9.1. The reader can simply look back at Figure 9.1.1, putting $\mathscr{A}$ in place of S and $y^\#$ in place of y_c.

We will need the notion that every affine space $\mathscr{A}$ has a unique translate $\mathscr{S}$, which is a subspace. For any $A_0 \in \mathscr{A}$, $\mathscr{S} = \{A - A_0 : A \in \mathscr{A}\}$ is independent of A_0.

THEOREM 11.3.1   [Dennis and Schnabel (1979)]   Let $s \ne 0$ and let $P_{\mathscr{A}}$ and $P_{\mathscr{S}}$ be the orthogonal projectors into $\mathscr{A}$ and $\mathscr{S}$. Let P be the $n \times n$ matrix whose jth column is

$$\left[P_{\mathscr{S}}\left(\frac{e_j s^T}{s^T s} \right) \right] s$$

and let $A \in \mathscr{A}$. If v is any solution to

$$\min_{v \in \mathbb{R}^n} \| Pv - (y^\# - As) \|_2 \qquad (11.3.8)$$

or equivalently to

$$\min_{v \in \mathbb{R}^n} \| P_{\mathscr{S}}\left(\frac{vs^T}{s^T s} \right)s - (y^\# - As) \|_2 ,$$

then

$$A_+ = A + P_{\mathscr{S}}\left(\frac{vs^T}{s^T s} \right) \qquad (11.3.9)$$

is the nearest to A of all the nearest matrices of $\mathscr{A}$ to $Q(y^\#, s)$. If the minimum is zero, then $A_+ \in \mathscr{A} \cap Q(y^\#, s)$.

Combining (11.3.8) and (11.3.9), we can express our least-change update as

$$A_+ = A + P_{\mathscr{S}}\left(\frac{P^+(y^\# - As)s^T}{s^T s} \right), \qquad (11.3.10)$$

where P^+ denotes the generalized inverse of P (see Section 3.6). Now we will use (11.3.10) to derive the sparse secant update and the symmetric sparse secant update that we discussed in Section 11.2.

To derive the sparse secant update using Theorem 11.3.1, we choose

$y^{\#} = y$, $\mathscr{A} = \mathscr{S} = \mathrm{SP}(Z)$ and $P_{\mathscr{A}} = P_{\mathscr{S}} = P_Z$ given by (11.2.5). Then

$$P_{\cdot j} = \left[P_Z\!\left(\frac{e_j s^T}{s^T s} \right) \right] s = \frac{s_i^T s}{s^T s}\, e_j,$$

so $P = (1/s^T s)\, \mathrm{diag}\,(s_j^T s_j)$ and (11.3.10) gives

$$A_+ = A + P_Z(\mathrm{diag}\,(s_j^T s_j))^+ (y - As)s^T$$

which is equivalent to (11.2.7).

To derive the symmetric sparse update, we choose $y^{\#} = y$, $\mathscr{A} = \mathscr{S} = \mathrm{SP}(Z) \cap \{A \in \mathbb{R}^{n \times n}: A = A^T\}$. It is an easy exercise to show that for any $M \in \mathbb{R}^{n \times n}$, $P_{\mathscr{A}}(M) = P_{\mathscr{S}}(M) = \frac{1}{2} P_Z(M + M^T)$. The reader can then show that in this instance, the matrix P in Theorem 11.3.1 is given by

$$P = \frac{1}{2s^T s}\, [\mathrm{diag}\,(s_j^T s_j) + P_Z(ss^T)], \tag{11.3.11}$$

so that (11.3.10) gives

$$A_+ = A + P_Z(vs^T + sv^T),$$
$$v = P^+(y - As). \tag{11.3.12}$$

It is a further exercise to show that (11.3.12) reduces to the symmetric secant update (9.1.3) when $\mathrm{SP}(Z) = \mathbb{R}^{n \times n}$ and $P_Z = I$. Notice, however, that in general the computation of update (11.3.12) requires v, which in turn requires the solution of a new sparse linear system at each iteration.

Finally, let us give the updates of Theorem 11.3.1 for the three examples of this section. It is an easy exercise to see that for the first example (11.3.2), the update is

$$A_+ = A + \begin{pmatrix} 0 & 0 \\ \dfrac{e_2^T(y^{\#} - As)}{s^T s}\, s_1 & \dfrac{e_2^T(y^{\#} - As)}{s^T s}\, s_2 \end{pmatrix} \tag{11.3.13}$$

and for the second (11.3.12), if $\gamma \triangleq [(s_2)^2 + (s_3)^2]^+$,

$$A_+ = A + \begin{pmatrix} 0 & \gamma e_1^T(y^{\#} - As)s_2 & \gamma e_1^T(y^{\#} - As)s_3 \\ 0 & \gamma e_2^T(y^{\#} - As)s_2 & \gamma e_2^T(y^{\#} - As)s_3 \\ 0 & \gamma e_3^T(y^{\#} - As)s_2 & \gamma e_3^T(y^{\#} - As)s_3 \end{pmatrix}. \tag{11.3.14}$$

We have already indicated in Chapter 10 that for function (11.3.4), the DFP analog is update (10.3.6), with $y^{\#}$ now given by (11.3.7) (see also Exercise 16 of Chapter 10).

11.4   ANALYZING LEAST-CHANGE SECANT METHODS

This section, based on Dennis and Walker (1981), presents two theorems giving necessary conditions on $y^{\#}$ and $\mathscr{A}$ for the local convergence of any secant method that is derived using Theorem 11.3.1. These theorems show that every least-change secant method we have presented in this book is locally and q-superlinearly convergent. They also indicate how secant updates for other problems with special structure should be constructed.

Let us assume below that $J(x)$ and $C(x)$ both satisfy Lipschitz conditions and that $J(x_*)$ is invertible.

It is useful for our purposes to distinguish between two kinds of least-change secant updates: those, like Broyden's method, which use the same unscaled or fixed-scale Frobenius norm at every iteration in calculating the least-change update, and those, like the DFP, which can be thought of as rescaling the problem anew at every iteration before calculating the update. We will state a theorem for each of these two types of methods, but for reasons of brevity we won't discuss methods like the BFGS, which are usually viewed in the least-change sense as projections of the inverse Hessian approximation. We will also restrict ourselves here to Frobenius and weighted Frobenius norms, although the Dennis-Walker paper is more general.

The theorems can be interpreted as saying roughly that least-change secant update methods are convergent if the space of approximants $\mathscr{A}$ is chosen reasonably and if the secant condition characterized by $y^\#$ is chosen properly and that they are superlinearly convergent if in addition $\mathscr{A}$ is chosen properly. Thus, the choice of secant condition is crucial to the convergence of the method. This fact has been somewhat obscured by the ease with which $y^\#$ can be chosen for the traditional problems or even for the more general problems of Section 11.3. The interaction of $\mathscr{A}$ and $y^\#$ in these cases is expressed by the always allowed choice

$$y^\# = P_{\mathscr{A}} \left[\left(\int_{t=0}^{1} J(x + ts)\, dt \right) - C(x_+) \right] s, \tag{11.4.1}$$

which reduces to the default choice

$$y^\# = y - C(x_+)s$$

if $A(x) \in \mathscr{A}$ for every x, as it always was in Section 11.3.

Theorem 11.4.1 is going to be stated with a rather minimal assumption on $\mathscr{A}$. It amounts to assuming that there is some $A_* \in \mathscr{A}$ for which the stationary quasi-Newton iteration

$$x_+ = x_c - (C(x_*) + A_*)^{-1} F(x_c)$$

would be locally q-linearly convergent. Surely this is little enough to require of $\mathscr{A}$, and if $A(x_*) \in \mathscr{A}$, then $A_* \triangleq A(x_*)$ will certainly satisfy this condition. The requirement on $y^\#$ is also pretty minimal; it should be thought of as requiring that the Frobenius distance from either $A(x_*)$ or $A(x_+)$ to $Q(y^\#, s)$ decreases like the maximum of $\| x_+ - x_* \|^p$ and $\| x_c - x_* \|^p$, for some $p > 0$. In other words, the size of the update to the model should get smaller as the solution is approached. As in Section 11.3, the theory includes the case when $\mathscr{A} \cap Q(y^{\#}, s)$ is empty. This is an important case for some applications that we comment on briefly at the end of this section.

We note that $\| \cdot \|$ is used to denote an arbitrary vector and subordinate matrix norm while $\| \cdot \|_F$ is still used to denote the Frobenius norm.

THEOREM 11.4.1 Let the hypotheses of Theorem 5.2.1 hold. Let $\mathscr{A}$, $\mathscr{S}$, $Q(y, s)$, and $P_{\mathscr{A}}$ be defined as in Theorem 11.3.1, let $P^{\perp}_{\mathscr{S} \cap Q(0, s)}$ denote the Euclidean projector orthogonal to $S \cap Q(0, s)$ and let $C: \mathbb{R}^n \longrightarrow \mathbb{R}^{n \times n}$. Let $A_* = P_{\mathscr{A}}[J(x_*) - C(x_*)]$ and $B_* = C(x_*) + A_*$ have the properties that B_* is invertible and that there exists an r_* for which

$$\| B_*^{-1} [J(x_*) - (C(x_*) + A_*)] \| \le r_* < 1. \tag{11.4.2}$$

Assume also that the choice rule $y^{\#}(s)$, $s = x_+ - x$, either is the default choice (11.4.1) or more generally has the property with $\mathscr{A}$ that there exists an $\alpha > 0$, $p > 0$ such that for any x, x_+ in a neighborhood of x_*, one has

$$\| P^{\perp}_{\mathscr{S} \cap Q(0, s)}(G - A_*) \| \le \alpha \sigma(x, x_+)^p \tag{11.4.3}$$

for at least one G that is a nearest point in $\mathscr{A}$ to $Q(y^{\#}(s), s)$, where

$$\sigma(x, x_+) = \max \{\| x - x_* \|, \| x_+ - x_* \|\}.$$

Under these hypotheses, there exist positive numbers ε, δ such that if $\| x_0 - x_* \| < \varepsilon$ and $\| A_0 - A_* \| < \delta$, the sequence defined by

$$x_{k+1} = x_k + (C(x_k) + A_k)^{-1} F(x_k)$$

$$A_{k+1} = (A_k)_+ \in \mathscr{A},$$

given by Theorem 11.3.1 exists and converges to x_* at least q-linearly with

$$\overline{\lim_{k \to \infty}} \frac{\| x_{k+1} - x_* \|}{\| x_k - x_* \|}$$

$$= \overline{\lim_{k \to \infty}} \left\| B_*^{-1} [J(x_*) - (C(x_*) + A_*)] \frac{x_k - x_*}{\| x_k - x_* \|} \right\| \le r_*.$$

Hence $\{x_k\}$ converges q-superlinearly to x_* if and only if

$$\lim_{k \to \infty} [J(x_*) - (C(x_*) + A_*)] \frac{x_k - x_*}{\| x_k - x_* \|} = 0.$$

In particular, $\{x_k\}$ converges q-superlinearly to x_* if

$$J(x_*) - C(x_*) = A(x_*) \in \mathscr{A}.$$

The proof of Theorem 11.4.1 is difficult, as the reader might suspect, since it implies the local q-superlinear convergence of every fixed-scale secant method we have discussed in this book. For example, for the sparse secant method in Section 11.2, $\mathscr{A} = \text{SP}(Z)$ and $C(x) \equiv 0$, so we have trivially that $A(x_*) = J(x_*) - C(x_*) = J(x_*) \in \mathscr{A}$, and (11.4.2) is true with $r_* = 0$. Furthermore $y^\# = y = F(x_+) - F(x_c)$ is the default choice (11.4.1). Thus, it is a trivial corollary of Theorem 11.4.1 that the sparse secant method is locally convergent, and since $A(x_*) \in \mathscr{A}$, is q-superlinearly convergent to x_* under standard assumptions. Virtually the same words suffice to prove the local q-superlinear convergence of the sparse symmetric secant method from Theorem 11.4.1. Of course, the convergence of the nonsparse secant and symmetric secant methods of Chapters 8 and 9 are just the special cases where the sparsity pattern has no zero elements.

All the standard methods of Chapters 8 and 9 use the default choice, and so the more general and mysterious condition (11.4.3) is not needed. On the other hand, it is needed in the fixed-scale or rescaled least-squares methods with $y^\# = J(x_+)^T F(x_+) - J(x)^T F(x_+)$. Dennis and Walker elaborate on this question, but we mention that for $\mathscr{S} = \{M: M = M^T\}$ it suffices to have

$$\| y^\# - A_* s \| \le \alpha \sigma(x, x_+)^p \| s \|.$$

For general $\mathscr{S}$, it suffices to have

$$y^\# - A_* s = Es \quad \text{for some } E \in \mathscr{S} \text{ with } \| P^\perp_{\mathscr{S} \cap Q(0, s)} E \| \le \alpha \sigma(x, x_+)^p.$$

(11.4.4)

In fact, we will see an example later for which one can more easily show the stronger result $\| E \| \le \alpha \sigma(x, x_+)^p$.

The statement of Theorem 11.4.2 on iteratively rescaled least-change secant methods, differs from the statement of Theorem 11.4.1 on fixed-scale methods in that the roles of y and $y^\#$ as functions of $s = x_+ - x$ must be clarified. The distinction is simple: the scaling matrices are from $Q(y, s)$ and the secant matrices are in $Q(y^\#, s)$. In short, y is trying to look like $J(x_*)s$ and $y^\#$ is trying to look like $A(x_*)s$.

For an affine space $\mathscr{A}$ and symmetric positive definite matrix W, it will be convenient to have the notation $P_{\mathscr{A}, W}$ to denote the orthogonal projection into $\mathscr{A}$ in the inner product norm $\| M \|_W = [\text{trace } (W^{-1} M W^{-1} M^T)]^{1/2}$ on $\mathbb{R}^{n \times n}$. If $W = JJ^T$ and M is symmetric, then it is useful to think of $\| M \|_W = \| J^{-1} M J^{-T} \|_F$.

THEOREM 11.4.2   Let the hypotheses of Theorem 5.2.1 hold. Let $\mathscr{A}$, $\mathscr{S}$, and $Q(y, s)$ be defined as in Theorem 11.3.1, let $P^\perp_{\mathscr{S} \cap Q(0, s), W}$ denote the projector orthogonal to $\mathscr{S} \cap Q(0, s)$ in the inner-product norm $\| M \|_W$, and let $C: \mathbb{R}^n \longrightarrow \mathbb{R}^{n \times n}$. Let $J(x_*)$ be symmetric and positive definite and let $\mathscr{S}$ have the property that for any s, y with $s^T y > 0$, the projection

$P_{\mathscr{S}, W}$ is independent of any particular symmetric positive definite $W \in Q(y, s)$. In addition, let $\mathscr{A}$ have the properties that

$$A_* = P_{\mathscr{A}, J(x_*)}[J(x_*) - C(x_*)] \quad \text{and} \quad B_* = A_* + C(x_*)$$

are such that B_* is invertible and there exists an r_* with

$$\| B_*^{-1}[J(x_*) - (C(x_*) + A_*)] \| \leq r_* < 1.$$

Also assume that there exist $\alpha_1, \alpha_2, \geq 0$, $p > 0$, such that the choice rule $y(s)$ obeys

$$\| y - J(x_*)s \| \leq \alpha_1 \sigma(x, x_+)^p \| s \|$$

and the choice rule $y^{\#}(s)$ either is the default choice

$$y^{\#} = P_{\mathscr{A}, W}\left[\left(\int_{t=0}^{1} J(x + ts)\, dt\right) - C(x_+)\right]s$$

for $W = J(x_*)$ or some $W \in Q(y, s)$, or has the property with $\mathscr{A}$ that for any x, x_+ in a neighborhood of x_*,

$$\| P_{\mathscr{S} \cap Q(0, s), W}^{\perp}(G - A_*) \|_W \leq \alpha_2 \sigma(x, x_+)^p \tag{11.4.5}$$

for every symmetric positive definite $W \in Q(y, s)$ and at least one G that in the W-norm is a nearest point in $\mathscr{A}$ to $Q(y^{\#}, s)$. Under these hypotheses, there exist positive constants ε, δ such that if $\| x_0 - x_* \| < \varepsilon$ and $\| A_0 - A_* \| < \delta$, the sequence defined by

$$x_{k+1} = x_k - (C(x_k) + A_k)^{-1}J(x_k),$$

$$A_{k+1} = (A_k)_+ \in \mathscr{A},$$

A_{k+1} the least-change secant update with respect to $\mathscr{A}$ and $Q(y_k^{\#}, s_k)$ in a W_k-norm, $W_k \in Q(y_k, s_k)$, exists and converges at least q-linearly to x_*. Furthermore,

$$\overline{\lim_{k \to \infty}} \frac{\| x_{k+1} - x_* \|}{\| x_k - x_* \|} = \overline{\lim_{k \to \infty}} \left\| B_*^{-1}[J(x_*) - (C(x_*) + A_*)] \frac{x_k - x_*}{\| x_k - x_* \|} \right\|$$

$$\leq r_* < 1.$$

Hence, $\{x_k\}$ converges q-superlinearly to x_* if and only if

$$\lim_{k \to \infty} [J(x_*) - (C(x_*) + A_*)] \frac{x_k - x_*}{\| x_k - x_* \|} = 0.$$

In particular, $\{x_k\}$ converges q-superlinearly to x_* if $(J(x_*) - C(x_*)) \in \mathscr{A}$.

The discussion of condition (11.4.3) following Theorem 11.4.1 applies to condition (11.4.5) as well. Using this, it is an exercise to apply Theorem 11.4.2 to prove the local q-superlinear convergence of the secant method of Dennis,

Gay, and Welsch given in Section 10.3 for solving the nonlinear least-squares problem. Of course, Theorem 11.4.2 proves the convergence of the DFP method as well. In general, Theorems 11.4.1–11.4.2 say that if you use the right kind of approximating space $\mathscr{A}$ and know how to choose the correct secant condition for that class of approximants, then the least-change principle is an excellent way to choose the secant update.

Finally, some interesting possibilities open up when we drop the requirement on $\mathscr{A}$ that $A(x) \in \mathscr{A}$ for every $x \in \mathbb{R}^n$. This may occur when we decide to approximate $A(x)$ by a sequence of matrices that have a simpler structure than $A(x)$. For example, in a secant version of the nonlinear Jacobi algorithm, the iteration is

$$x_{k+1} = x_k - A_k^{-1} F(x_k),$$

where each A_k is diagonal even though $J(x)$ is not diagonal; that is, $\mathscr{A} = \{\text{diagonal matrices} \in \mathbb{R}^{n \times n}\}$ but $J(x) \notin \mathscr{A}$. If $F(x_*) = 0$ and $\| I - (\text{diag } (J(x_*))^{-1})J(x_*) \| < 1$, then Theorem 11.4.1 says that given a proper choice rule for $y^\#$, the method will still be locally convergent. Theorem 11.4.1 also suggests that we choose $y_i^\#$ so that it approximates

$$((\text{diag } (J(x_+)))s)_i = \frac{\partial f_i(x_+)}{\partial x_i} s_i \cong f_i(x_+ + s_i e_i) - f_i(x_+) \triangleq y_i^\#$$

if $s_i \neq 0$, and $y_i^\# = 0$ if $s_i = 0$. It is an easy exercise to show that (11.4.4) with

$$E = \begin{cases} \dfrac{y_i^\#}{s_i} & \text{if } s_i \neq 0, \\ 0, & \text{if } s_i = 0, \end{cases}$$

holds for this choice of y_i, and so this method is q-linearly convergent under the proper assumptions.

The important thing to notice in the above example is that the choice of $y^\#$ is not the same as the default choice (11.3.5), which would yield $y^\# = y = F(x_+) - F(x)$, since $C(x_*) = 0$. In general, the default rule (11.3.5) is unsatisfactory when $A(x_*) \notin \mathscr{A}$, because it causes $y_k^\#$ to approximate $A(x_*)s_k$ when we want it to approximate $(P_{\mathscr{A}}(A(x_*)))s_k = A_* s_k$. There are other practical examples where the subspace of approximants $\mathscr{A}$ has a simpler structure than $\{A(x)\}$, and in many of these selecting an appropriate choice rule for $y^\#$ is particularly difficult (see, for example, Exercise 19). If we could find a simple idea like the default condition (11.3.5) to choose the $y^\#$'s properly, then the Dennis-Walker theorems would tell us that the methods would be locally convergent and that the speed of convergence would depend on how near $A(x_*)$ is to $\mathscr{A}$. This is an area of current research. [See Dennis-Walker (1983).]

11.5   EXERCISES

1. Let $F: \mathbb{R}^n \longrightarrow \mathbb{R}^n$, and let $J(x) = F'(x)$ have the property that $J(x)_{ij} = 0$ if $|i - j| > m$, where $m < n/2$. Using the techniques of Section 11.1, how many evaluations of $F(x)$ are required to calculate a finite-difference approximation to $J(x)$?

2. [This exercise is taken from Powell and Toint (1979).] Let $J(x) \in \mathbb{R}^{5 \times 5}$ have the nonzero structure

$$J(x) = \begin{bmatrix} x & & & & x \\ & x & & & x \\ & & x & & x \\ & & & x & x \\ x & x & x & x & x \end{bmatrix}.$$

Show that the techniques of Section 11.1 require the full five evaluations of $F(x)$ in addition to $F(x_c)$ to approximate $J(x)$. However, if $J(x)$ is symmetric for all x, show that $J(x)$ can be approximated using only two additional evaluations of $F(x)$.

3. Let SP(Z) be defined by (11.2.3). Show that the problem

$$\min_{B \in R^{n \times n}} \| B - A \|_F \quad \text{subject to} \quad B \in \text{SP}(Z)$$

is solved by $B = P_Z(A)$, P_Z defined by (11.2.5).

4. Using the Cauchy-Schwarz inequality, show that the solution to (11.2.8) is (11.2.9).

5. Let $F: \mathbb{R}^n \longrightarrow \mathbb{R}^n$ be continuously differentiable in an open convex region $D \subset \mathbb{R}^n$, $x, x_+ \in D$, $s = x_+ - x$, $y = F(x_+) - F(x)$, s_i defined by (11.2.6). Prove that if $s_i = 0$, then $y_i = 0$. [*Hint*: Use equation (4.1.8).]

6. Show by example that the inverse of a sparse matrix may have no zero elements. For example, consider $A \in \mathbb{R}^{n \times n}$ defined by

$$A_{ij} = \begin{cases} 4, & i = j, \\ 1, & |i - j| = 1, \\ 0, & |i - j| > 1. \end{cases}$$

(This matrix arises in calculating cubic splines).

7. Let $A \in \mathbb{R}^{n \times n}$ be tridiagonal, symmetric, and positive definite. Construct an algorithm and data structure to solve the linear system $Ax = b$ using $O(n)$ arithmetic operations and $O(n)$ storage. Why is it important that A is positive definite?

8. Calculate $\nabla \rho(R(x))$ and $\nabla^2 \rho(R(x))$ for the data-fitting functions (10.4.4) and (10.4.5). Use these to calculate $\nabla f(x)$ and $\nabla^2 f(x)$ in each case.

9. Show that the values of $y^\#$ obtained in Section 11.3 for functions (11.3.2) and (11.3.3) are equivalent to the default choice (11.3.5).

10. Let $f(x)$ be the nonlinear least-squares function, (11.3.4) with $\rho(R(x)) = \frac{1}{2} R(x)^T R(x)$. Show by example that the default choice of $y^\#$, (11.3.6), and the choice used in Section 10.3, (11.3.7), may differ.

11. Work through the proof of Theorem 11.3.1 using Dennis and Schnabel (1979). Alternatively, see Powell (1981) or Griewank (1982).

12. Let $\mathscr{A} = \text{SP}(Z) \cap \{A \in \mathbb{R}^{n \times n} : A = A^T\}$. Show that $P_{\mathscr{A}}(M) = \frac{1}{2} P_Z(M + M^T)$.

13. Using Theorem 11.3.1 and Exercise 12, complete the derivation of the least-change sparse symmetric secant update. Also, show that P given by (11.3.11) is positive definite if $s_i \neq 0$ for all i. [See Dennis and Schnabel (1979).]

14. Show that the sparse symmetric secant update (11.3.12) reduces to the symmetric secant update (9.1.3) when $\text{SP}(Z) = \mathbb{R}^{n \times n}$ so that $P_Z = I$.

15. Using Theorem 11.3.1, derive the least-change updates (11.3.13) and (11.3.14) for functions (11.3.2) and (11.3.3), respectively.

16. Read Dennis and Walker (1981), and then try to give proofs for Theorems 11.4.1–11.4.2. (These theorems are slight simplifications of the theorems in the paper.)

17. Use Theorem 11.4.2 to prove the local q-superlinear convergence of the secant method of Dennis, Gay, and Welsch for nonlinear least squares,

$$x_{k+1} = x_k + (J(x_k)^T J(x_k) + A_k)^{-1} J(x_k)^T R(x_k),$$

where A_k is updated by (10.3.6), $y_k^\#$ is given by (10.3.3).

18. Show that the nonlinear Jacobi secant method described in Section 11.4 is q-linearly convergent. [*Hint*: Show $\| E \| \leq \alpha \sigma(x, x_+)^p$.]

19. In an algorithm by Dennis and Marwil (1982), a sparse $J(x_k)$ is approximated for a finite number of iterations by $L_0 U_k$, where $J(x_0) = L_0 \cdot U_0$ and $U_k \in \{$matrices with the sparsity pattern of $U_0\}$. Then when the algorithm decides that the Jacobian has become too poor, $J(x_{k+1}) = L(x_{k+1})U(x_{k+1})$ is recalculated using finite differences, and the process is restarted. What does Theorem 11.4.1 suggest that $y^\#$ should be for this algorithm? How could you implement this choice? What criterion might you use to decide when to restart? Discuss the incorporation of pivoting.

20. In an algorithm by Johnson and Austria (1983), the LU factorization is updated to satisfy $LUs = y$ by updating U and L^{-1} to satisfy $Us - L^{-1}y = 0$ a row at a time. Remember that L^{-1} is also lower triangular and see if you can derive from this hint the clever Johnson and Austria update before you read their paper.

Appendix A

A Modular System of Algorithms
for Unconstrained Minimization
and Nonlinear Equations *

CONTENTS

* by Robert B. Schnabel

PREFACE

The appendix to this book has several diverse purposes that are discussed in Section I.1 of Appendix A. The following comments concern how the appendix may be used to create class projects. They should be read after reading sections I.1 and I.2 of Appendix A.

I have used the pseudo-code in appendix A, in various forms, to create class projects for my graduate optimization class each year for the past five years. The students, usually working in groups of two, select a specific method for unconstrained optimization or nonlinear equations, and code a complete routine using the pseudo-code in Appendix A. Usually each group codes only one global strategy (line search, hookstep, or dogleg) but several derivative evaluation options (e.g. finite difference and secant approximation). They create and debug their code as the semester progresses, and then run it on several test problems, such as the problems in appendix B. I have found these projects to be the most valuable part of a course based on this book. Even though the pseudo-code is completely detailed, the students must understand it thoroughly to implement it successfully and to remove the inevitable coding errors. Watching the completed algorithms perform on test problems also is very instructive. This is especially true if the students print out details of the step selection process, for example the values attempted for the line search parameter at each iteration, and answer questions based on this data. ("Is the cubic backtrack ever used? How often are the default backtracks taken?) It also is instructive to compare finite difference and secant methods on the same problems.

The project can be pursued throughout the semester. For example, for unconstrained minimization, the groups can code the machineps and Cholesky backsolve routines towards the start of the semester, and the finite difference derivative and model Hessian / perturbed Cholesky decomposition routines after they are covered in Chapter 5. After Chapter 6 is studied, they can select a global method, code it and the driver program. The addition of the stopping routines from Chapter 7 gives a complete finite difference routine which the students can test. It can then be changed to a BFGS method after secant updates are covered in Chapters 8-9. Similar remarks apply to a nonlinear equations project. Two simplifications may be useful for class projects: omitting the scaling matrices (assuming that $D_x = D_F = I$ in all cases), and using the simpler to understand but less efficient matrix storage scheme that is the default in the pseudo-code (see Guidelines 3 and 6). Omitting scaling has virtually no effect on test problems since almost all of the standard ones are well scaled. It is easiest to use the default tolerances given in Guidelines 2 and 5.

My use of this pseudo-code for class projects over the years led to several changes. Originally, I used strictly PASCAL control structures, with BEGINs and ENDs. Students unfamiliar with block structured languages had difficulty following the structure, so I switched to the hybrid PASCAL / numbering scheme found in the present pseudo-code. The switch met with indifference from the block-structured students and was a big help to students familiar only with FORTRAN. A second change was to expand fully in the pseudo-code products involving matrices. Originally the pseudo-code contained statements like $t \leftarrow H * s$ where H is symmetric and only its upper triangle is stored, but many errors occurred due to the triangular matrix storage. My students find the pseudo-code in its present form easy to use, whether they are coding in FORTRAN or in a block-structured language. Many of them have more difficulty debugging and testing a software system of this size; the comments at the start of Section 7.3 in the text are aimed at this problem.

Unfortunately, it is inevitable that the pseudo-code will contain errors. A few of the algorithms have not been used by my students and may contain algorithmic errors. The rest of the algorithms may contain typographic errors. I have tried to reduce the number of errors by preparing the appendix myself, using the UNIX text-editing facilities on the VAX 11/780 system of the Department of Computer Science at the University of Colorado. I will maintain a computer-generated list of errata, and distribute it to anyone who requests it by sending a stamped, self-addressed envelope. Any instructor using this pseudo-code for a class project is encouraged to send for the errata.

Another possibility for class projects is to obtain a production code for nonlinear equations or unconstrained minimization, and have the students run it on test problems and perhaps alter portions of it if the source is available. The code of Schnabel, Weiss and Koontz that corresponds closely to the pseudo-code for unconstrained minimization in this appendix may be used for this purpose. It is available through the author.

Robert B. Schnabel

SECTION I.   DESCRIPTION

I.1.   PURPOSE OF THE MODULAR SYSTEM OF ALGORITHMS

Appendix A contains a *modular system* of *fully detailed algorithms* for solving both unconstrained minimization problems and systems of nonlinear equations. By *modular*, we mean that each distinct functional component of these methods is given in one or more separate modules. For example, the stopping check, derivative calculation, matrix factorization, and step selection strategy ("global method") each correspond to one or more separate modules. The word *system* refers to the fact that several alternative modules are provided for several parts of the minimization or nonlinear equations process, so that a large number of different methods may be obtained by using different combinations of these modules. In addition, the methods for unconstrained minimization and nonlinear equations share several components. For example, there are three alternative global methods (line search, dogleg, and hookstep) that can be used interchangeably along with any of the other parts of the algorithms for either problem. There are also several alternatives for calculation or approximation of derivatives (analytic, finite differences, secant approximation). Finally, by *fully detailed algorithms* we mean that each module is completely specified, including storage and any tolerances or heuristic decisions that usually are omitted in high level descriptions.

There are five main reasons why we included these algorithms in this book:

1) *To provide a completely detailed description of any method for unconstrained minimization or nonlinear equations discussed in the text.* In the text, we present algorithms to the level of detail we consider appropriate for a thorough introduction to the subject. Readers who desire the remaining details can find them in this appendix.

2) *To show that the modular framework for quasi-Newton algorithms that we refer to throughout the text (Alg. 6.1.1) can be carried to the implementation level.* This substantiates a basic premise of the text, that all the approaches we discuss for both problems are variations on one framework.

3) *To provide a mechanism for class projects.* We have found that implementing one or more particular methods from this system is an excellent way to reinforce the material presented in the text. Class projects are discussed further in the preface to the appendix.

4) *To encourage the use of structured and controlled software environments in testing new methods for optimization and related problems.* If production optimization codes were written in a modular manner such as we describe, then many new methods could be developed and tested by changing one or a few modules. This would provide an easy and controlled testing environment. The code of Schnabel, Weiss and Koontz mentioned below has been used in this manner.

5) *To aid practitioners who need to develop their own codes.* We emphasize that one should use codes from numerical software libraries

whenever possible. However, the special nature of a class of applied problems sometimes necessitates the development of specialized codes, either from scratch or adapted from existing routines. The algorithms in this appendix may serve as a refence in such cases.

A production code for unconstrained minimization that corresponds closely but not exactly to the algorithms in this appendix has been developed by Schnabel, Weiss and Koontz [1982]. The purpose of this appendix was not to create this code, because several software libraries contain good codes for unconstrained minimization, including the Harwell, IMSL, MINPACK and NAG libraries. The code was developed to test the algorithms in this appendix and to provide the research environment mentioned above. It is generally available through the author and may be used for class projects or to supplement this text.

The main reason for providing algorithms in the pseudo-code form that follows instead of an actual coded version (e.g. in FORTRAN) is to make them easier to read. The pseudo-code algorithms differ from algorithms in a computer language in that mathematical notation has been retained, for example Greek letters (λ), summation notation ($\sum_{i=j+1}^{n} H[i,j]$), inner products of vectors ($v^T w$) and norms ($\|x\|_2$). The pseudo-code we use was developed so that it translates readily either into FORTRAN or into block structured languages like PASCAL or ALGOL, and so that it can be implemented in most computing environments. It is related to the pseudo-code used by Vandergraft [1978].

The remainder of Section I as well as Sections II and III of this appendix primarily will be of interest to readers who wish to implement algorithms contained herein. Section I.2 explains how the parts of this appendix can be used together to create one, several, or the entire system of algorithms for solving unconstrained minimization or nonlinear equations problems. This includes describing the organization of Sections II and III, which contain more detailed instructions for using this appendix to create unconstrained minimization or nonlinear equations algorithms, respectively. Section I.3 discusses the conversion of the individual modules of this system into computer code, including the effects of computer environment. Sections I.4 and I.5 briefly discuss two more specific considerations in coding these algorithms.

Section IV contains the fully detailed individual modules for the various parts of our methods. It should be of interest to all readers of this book, as an accompaniment to the discussion of the methods in the body of the text. The modules are numbered to correspond to the text: the first two digits of the three digit algorithm number refer to the section in the text where the method is covered. (For example, Algorithms A7.2.1 through A7.2.4 are stopping algorithms discussed in Section 7.2 of the text.) The modules are also given short descriptive names (e.g. Alg. A7.2.1 is UMSTOP) that are used within the pseudo-code for increased readability.

I.2.   ORGANIZATION OF THE MODULAR SYSTEM OF ALGORITHMS

The organization of this system of algorithms for unconstrained minimization and nonlinear equations follows the quasi-Newton framework presented in Section 6.1 of the text. The methods for both problems are divided into a number of separate functional components (e.g. stopping check, derivative calculation, global method), and each component is implemented by one or more modules. In some cases (derivative calculation or approximation, and global method), a number of alternatives are given for one component, and any one may be selected. A complete method for unconstrained minimization or nonlinear equations is created by selecting and coding algorithms for each of the separate components, and joining them together with a driver program.

In the text, Alg. 6.1.1 served as an indication of the required components as well as the driver program, but here more detail is required. The Description sections of Algorithms D6.1.1 and D6.1.3 (in Sections II.1 and III.1 below) list the steps of our methods for unconstrained minimization and nonlinear equations, respectively. Some of these steps are simple lines of code in the driver algorithm, while others are the components mentioned above that correspond to one or more individual modules. Sections II.2 and III.2 list these separate components, and the modules that are used to implement them, including alternatives if there are any.

The Algorithm sections of Algorithms D6.1.1 and D6.1.3 contain the pseudo-code for the driver algorithms one would use to implement the *entire system* of algorithms for unconstrained minimization or nonlinear equations, with all the alternatives for global method and derivative calculation. The driver for the code of Schnabel, Weiss and Koontz [1982] is similar to this. To implement one (or a few) specific method rather than the whole system, a pared down driver suffices. Algorithms D6.1.2 and D6.1.4 are examples of drivers that implement one specific method for unconstrained minimization and nonlinear equations, respectively. Drivers for other specific methods can be created similarly by selecting the appropriate parts from Algorithms D6.1.1 or D6.1.3.

Each main driver algorithm (D6.1.1 and D6.1.3) is supplemented by the list of modules mentioned above, and by three guidelines. Guideline 1 contains suggestions for choosing among the algorithmic options provided for the global method and for derivative calculation, for unconstrained minimization. It is intended as advice to a user of this system and as information to students and other interested readers. Guideline 2 discusses the choice of the stopping tolerances and all other tolerances used in the unconstrained minimization algorithms. In a production implementation, these tolerances might be selected by the user or given values in the code; in a class project, the student will select them based on Guideline 2. Guideline 3 discusses in some detail the matrix storage required by the unconstrained minimization code. We found efficient storage allocation to be the most difficult aspect of these algorithms to specify completely in a pseudo-code description intended mainly for easy readability. Guideline 3 together with the Storage Considerations sections of the individual modules attempts to remedy this situation.

Guidelines 4-6 contain the analogous information to Guidelines 1-3, for nonlinear equations. In cases where the information is repetitious, the nonlinear equations guidelines refer to the corresponding unconstrained minimization guidelines.

I.3.   ORGANIZATION OF THE INDIVIDUAL MODULES

The structure of each module in Section IV, and of the driver modules D6.1.1-4, is :
Purpose
Input Parameters
Input-Output Parameters
Output Parameters
Meaning of Return Codes
Storage Considerations
Additional Considerations
Description
Algorithm

The **Purpose** section gives a one or two sentence summary of the purpose of the module. The three **Parameter** sections list the names, data types, and occasional additional information about all the parameters to the module. When the module is called by other modules in the system, the order of the arguments in the calling statement corresponds to the order of the parameters in these sections. The **Return Codes** section is provided for algorithms like stopping checks that return an integer code to the algorithm that calls them; for all other algorithms this section is omitted.

The **Storage Considerations** sections of the modules, together with Guidelines 3 and 6, specify the matrix and vector storage requirements of these algorithms. Alternative implementations are discussed that achieve more efficient storage while diminishing the readability of the code. Item 1 of the Storage Considerations section of each individual module lists any local variables that are vectors or matrices. All variables in the module not listed in this item or in the parameter lists are scalar local variables (real, integer, or Boolean). The code is constructed so that all communication between routines is via parameters and no global variables (i.e. FORTRAN COMMON) are required. The single exception is two variables that must be passed as global variables between algorithms NEFN and D6.1.3 for nonlinear equations, as described in these algorithms.

The **Additional Considerations** section is included occasionally for an additional comment about the module. The **Description** section is intended to make clear the basic structure of the algorithm that follows. It may oversimplify some details of the algorithm. For very simple algorithms, this section may be omitted.

The **Algorithm** section contains the fully detailed pseudo-code for the module. The format of the pseudo-code is intended to be self-evident to all readers, and is similar to that used by Vandergraft [1978]. It is based on PASCAL control structures described briefly below; they were selected because their function is obvious from their wording. (PASCAL control structures are very similar to those in ALGOL, PL/1 and FORTRAN 77.) Rather than use PASCAL BEGINs and ENDs, a numbering scheme is used that indicates sequence number and nesting level. In our experience, it makes the pseudo-code easier to follow for readers unfamiliar with block structured languages, and has the additional advantage of giving each statement a unique number.

The following six control structures are used. A reader familiar with block structured languages should glance at these only to notice the sample numbering schemes.

i) **IF-THEN** statement
    1. IF (*condition i*) THEN
        1.1 first statement inside IF-THEN
          ...
        1.5 last statement inside IF-THEN (5 is sample length)
    2. next statement
meaning : If *condition i* is true, execute statements 1.1-5 and then proceed to statement 2, otherwise proceed directly to statement 2.

ii) **IF-THEN-ELSE** statement
    4. IF (*condition ii*)
    4T. THEN
        4T.1 first statement in THEN block
          ...
        4T.3 last statement in THEN block
    4E. ELSE
        4E.1 first statement in ELSE block
          ...
        4E.7 last statement in ELSE block
    5. next statement
meaning : If *condition ii* is true, execute statements 4T.1-3, otherwise execute statements 4E.1-7. Then proceed to statement 5.

iii) **Multiple branch** statement
    2. last statement
    3a. IF (*condition iiia*) THEN
        statements 3a.1 through 3a.4
    3b. ELSEIF (*condition iiib*) THEN
        statement 3b.1
    3c. ELSE
        statements 3c.1 through 3c.12
    4. next statement
meaning : If *condition iiia* is true, execute statements 3a.1-4 and proceed to statement 4; else if *condition iiib* is true, execute statement 3b.1 and proceed to statement 4; otherwise execute statements 3c.1-12 and proceed to statement 4.
This control structure with its characteristic labelling is included because it indicates the conceptual structure in several of our algorithms better than an IF-THEN-ELSE statement whose ELSE clause is itself an IF-THEN-ELSE. The number of alternatives can be increased to any number desired; for example, Algorithm A7.2.1 has statements 2a through 2f.

iv) **FOR** loop (equivalent to FORTRAN DO loop)
    6.2 FOR *integer variable* =*integer value*#1 TO *integer value*#2 DO
        statements 6.2.1 through 6.2.9
    6.3 next statement
meaning :
    a. *integer variable* ← *integer value*#1
    b. if *integer variable* > *integer value*#2, go to next statement (6.3), otherwise go to step c.
    c. execute statements 6.2.1-9.
    d. *integer variable* ← (*integer variable* +1)
    e. go to step b.
note that the loop is never executed if *integer value*#1 >

integer value#2; it is executed once if these two values are equal.

a FOR ... DOWNTO statement also is used occasionally; its meaning differs from the above only in that the > is changed to a < in step b, and the + is changed to a - in step d.

v) WHILE loop

7.3.2 WHILE (*condition v*) DO
        statements 7.3.2.1 through 7.3.2.8
7.3.3 next statement

meaning :

a. if *condition v* is true, go to step b, otherwise go to next statement (7.3.3).

b. execute statements 7.3.2.1-9.

c. go to step a.

vi) REPEAT loop

2.3 REPEAT
        statements 2.3.1 through 2.3.6
2.3U UNTIL (*condition vi*)
2.4 next statement

meaning :

a. execute statements 2.3.1-6.

b. if *condition vi* is true go to next statement (2.4), otherwise go to step a.

note that the body of a repeat loop is always executed at least once but the body of a while loop may be executed zero times.

The *condition* in the above statements can either be a simple condition ($s^2 > x - 2$), a compound condition (($i = 2$) AND (($b < 1$) OR ($b > 3$))), or a Boolean variable. Compound conditions use the operators AND, OR, and NOT and follow normal precedence rules. Boolean variables may be assigned the values TRUE and FALSE; if *boo* is a boolean variable then *boo* and NOT *boo* are possible conditions. Boolean variables may be part of compound conditions as well.

Since these algorithms contain no input or output statements (see Section I4) there are only two other types of statements besides the six given above, an assignment statement and a procedure call statement. (One GO TO is used in Alg. D6.1.3., and one RETURN statement is used in each driver algorithm.) An assignment statement is written

variable ← expression

read, "variable gets value of expression". The operator = is reserved for conditions and FOR loops. Procedure calls are written

CALL *modulename* (*list of arguments*).

The module name is the descriptive name (e.g. UMSTOP); the module number (A7.2.1) is always given as an accompanying comment. The arguments correspond, in order, to the Input Parameters, Input-Output Parameters and Output Parameters of the module being called. If the body of one of the control statements consists of only one assignment statement or procedure call, it is usually not given a separate number, for example:

3.2 FOR $i = 1$ TO n DO
        $x[i] ← 1$
3.3 next statement

Elements of vectors and matrices are denoted as in PASCAL: the ith element of a vector x is $x[i]$, the i,j element of a matrix M is $M[i,j]$. Scalar variables may have type integer, real or Boolean, their type is implied by their use in the algorithm. Variable names are sometimes Greek letters when this corresponds to the usage in the text; in a coded version these could be converted to the English equivalent (i.e α becomes *alpha*). Variable names are descriptive and some are longer than allowed by standard FORTRAN. All variable names are italicized.

The precision of real variables is not specified. These methods usually should have at least 10-15 base 10 digits of accuracy available for real numbers. Thus on a CDC, Cray, or similar machine, single precision is sufficient, but on an IBM, DEC or similar machine, real variables should be double precision. The only machine dependent constants in the code are functions of the machine epsilon, *macheps*, which is calculated at the beginning of any method for unconstrained minimization or nonlinear equations.

The main feature that makes the pseudo-code different from and more readable than actual computer language code is the use of mathematical and vector notation. The main instances are summation notation, max and min operators, exponents, assignments of vectors to vectors, vector products, norms of vectors, and multiplication of vectors by diagonal matrices (see Section I5). Coding these operations sometimes will require the introduction of additional variables. Operations involving non-diagonal matrices generally are specified componentwise. The reason is that too much confusion arose when our students coded statements like $t \leftarrow H * s$ where $H \in R^{n \times n}$ is a symmetric matrix and only the upper triangle of H is stored. The matrix form of the statement is always provided in an accompanying comment.

Comments are denoted using PASCAL notation: (* starts the comment and the next *) ,not necessarily on the same line, ends it. The RETURN statements that would be required in FORTRAN, but not in a block structured language, are indicated in comments.

I.4. INPUT AND OUTPUT

This system of algorithms does not contain any input (read) or output (write, print) statements. It communicates to the outside world solely via the parameter list of the main driver program. This is the standard and desirable way to receive input, and there should be no need to insert any input statements into the algorithms. However, most implementors will want to insert output statements.

A production code usually has several levels of printed output available, including no printed output, sometimes desirable when the code is imbedded inside another routine, and a small amount of printed output, say the method used, the function (and gradient) values at the initial and final points, a termination message, and the number of iterations and function and derivative evaluations used. Another option usually is to also print the function (and gradient) values at the end of each iteration. For debugging or a class project, it is useful to have the additional option of printing information about the step selection process, for example, the values of λ attempted at each iteration in a line search method, the successive values of the trust radius in a trust region method, some details of the μ iteration in a hookstep method. The level of printed output can be controlled by adding an input parameter to the main driver, say *printcode*; the output statements in the code are made conditional on the value of *printcode*.

The output parameters of the main drivers also are not completely specified. They must include the termination code, and the final value of the independent variable x. They may include the final function vector (for non-linear equations) or function and gradient values (for minimization) as well. For nonlinear equations, the Jacobian or its approximation is calculated at the final point and is available for output. For minimization, the Hessian or its approximation is not calculated at the final point; a user desiring this information must modify the algorithms slightly.

I.5.   MODULES FOR BASIC ALGEBRAIC OPERATIONS

Several algebraic operations appear frequently enough in the pseudo-code that an implementor of these algorithms may wish to provide special subroutines to perform them. Some possibilities are max and min, vector product $(v^T w)$, l_2 norm of a vector, and operations involving the diagonal scaling matrices D_x and D_F. When coding in FORTRAN, it may be convenient to use the Basic Linear Algebra Subroutines (BLAS) to implement some of these; see Lawson, Hanson, Kincaid and Krogh [1979].

Algorithms A6.4.1-5 and A8.3.1-2 contain a number of expressions of the form $D_x^k v$ or $\|D_x^k v\|_2$, where k is +2 or ±1, and $v \in R^n$. In all cases, a comment in the pseudo-code reminds the reader that the diagonal scaling matrix D_x is not stored as such, rather the vector $S_x \in R^n$ is stored, where $D_x = \text{diag}((S_x)_1,...,(S_x)_n)$. Thus the ith component of $D_x^k v$ is $(S_x[i])^k * v[i]$. If one is implementing scaling, it may be convenient to write a subroutine that, given k, S_x, and v, returns $D_x^k v$, and another that returns $\|D_x^k v\|_2$.

SECTION II

DRIVER MODULES AND GUIDELINES FOR
UNCONSTRAINED MINIMIZATION

II.1   ALGORITHM D6.1.1. (UMDRIVER)

DRIVER FOR A MODULAR SYSTEM OF ALGORITHMS
FOR UNCONSTRAINED MINIMIZATION

Purpose : Attempt to find a local minimizer x_* of $f(x) : R^n \to R$ starting from x_0, using one of a variety of algorithms.

Input Parameters : $n \in Z$ $(n \geq 1)$, $x_0 \in R^n$, FN (the name of a user-supplied function $f : R^n \to R$ that is at least twice continuously differentiable) plus optionally :

   i) GRAD, HESS -- names of user-supplied functions for evaluating $\nabla f(x)$ and $\nabla^2 f(x)$, respectively.

   ii) $global \in Z$, $analgrad \in$ Boolean, $analhess \in$ Boolean, $cheapf \in$ Boolean, $factsec \in$ Boolean -- parameters used in choosing algorithmic options. These are explained in Guideline 1. (Parameters that are not input are given values by algorithm UMINCK).

   iii) $typx \in R^n$, $typf \in R$, $fdigits \in Z$, $gradtol \in R$, $steptol \in R$, $maxstep \in R$, $itnlimit \in Z$, $\delta \in R$ -- tolerances and constants used by the algorithm. These are explained in Guideline 2. (Tolerances or constants that are not input are given values by algorithm UMINCK).

   iv) $printcode \in Z$ -- see Section I.4 of this Appendix.

Input-Output Parameters : none (see however item 2 under Storage Considerations)

Output Parameters : $x_f \in R^n$ (the final approximation to x_*), $termcode \in Z$ (indicates which termination condition caused the algorithm to stop) plus optionally :

   $f_c \in R$ $(= f(x_f))$, $g_c \in R^n$ $(= \nabla f(x_f))$, other output parameters as desired by the implementor (see Section I.4)

Meaning of Return Codes :

   $termcode < 0$: algorithm terminated due to input error, see Alg. UMINCK.

   $termcode > 0$: algorithm terminated for reason given in Alg. A7.2.1.

Storage Considerations :
1) Matrix and vector storage requirements are discussed in Guideline 3. Be sure to read item 5 in Guideline 3 if you are implementing these algorithms in FORTRAN.
2) The names x_0 and x_f are used in this driver for clarity only. In the implementation, x_0 and x_f can both share storage with x_c.

Additional Considerations :
1) The List of Modules for Unconstrained Minimization shows which modules correspond to each step in the Description below. The choice between alternatives in the list is determined by the algorithmic option parameters described in Guideline 1.
2) The Algorithm below is used for implementing the *entire system* of unconstrained minimization algorithms. A simpler driver can be used to implement a *particular* algorithm for unconstrained minimization, as is illustrated by Algorithm D6.1.2.
3) On some computing systems, it will be necessary to supply dummy routines GRAD and HESS, for the case when the user doesn't supply routines for evaluating $\nabla f(x)$ or $\nabla^2 f(x)$, respectively. These dummy routines are required solely to allow compilation of the program.

Description :

Initialization :
1) Calculate machine epsilon.
2a) Check input for reasonableness, assign default values to optional parameters and tolerances that weren't input to user. (Return *termcode* to driver.)
2b) If *termcode* < 0, terminate algorithm, returning $x_f = x_0$ and *termcode*. If *termcode* = 0, set *itncount* $\leftarrow$ 0 and continue.
3) Calculate $f_c = f(x_0)$.
4) Calculate or approximate $g_c = \nabla f(x_0)$.
5a) Decide whether to stop. (Return *termcode* to driver).
5b) If *termcode* > 0, terminate algorithm, returning $x_f = x_0$ and *termcode*. If *termcode* = 0, continue.
6) Calculate or approximate $H_c = \nabla^2 f(x_0)$.
7) Set $x_c = x_0$.

Iteration :
1) *itncount* $\leftarrow$ *itncount* + 1
2) Form positive definite model Hessian H_c ($= H_c + \mu I$, $\mu \geq 0$, $\mu = 0$ if H_c is already safely positive definite) and its Cholesky decomposition $H_c = L_c L_c^T$, L_c lower triangular.
3) Solve $(L_c L_c^T) s_N = -g_c$.
4) Choose $x_+ = x_c + s_c$ where $s_c = s_N$ or a step chosen by the global strategy. $f_+ = f(x_+)$ is calculated during this step.
5) Calculate or approximate $g_+ = \nabla f(x_+)$.
6a) Decide whether to stop. (Return *termcode* to driver).
6b) If *termcode* > 0, terminate algorithm, returning $x_f = x_+$ and *termcode*. If *termcode* = 0, continue.
7) Calculate or approximate $H_c = \nabla^2 f(x_+)$.
8) Set $x_c \leftarrow x_+$, $f_c \leftarrow f_+$, $g_c \leftarrow g_+$ and return to step 1 of the iteration section.

Algorithm :

(* Initialization section : *)
1. CALL MACHINEPS (*macheps*)
2. CALL UMINCK (parameter list selected by implementor)
   (* see UMINCK description *)
3. IF *termcode* < 0 THEN
   3.1 $x_f \leftarrow x_0$
   3.2 RETURN from Algorithm D6.1.1   (* write appropriate message, if desired *)
4. *itncount* ← 0
5. CALL FN (n, x_0, f_c)
6. IF *analgrad*
6T. THEN CALL GRAD (n, x_0, g_c)
6E. ELSE CALL FDGRAD $(n, x_0, f_c, FN, S_x, \eta, g_c)$ (* Alg. A5.6.3 *)
7. CALL UMSTOP0 $(n, x_0, f_c, g_c, S_x, typf, gradtol, termcode, consecmax)$
   (* Alg. A7.2.2 *)
8. IF *termcode* > 0
8T. THEN $x_f \leftarrow x_0$
8E. ELSE (* calculate initial Hessian or approximation *)
   8E.1a IF *analhess* THEN
          CALL HESS (n, x_0, H_c)
   8E.1b ELSEIF *analgrad* AND *cheapf* THEN
          CALL FDHESSG $(n, x_0, g_c, GRAD, S_x, \eta, H_c)$ (* Alg. A5.6.1 *)
   8E.1c ELSEIF *cheapf* THEN
          CALL FDHESSF $(n, x_0, f_c, FN, S_x, \eta, H_c)$ (* Alg. A5.6.2 *)
   8E.1d ELSEIF *factsec* THEN
          CALL INITHESSFAC $(n, f_c, typf, S_x, L_c)$ (* Alg. A9.4.4 *)
          (* note: *factsec* must be false if *global* =2 *)
   8E.1e ELSE (* *factsec* = FALSE *)
          CALL INITHESSUNFAC $(n, f_c, typf, S_x, H_c)$ (* Alg. A9.4.3 *)
9. $x_c \leftarrow x_0$

(* Iteration section : *)
10. WHILE *termcode* = 0 DO
    10.1 *itncount* ← *itncount* + 1
    10.2 IF NOT *factsec* THEN
              CALL MODELHESS $(n, S_x, macheps, H_c, L_c)$ (* Alg. A5.5.1 *)
    10.3 CALL CHOLSOLVE (n, g_c, L_c, s_N) (* Alg.A3.2.3 *)
    10.4a IF *global* = 1 THEN
              CALL LINESEARCH $(n, x_c, f_c, FN, g_c, s_N, S_x, maxstep, steptol,$
              $retcode, x_+, f_+, maxtaken)$ (* Alg. A6.3.1 *)
    10.4b ELSEIF *global* = 2 THEN
              CALL HOOKDRIVER $(n, x_c, f_c, FN, g_c, L_c, H_c, s_N, S_x, maxstep,$

steptol, itncount, macheps, $\delta, \mu, \delta prev,$ $\varphi, \varphi',$ *retcode,* $x_+,$ $f_+,$
maxtaken) (*Alg. A6.4.1 *)

10.4c ELSEIF *global* = 3 THEN

CALL DOGDRIVER $(n, x_c, f_c,$ FN, $g_c, L_c, s_N, S_x,$ *maxstep, steptol,*
$\delta,$ *retcode,* $x_+, f_+,$ *maxtaken*)   (* Alg. A6.4.3 *)

10.4d ELSE (* *global* = 4 *)

CALL LINESEARCHMOD $(n, x_c, f_c,$ FN, $g_c,$ *analgrad,* GRAD, $s_N, S_x,$
maxstep, steptol, retcode, $x_+, f_+, g_+,$ *maxtaken*)   (* Alg.
A6.3.1mod *)

10.5 IF *global* $\neq$ 4 THEN

10.5.1 IF *analgrad*

10.5.1T THEN CALL GRAD (n, x_+, g_+)

10.5.1E ELSE CALL FDGRAD $(n, x_+, f_+,$ FN, $S_x, \eta, g_+)$ (* Alg. A5.6.3
*)

(* if central difference switch is used, perhaps call CDGRAD $(n, x_+,$
FN, $S_x, \eta, g_+)$ instead -- see Alg. A5.6.4 *)

10.6 CALL UMSTOP $(n, x_c, x_+, f_+, g_+, S_x, typf,$ *retcode, gradtol, step-*
tol, itncount, itnlimit, maxtaken, consecmax, termcode)   (* Alg.
A7.2.1 *)

10.7 IF *termcode* > 0

10.7T THEN $x_f \leftarrow x_+$

10.7E ELSE (* calculate next Hessian or approximation *)

10.7E.1a IF *analhess* THEN

CALL HESS (n, x_+, H_c)

10.7E.1b ELSEIF *analgrad* AND *cheapf* THEN

CALL FDHESSG $(n, x_+, g_+,$ GRAD, $S_x, \eta, H_c)$ (* Alg. A5.6.1 *)

10.7E.1c ELSEIF *cheapf* THEN

CALL FDHESSF $(n, x_+, f_+,$ FN, $S_x, \eta, H_c)$ (* Alg. A5.6.2 *)

10.7E.1d ELSEIF *factsec* THEN

CALL BFGSFAC $(n, x_c, x_+, g_c, g_+,$ *macheps,* $\eta,$ *analgrad,* $L_c)$
(* Alg. A9.4.2 *)

10.7E.1e ELSE (* *factsec* = FALSE *)

CALL BFGSUNFAC $(n, x_c, x_+, g_c, g_+,$ *macheps,* $\eta,$ *analgrad,*
$H_c)$ (* Alg. A9.4.1 *)

10.7E.2 $x_c \leftarrow x_+$

10.7E.3 $f_c \leftarrow f_+$

10.7E.4 $g_c \leftarrow g_+$

(* END, WHILE loop 10 *)

(* write appropriate termination message, if desired *)

(* RETURN from Algorithm D6.1.1 *)

(* END, Algorithm D6.1.1 *)

II.2. LIST OF MODULES FOR UNCONSTRAINED MINIMIZATION

The steps in the Description section of Driver D6.1.1 correspond to the following modules. The steps in the Description that are not listed below correspond to lines of code in the driver algorithm.

Initialization
- Step 1 -- MACHINEPS
- Step 2a -- UMINCK
- Step 3 -- FN
- Step 4 -- GRAD or FDGRAD
- Step 5a -- UMSTOP0
- Step 6 -- HESS or (FDHESSG / FDJAC) or FDHESSF or INITHESSUNFAC or INITHESSFAC

Iteration
- Step 2 -- (MODELHESS / CHOLDECOMP)
- Step 3 -- CHOLSOLVE
- Step 4 -- LINESEARCH or LINESEARCHmod or (HOOKDRIVER / HOOKSTEP / TRUSTREGUP) or (DOGDRIVER / DOGSTEP / TRUSTREGUP)
- Step 5 -- GRAD or FDGRAD (or perhaps CDGRAD if using central difference switching option)
- Step 6a -- UMSTOP
- Step 7 -- HESS or (FDHESSG / FDJAC) or FDHESSF or BFGSUNFAC or (BFGSFAC / QRUPDATE)

When a step above contains two or more modules or groups of modules separated by the word "or", these are alternative choices selected by the algorithmic option parameters. A given method would use only one choice throughout the algorithm for each such step.

When two or more modules are listed above in parentheses separated by slashes (/), the first module in the parenthesized group is called by the driver, the remaining module(s) in the group are called by the first module. For example, FDHESSG calls FDJAC. In addition, the following modules call the following service and function evaluation modules :
  HOOKSTEP calls CHOLDECOMP and LSOLVE
  LINESEARCH calls FN
  LINESEARCHmod calls FN and (GRAD or FDGRAD (or CDGRAD))
  TRUSTREGUP calls FN

The following modules also call their following subalgorithms :
  CHOLSOLVE calls LSOLVE and LTSOLVE
  QRUPDATE calls JACROTATE

II.3.   GUIDELINE 1

CHOOSING ALGORITHMIC OPTIONS
FOR UNCONSTRAINED MINIMIZATION

The following are the algorithmic options available when using this system of algorithms for unconstrained minimization, and an indication of how to choose them.

global -- A positive integer designating which global strategy to use at each
   iteration, as follows
   =1 Line search (Alg. A6.3.1)
   =2 Hookstep trust region (Alg. A6.4.1)
   =3 Dogleg trust region (Alg. A6.4.3)
   =4 Modified line search (Alg. A6.3.1mod)
On some problems, there may be rather large variations in the accuracy
and efficiency obtained by interchanging methods 1, 2, and 3 while keeping all other parameters and tolerances unchanged. However, no method seems clearly superior or inferior to the others in general, and each is likely to be best on some problems. The first three possibilities are provided for comparison, and to allow a user to select the best method for a particular class of problems. Method 4 is a slight augmentation of method 1 that may be preferred when secant approximations to the Hessian are used; in our experience, the results using methods 1 and 4 are very similar.
Suggested default value : *global* = 1 (because this probably is the easiest
   method to understand and to code)

analgrad -- A Boolean variable that is TRUE if a routine that computes the
   analytic gradient $\nabla f(x)$ has been supplied by the user, and is FALSE otherwise. If analytic gradients are not supplied, the system will approximate the gradient at each iteration by finite differences. The algorithms in this system usually will be able to find a more accurate approximation to the minimizer, and sometimes may perform more efficiently, using analytic rather than finite difference gradients. Therefore it is important to supply a routine that evaluates the analytic gradient if it is readily available.
Suggested default value : *analgrad* = FALSE

analhess -- A Boolean variable that is TRUE if a routine that computes the
   analytic Hessian matrix $\nabla^2 f(x)$ has been supplied by the user, and is FALSE otherwise. If analytic Hessians are not supplied, the system will approximate the Hessian at each iteration by finite differences or secant approximations depending on the value of *cheapf*. The algorithms in this system may perform more efficiently if analytic Hessians are provided; a routine to evaluate the analytic Hessian should be supplied if it is readily available.
Suggested default value : *analhess* = FALSE

cheapf -- A Boolean variable that is TRUE if the objective function $f(x)$ is
   inexpensive to evaluate and is FALSE otherwise. If *analhess* = TRUE, *cheapf* is disregarded. If *analhess* = FALSE, *cheapf* is used to choose the method of Hessian approximation: finite difference Hessians are used if *cheapf* = TRUE, secant approximations to the Hessian (BFGS

updates) are used if *cheapf* = FALSE. Methods using finite difference Hessians usually (but not always) will require fewer iterations than methods using secant approximations on the same problem. Methods using secant approximations usually will require fewer objective function evaluations than methods using finite difference Hessians (including the function evaluations used for finite differences), especially for problems with medium or large values of n. Thus if the cost of evaluating $f(x)$ is appreciable or if secant approximations are desired for any other reason, *cheapf* should be set FALSE, otherwise *cheapf* should be set TRUE.
Suggested default value : *cheapf* = FALSE

factsec -- A Boolean variable that only is used if secant approximations to the Hessian are used, i.e. if *analhess* = FALSE and *cheapf* = FALSE, and is disregarded in all other cases. When it is used, the value of *factsec* only affects the algorithmic overhead of the algorithm, the results and sequence of iterates are unaffected. If *factsec* = TRUE, the system uses a factored secant approximation, updating an LL^T factorization of the Hessian approximation and requiring $O(n^2)$ operations per iteration; if *factsec* = FALSE, it uses an unfactored approximation, requiring $O(n^3)$ operations per iteration. (See Section 9.4 of the text for details.) If *global* = 2, *factsec* must be set FALSE because the hookstep algorithm as written could overwrite the Cholesky factor L of the Hessian approximation. If *global* = 1, 3 or 4, a production code should use *factsec* = TRUE because this is more efficient. In all cases, *factsec* = FALSE may be preferred for class projects as this version is easier to understand.
Suggested default value : *factsec* = FALSE for class projects or when *global* = 2, *factsec* = TRUE otherwise

II.4.   GUIDELINE 2

CHOOSING TOLERANCES FOR UNCONSTRAINED MINIMIZATION

The following are the scaling, stopping and other tolerances used for unconstrained minimization, and an indication of appropriate values for them.

typx -- An n dimensional array whose i^{th} component is a positive scalar specifying the typical magnitude of $x[i]$. For example, if it is anticipated that the range of values for the iterates x will be

$$x_1 \in [-10^{10}, 10^{10}], \; x_2 \in [-10^2, -10^4], \; x_3 \in [-6 * 10^{-6}, 9 * 10^{-6}]$$

then an appropriate choice would be

$$typx = (10^{10}, 10^3, 7 * 10^{-6}).$$

typx is used to determine the scale vector S_x and diagonal scaling matrix D_x used throughout the code; $D_x = \text{diag}((S_x)_1, ..., (S_x)_n)$ where $S_x[i] = 1/ typx[i]$. It is important to supply values of *typx* when the magnitudes of the components of x are expected to be very different; in this case, the code may work better with good scale information than with $D_x = I$. If the magnitudes of the components of x are similar, the

code will probably work as well with $D_x = I$ as with any other values.
Suggested default value : $typx[i] = 1, i = 1, \cdots, n$

typf -- A positive scalar estimating the magnitude of $f(x)$ near the minimizer x_*. It is used only in the gradient stopping condition given below. If too large a value is provided for *typf*, the algorithm may halt prematurely. In particular, if $f(x_0)$ is $>> f(x_*)$, *typf* should be approximately $|f(x_*)|$.
Suggested default value : $typf = 1$

fdigits -- A positive integer specifying the number of reliable digits returned by the objective function FN. For example, if FN is the result of an iterative procedure (quadrature, a p.d.e. code) expected to provide five good digits to the answer, *fdigits* should be set to 5. If FN is expected to provide within one or two of the full number of significant digits available on the host computer, *fdigits* should be set to -1. *fdigits* is used to set the parameter η that is used in the code to specify the relative noise in $f(x)$; the main use of η is in calculating finite difference step sizes. η is set to *macheps* if *fdigits* $= -1$, to $\max\{macheps, 10^{-fdigits}\}$ otherwise. If $f(x)$ is suspected to be noisy but the approximate value of *fdigits* is unknown, it should be estimated by the routine of Hamming [1973] given in Gill, Murray and Wright [1981].
Suggested default value : *fdigits* = -1

gradtol -- A positive scalar giving the tolerance at which the scaled gradient is considered close enough to zero to terminate the algorithm. The algorithm is stopped if

$$\max_{1 \le i \le n}\left\{ \frac{|\nabla f(x_+)[i]| * \max\{|x[i]|, typx[i]\}}{\max\{|f|, typf\}} \right\} \le gradtol$$

(see Section 7.2). This is the primary stopping condition for unconstrained minimization, and *gradtol* should reflect the user's idea of what constitutes a solution to the problem. If the gradient is approximated by finite differences, $\sqrt{\eta}$ is likely to be the smallest value of *gradtol* for which the above condition can be satisfied.
Suggested default value : $gradtol = macheps^{1/3}$

steptol -- A positive scalar giving the tolerance at which the scaled distance between two successive iterates is considered close enough to zero to terminate the algorithm. The algorithm is stopped if

$$\max_{1 \le i \le n}\left\{ \frac{|x_+[i] - x_c[i]|}{\max\{|x_+[i]|, typx[i]\}} \right\} \le steptol$$

(see Section 7.2). *steptol* should be at least as small as 10^{-d} where d is the number of accurate digits the user desires in the solution x_*. The algorithm may terminate prematurely if *steptol* is too large, especially when secant updates are being used. For example, *steptol* = $macheps^{1/3}$ is often too large when secant methods are used.
Suggested default value : $steptol = macheps^{2/3}$

maxstep -- A positive scalar giving the maximum allowable scaled steplength $\|D_x(x_+ - x_c)\|_2$ at any iteration. *maxstep* is used to prevent steps that would cause the optimization algorithm to overflow or leave the domain of interest, as well as to detect divergence. It should be chosen small

enough to prevent the first two of these occurrences but larger than any anticipated reasonable stepsize. The algorithm will halt if it takes steps of length *maxstep* on five consecutive iterations.
Suggested default value : $maxstep = 10^3 * \max\{\|D_x x_0\|_2, \|D_x\|_2\}$

itnlimit -- A positive integer specifying the maximum number of iterations that may be performed before the algorithm is halted. Appropriate values depend strongly on the dimension and difficulty of the problem, and on the cost of evaluating the nonlinear function.
Suggested default value : *itnlimit* $= 100$

δ -- A positive scalar giving the initial trust region radius (see Section 6.4). It is used only when *global* $= 2$ or 3 (hookstep or dogleg methods), ignored when *global* $= 1$ or 4 (line search). The value of δ should be what the user considers a reasonable scaled steplength for the first iteration, and should obey $\delta \leq maxstep$. If no value of δ is supplied by the user or if $\delta=-1$, the length of the initial scaled gradient is used instead. In all cases, the trust radius is adjusted automatically thereafter.
Suggested default value : $\delta = -1$

II.5.   GUIDELINE 3

STORAGE CONSIDERATIONS FOR UNCONSTRAINED MINIMIZATION

The following are some storage considerations relevant to the entire system of algorithms for unconstrained minimization. In addition, many individual modules contain a Storage Considerations section giving comments specific to that module.

1) The entire system of algorithms for unconstrained minimization uses two $n \times n$ matrices, L and H. However, only the lower triangle of L and the upper triangle of H, including the main diagonals of both, actually are used. Thus the entire system of algorithms could be implemented using $n^2 + O(n)$ storage locations. We have written it using separate matrices for L and H, thus wasting $n^2 - n$ locations, because this makes the pseudo-code easier to follow. However, the system also is written so that with only a few minor modifications to a few modules, discussed in item 2 below, it uses only $n^2 + O(n)$ storage locations. For class projects, the maximum problem dimension n probably will be small, so the system probably should be implemented as written to make it easy to understand. Any production implementation certainly should make the modifications given in item 2.

2) To make the entire system of algorithms for unconstrained minimization use only one $n \times n$ matrix, minor modifications are required to Algorithms A5.5.1-2, A6.4.1-2, A6.4.5, A9.4.1, and the driver D6.1.1 (or the pared down driver being used). Basically, they consist of combining L and H into one matrix, with the main diagonal of H sometimes stored in a separate n-vector *hdiag*, and changing the code, parameter lists and calling sequences accordingly. The modifications are :

Algorithm A5.5.1
i) Change output parameter $L \in R^{n \times n}$ to output parameter *hdiag* $\in R^n$
ii) Between lines 11 and 12, insert the statement
$11\frac{1}{2}$. FOR $i = 1$ TO n DO
$hdiag[i] \leftarrow H[i,i]$
iii) Change lines 12 and 13.8 to
CALL CHOLDECOMP $(n,\ hdiag,\ maxo\, f\!f\!l,\ macheps,\ H,\ maxadd)$
iv) On lines 13.3.2, 13.3.3 and 13.7, change all occurrences of $H[i,i]$ to
$hdiag[i]$
v) Replace statement 14.1 by
14.1 FOR $j = i+1$ TO n DO
$H[i,j] \leftarrow H[i,j] * S_x[i] * S_x[j]$
14.2 $hdiag[i] \leftarrow hdiag[i] * S_x[i]^2$
vi) On line 15.1, change both instances of $L[i,j]$ to $H[i,j]$

Algorithm A5.5.2
i) Change input parameter $H \in R^{n \times n}$ to input parameter *hdiag* $\in R^n$
ii) On line 2.1, change $H[i,i]$ to $hdiag[i]$
iii) On line 4.1, change $H[j,j]$ to $hdiag[j]$
iv) On line 4.3.1, change $H[j,i]$ to $L[j,i]$

Algorithm A6.4.1
i) Change input parameter $H \in R^{n \times n}$ to input parameter *hdiag* $\in R^n$
ii) On lines 5.1 and 5.3, change the parameter H to *hdiag*

Algorithm A6.4.2
i) Change input parameter $H \in R^{n \times n}$ to input parameter *hdiag* $\in R^n$
ii) On lines 3E.8.2 and 3E.8.5, change all occurrences of $H[i,i]$ to
$hdiag[i]$
iii) On line 3E.8.3, change the parameter H to *hdiag*

Algorithm A6.4.5
i) Change input parameter $H \in R^{n \times n}$ to input parameter *hdiag* $\in R^n$
ii) On line 9c.2T.1.1, change $H[i,i]$ to $hdiag[i]$

Algorithm A9.4.1
i) Add input parameter *hdiag* $\in R^n$ as the final input parameter.
ii) Before line 1, insert the statement
$\frac{1}{2}$. FOR $i = 1$ TO n DO
$H[i,i] \leftarrow hdiag[i]$

Driver D6.1.1
i) On line 10.2, change the parameter L_c to *hdiag*
ii) On line 10.4b, change the parameter H_c to *hdiag*
iii) On line 10.7E.1e, insert the parameter *hdiag* between *analgrad* and
H_c
iv) Change all remaining instances of L_c in the parameter lists to H_c.
These occur on lines 8E.1d, 10.3, 10.4b, 10.4c, and 10.7E.1d.

3) It is possible to reduce the storage requirement further, to $\frac{1}{2}n^2+O(n)$
locations, in one special case : when factored secant approximations to the
Hessian are used. H never is used in this case; the only matrices used are
the lower triangular L, and the upper triangular R used by Alg. A3.4.1, and
they may share storage. In most languages, to actually use only $(n^2+n)/2$
locations to store a triangular matrix, the matrix would have to be

implemented as a vector with $(n^2+n)/2$ elements, for example with $L[i,j] = v[((i^2-i)/2)+j] = R[j,i]$, $1 \le j \le i \le n$. The remaining details of this conversion are left as an exercise.

For all the other methods, n^2 storage is required, because they all use Alg. A5.5.1 which requires L and H to be known simultaneously. The hookstep algorithms and any algorithms using unfactored secant updates also require both L and H.

4) The vector storage required by each module is the sum of the vectors in its parameter list, plus any vectors used as local variables. Item 1 of each module's Storage Considerations section lists any vectors used as local variables. The implementor should determine the vector storage required by the driver.

5) When implementing these algorithms in FORTRAN, all modules containing matrices as parameters may require an additional parameter *ndim* giving the actual row dimension of the matrices. This is required whenever the matrix dimensions in the modules are given as variables instead of as fixed integers. For example in a production implementation, any matrices used usually will be passed by the user as an additional parameter to the main driver. They will be declared in the user's program with some fixed dimensions, say $H(50,50)$, where the matrix dimension (50) is greater than or equal to the problem dimension n. All matrices in the driver and the modules of the system will then be declared as $H(ndim,n)$ or $L(ndim,n)$ where *ndim* is *an additional parameter* passed by the user to the driver and by the system to each module containing a matrix parameter. Here *ndim* contains the number of rows in the matrix in the users program, *ndim*=50, while n may be any number between 1 and 50. If instead the matrix dimensions in the modules are given as (n,n), the algorithms *will not work correctly*. The reason is that FORTRAN stores matrices by columns, and the compiler must know the number of locations reserved for each column, i.e. the number of rows, to find its place in the matrix correctly.

In any FORTRAN implementation, these considerations apply to any modules where matrix dimensions are given as variables. The dimensions must be $(ndim,n)$ where *ndim* is an input parameter to the module containing the number of rows in the matrix in whatever subroutine or main program it is declared with fixed integer dimensions. (In a class implementation of our system, this could be the driver.) An alternative for class projects is to declare all matrices in all modules with the same fixed integer dimensions, say (50,50). In this case, the extra parameter *ndim* is not needed and no modifications to our pseudo-code are required, but the system will only work for $n \le 50$.

In most strongly-typed languages, the additional parameter *ndim* would not be required, and our pseudo-code can be used as is. For example in PASCAL, one could declare a data type

MATRIX = ARRAY [1..50, 1..50] OF REAL

in the calling program. Again the matrix dimension (50) would be greater than or equal to the largest problem dimension n to be used. Then all matrices in all procedures in the system simply would be given the data type MATRIX. To run the system with $n > 50$, only the declaration of MATRIX would have to be changed. The disadvantage of this scheme is that the same word MATRIX must be used by all procedures in the system and by the calling program.

II.6   ALGORITHM D6.1.2. (UMEXAMPLE)

DRIVER FOR AN UNCONSTRAINED MINIMIZATION ALGORITHM USING LINE SEARCH, FINITE DIFFERENCE GRADIENTS, AND FACTORED SECANT APPROXIMATIONS TO THE HESSIAN

Purpose : Attempt to find a local minimizer x_* of $f(x) : R^n \to R$ starting from x_0, using a line search algorithm, finite difference gradients, and BFGS Hessian approximations.

Input Parameters : $n \in Z$ $(n \geq 1)$, $x_0 \in R^n$, FN (the name of a user-supplied function $f : R^n \to R$)

plus optionally : $typx \in R^n$, $typf \in R$, $fdigits \in Z$, $gradtol \in R$, $steptol \in R$, $maxstep \in R$, $itnlimit \in Z$, $printcode \in Z$ (for further information see Algorithm D6.1.1)

Input-Output Parameters : none (see item 2 under Storage Considerations, Alg. D6.1.1)

Output Parameters : $x_f \in R^n$ (the final approximation to x_*), $termcode \in Z$ (indicates which termination condition caused the algorithm to stop)

plus optionally : $f_c \in R$ $(= f(x_f))$, $g_c \in R^n$ $(= \nabla f(x_f))$, other output parameters as desired by the implementor (see Section I.4)

Meaning of Return Codes, Storage Considerations :

same as Algorithm D6.1.1

Algorithm :

(* Initialization section : *)
1. CALL MACHINEPS ($macheps$)
2. CALL UMINCK (parameter list selected by implementor)
    (* see UMINCK description *)
3. IF $termcode < 0$ THEN
    3.1 $x_f \leftarrow x_0$
    3.2 RETURN from Algorithm D6.1.2   (* write appropriate message, if desired *)
4. $itncount \leftarrow 0$
5. CALL FN (n, x_0, f_c)
6. CALL FDGRAD $(n, x_0, f_c, FN, S_x, \eta, g_c)$ (* Alg. A5.6.3 *)
7. CALL UMSTOP0 $(n, x_0, f_c, g_c, S_x, typf, gradtol, termcode, consecmax)$
    (* Alg. A7.2.2 *)
8. IF $termcode > 0$
8T. THEN $x_f \leftarrow x_0$
8E. CALL INITHESSFAC $(n, f_c, typf, S_x, L_c)$ (* Alg. A9.4.4 *)
9. $x_c \leftarrow x_0$

(* Iteration section : *)

10. WHILE $termcode$ = 0 DO

    10.1 $itncount \leftarrow itncount + 1$

    10.2 CALL CHOLSOLVE (n, g_c, L_c, s_N) (* Alg.A3.2.3 *)

    10.3 CALL LINESEARCH $(n, x_c, f_c,$ FN, $g_c, s_N, S_x, maxstep, steptol,$ $retcode, x_+, f_+, maxtaken)$ (* Alg. A6.3.1 *)

    10.4 CALL FDGRAD $(n, x_+, f_+,$ FN, $S_x, \eta, g_+)$ (* Alg. A5.6.3 *)

    10.5 CALL UMSTOP $(n, x_c, x_+, f_+, g_+, S_x, typf, retcode, gradtol, step$-$tol, itncount, itnlimit, maxtaken, consecmax, termcode)$ (* Alg. A7.2.1 *)

    10.6 IF $termcode > 0$

    10.6T THEN $x_f \leftarrow x_+$

    10.6E ELSE (* calculate next Hessian approximation *)

        10.6E.1 CALL BFGSFAC $(n, x_c, x_+, g_c, g_+, macheps, \eta, analgrad, L_c)$ (* Alg. A9.4.2 *)

        10.6E.2 $x_c \leftarrow x_+$

        10.6E.3 $f_c \leftarrow f_+$

        10.6E.4 $g_c \leftarrow g_+$

    (* END, WHILE loop 10 *)

(* write appropriate termination message, if desired *)

(* RETURN from Algorithm D6.1.2 *)

(* END, Algorithm D6.1.2 *)

SECTION III

**DRIVER MODULES AND GUIDELINES FOR
NONLINEAR EQUATIONS**

III.3   ALGORITHM D6.1.3. (NEDRIVER)

**DRIVER FOR A MODULAR SYSTEM OF ALGORITHMS
FOR NONLINEAR EQUATIONS**

Purpose : Attempt to find a root x_* of $F(x) : R^n \to R^n$ starting from x_0, using one of a variety of algorithms.

Input Parameters : $n \in Z$ $(n \geq 1)$, $x_0 \in R^n$, FVEC (the name of a user-supplied function $F: R^n \to R^n$ that is a least once continuously differentiable)
plus optionally :
   i) JAC -- name of a user-supplied function for evaluating $F'(x)$. $(F'(x)$ is denoted hereafter by $J(x)$.)
   ii) $global \in Z$, $analjac \in$ Boolean, $cheapF \in$ Boolean, $factsec \in$ Boolean -- parameters used in choosing algorithmic options. These are explained in Guideline 4. (Parameters that are not input are given values by algorithm NEINCK).
   iii) $typx \in R^n$, $typF \in R^n$, $Fdigits \in Z$, $fvectol \in R$, $steptol \in R$, $mintol \in R$, $maxstep \in R$, $itnlimit \in Z$, $\delta \in R$ -- tolerances and constants used by the algorithm. These are explained in Guideline 5. (Tolerances or constants that are not input are given values by algorithm NEINCK).
   iv) $printcode \in Z$ -- see Section I.4 of this Appendix.

Input-Output Parameters : none (see however item 2 under Storage Considerations)

Output Parameters : $x_f \in R^n$ (the final approximation to x_*), $termcode \in Z$ (indicates which termination condition caused the algorithm to stop)
plus optionally :
   $FV_+ \in R^n$ $(= F(x_f))$, $J_c \in R^{n \times n}$ $(= J(x_f))$, other output parameters as desired by the implementor (see Section I.4)

Meaning of Return Codes :
   $termcode < 0$: algorithm terminated due to input error, see Alg. NEINCK.
   $termcode > 0$: algorithm terminated for reason given in Alg. A7.2.1.

Storage Considerations :
1) Matrix and vector storage requirements are discussed in Guideline 3 and 6. Be sure to read item 5 in Guideline 3 if you are implementing these algorithms in FORTRAN.
2) The names x_0 and x_f are used in this driver for clarity only. In the implementation, x_0 and x_f can both share storage with x_c.
3) Two global variables (FORTRAN COMMON), S_F and FV_+, must be shared between this driver and Algorithm NEFN. See Algorithm NEFN for further information.

Additional Considerations :
1) The List of Modules for Nonlinear Equations shows which modules correspond to each step in the Description below. The choice between alternatives in the list is determined by the algorithmic option parameters described in Guideline 4.
2) The Algorithm below is used for implementing the *entire system* of nonlinear equations algorithms. A simpler driver can be used to implement a *particular* algorithm for nonlinear equations, as is illustrated by Algorithm D6.1.4.
3) On some computing systems, it will be necessary to a supply dummy routine JAC for the case when the user doesn't supply a routine for evaluating $F'(x)$. This dummy routine is required solely to allow compilation of the program.

Description :

Initialization :
1) Calculate machine epsilon.
2a) Check input for reasonableness, assign default values to optional parameters and tolerances that weren't input to user. (Return *termcode* to driver.)
2b) If *termcode* < 0, terminate algorithm, returning $x_f = x_0$ and *termcode*. If *termcode* = 0, set *itncount* ← 0 and continue.
3) Calculate $FV_+ = F(x_0)$, $f_c = \frac{1}{2}\|D_F\,FV_+\|_2^2$.
4a) Decide whether to stop. (Return *termcode* to driver).
4b) If *termcode* > 0, terminate algorithm, returning $x_f = x_0$ and *termcode*. If *termcode* = 0, continue.
5) Calculate $J_c = J(x_0)$ or approximate it by finite differences.
6) Calculate $g_c = J_c^T\,D_F^2\,FV_+$.
7) Set $x_c = x_0$, $FV_c = FV_+$, and *restart* = TRUE.

Iteration :
1) *itncount* ← *itncount* + 1
2) Compute Newton step $s_N = -J_c^{-1}\,F_c$, or a variation if J_c is singular or ill-conditioned. Also compute associated minimization model if a trust region global algorithm is being used.
3) Choose $x_+ = x_c + s_c$ where $s_c = s_N$ or a step chosen by the global strategy. $F_+ = F(x_+)$ and $f_+ = \frac{1}{2}\|D_F\,FV_+\|_2^2$ are calculated during this step.
4) If J_c is being calculated by secant approximations and the algorithm has bogged down, reset J_c to $J(x_c)$ using finite differences, recalculate g_c, and restart the iteration.
5) Calculate or approximate $J_c = J(x_+)$.
6) Calculate $g_c = J_c^T\,D_F^2\,FV_+$.

7a) Decide whether to stop. (Return *termcode* to driver).
7b) If *termcode* > 0, terminate algorithm, returning $x_f = x_+$ and *termcode*.
If *termcode* = 0, continue.
8) Set $x_c \leftarrow x_+$, $f_c \leftarrow f_+$, $FV_c \leftarrow FV_+$ and return to step 1 of the iteration section.

Algorithm :

(* Initialization section : *)
1. CALL MACHINEPS (*macheps*)
2. CALL NEINCK (parameter list selected by implementor)
    (* see NEINCK description *)
3. IF *termcode* < 0 THEN
    3.1 $x_f \leftarrow x_0$
    3.2 RETURN from Algorithm D6.1.3   (* write appropriate message, if desired *)
4. *itncount* $\leftarrow$ 0
5. CALL NEFN (n, x_0, f_c)
    (* NEFN also calculates $FV_+ = F(x_0)$ and communicates it to the driver via a global parameter *)
6. CALL NESTOP0 $(n, x_0, FV_+, S_F, fvectol, termcode, consecmax)$ (* Alg. A7.2.4 *)
7. IF *termcode* > 0
7T. THEN $x_f \leftarrow x_0$
7E. ELSE (* calculate initial Jacobian *)
    7E.1 IF *analjac*
    7E.1T THEN CALL JAC (n, x_0, J_c)
    7E.1E ELSE CALL FDJAC $(n, x_0, FV_+, \text{FVEC}, S_x, \eta, J_c)$ (* Alg. A5.4.1 *)
    (* $g_c \leftarrow J_c^T D_F^2 FV_+$: *)
    7E.2 FOR i = 1 TO n DO
    $$g_c[i] \leftarrow \sum_{j=1}^{n} J_c[j,i] * FV_+[j] * S_F[j]^2$$
    7E.3 $FV_c \leftarrow FV_+$
8. $x_c \leftarrow x_0$
9. *restart* $\leftarrow$ TRUE

(* Iteration section : *)
10. WHILE *termcode* = 0 DO
    10.1 *itncount* $\leftarrow$ *itncount* + 1
    10.2 IF *analjac* OR *cheapF* OR (NOT *factsec*)
    10.2T THEN CALL NEMODEL $(n, FV_c, J_c, g_c, S_F, S_x, macheps, global, M, H_c, s_N)$ (* Alg. A6.5.1 *)
    10.2E ELSE CALL NEMODELFAC $(n, FV_c, g_c, S_F, S_x, macheps, global, restart, M, M2, J_c, H_c, s_N)$ (* Alg. A6.5.1fac *)
    10.3a IF *global* = 1 THEN

> CALL LINESEARCH $(n, x_c, f_c, \text{FVEC}, g_c, s_N, S_x, maxstep, steptol, retcode, x_+, f_+, maxtaken)$ (* Alg. A6.3.1 *)

10.3b ELSEIF $global = 2$ THEN

> CALL HOOKDRIVER $(n, x_c, f_c, \text{FVEC}, g_c, M, H_c, s_N, S_x, maxstep, steptol, macheps, itncount, \delta prev, \delta, \mu, \varphi, \varphi', retcode, x_+, f_+, maxtaken)$ (* Alg. A6.4.1 *)

10.3c ELSE   (* $global = 3$ *)

> CALL DOGDRIVER $(n, x_c, f_c, \text{FVEC}, g_c, M, s_N, S_x, maxstep, steptol, \delta, retcode, x_+, f_+, maxtaken)$ (* Alg. A6.4.3 *)

10.4 IF $(retcode = 1)$ AND (NOT $restart$) AND (NOT $analjac$) AND (NOT $cheapF$)

10.4T THEN   (* secant method restart : recalculate J_c and redo iteration *)

> 10.4T.1 CALL FDJAC $(n, x_c, FV_c, \text{FVEC}, S_x, \eta, J_c)$ (* Alg. A5.4.1 *)
>
> 10.4T.2 $g_c \leftarrow J_c^T * D_F^2 * FV_c$
>
> (* same as statement 7E.2 except substitute FV_c for FV_+ *)
>
> 10.4T.3 IF $(global = 2)$ OR $(global = 3)$ THEN $\delta \leftarrow -1$
>
> 10.4T.4 $restart \leftarrow$ TRUE

10.4E ELSE   (* complete the iteration *)

> 10.4E.1a IF $analjac$ THEN
>
> > CALL JAC (n, x_+, J_c)
>
> 10.4E.1b ELSEIF $cheapF$ THEN
>
> > CALL FDJAC $(n, x_+, FV_+, \text{FVEC}, S_x, \eta, J_c)$ (* Alg. A5.4.1 *)
>
> 10.4E.1c ELSEIF $factsec$ THEN
>
> > CALL BROYFAC $(n, x_c, x_+, FV_c, FV_+, \eta, S_x, S_F, J_c, M, M2)$ (* Alg. A8.3.2 *)
>
> 10.4E.1d ELSE   (* $factsec =$ FALSE *)
>
> > CALL BROYUNFAC $(n, x_c, x_+, FV_c, FV_+, \eta, S_x, J_c)$ (* Alg. A8.3.1 *)
>
> 10.4E.2 IF $factsec$
>
> 10.4E.2T THEN   (* calculate g_c using QR factorization *)
>
> $(* g_c \leftarrow J_c\, D_F\, FV_+ = Q^T\, D_F\, FV_+ : *)$
>
> > 10.4E.2T.1 FOR $i = 1$ TO n DO
> >
> > $$g_c[i] \leftarrow \sum_{j=1}^{n} J_c[i,j] * FV_+[j] * S_F[j]$$
>
> $(* g_c \leftarrow R^T\, g_c : *)$
>
> > 10.4E.2T.2 FOR $i = n$ DOWNTO 1 DO
> >
> > $$g_c[i] \leftarrow \sum_{j=1}^{i} M[j,i] * g_c[j]$$
>
> 10.4E.2E ELSE
>
> $$g_c \leftarrow J_c^T * D_F^2 * FV_+$$
>
> (* same as statement 7E.2 *)
>
> 10.4E.3 CALL NESTOP $(n, x_c, x_+, FV_+, f_+, g_c, S_x, S_F, retcode, fvectol, steptol, itncount, itnlimit, maxtaken, analjac, cheapF,$

$mintol$, $consecmax$, $termcode$)   (* Alg. A7.2.3 *)

10.4E.4a IF ($termcode$ = 2) AND (NOT $restart$) AND (NOT $analjac$)
AND (NOT $cheapF$) THEN   (* restart *)
    GO TO 10.4T.1

10.4E.4b ELSEIF $termcode$ > 0 THEN
    10.4E.4bT.1 $x_f \leftarrow x_+$
    10.4E.4bT.2 IF termcode = 1 THEN $FV_+ \leftarrow FV_c$

10.4E.4c ELSE $restart \leftarrow$ FALSE

10.4E.5 $x_c \leftarrow x_+$

10.4E.6 $f_c \leftarrow f_+$

10.4E.7 $FV_c \leftarrow FV_+$

(* END, IFTHENELSE 10.4 and WHILE loop 10 *)

(* write appropriate termination message, if desired *)

(* RETURN from Algorithm D6.1.3 *)

(* END, Algorithm D6.1.3 *)

III.2.   LIST OF MODULES FOR NONLINEAR EQUATIONS

The steps in the Description section of Driver D6.1.3 correspond to the following modules. The steps in the Description that are not listed below correspond to lines of code in the driver algorithm.

Initialization
    Step 1 -- MACHINEPS
    Step 2a -- NEINCK
    Step 3 -- (NEFN / FVEC)
    Step 4a -- NESTOP0
    Step 5 -- JAC or FDJAC

Iteration
    Step 2 -- (NEMODEL / QRDECOMP / CONDEST / QRSOLVE / CHOLDECOMP /
        CHOLSOLVE) or (NEMODELfac / QRDECOMP / QFORM / CONDEST /
        RSOLVE / CHOLDECOMP / CHOLSOLVE)
    Step 3 -- LINESEARCH or LINESEARCHmod or (HOOKDRIVER / HOOKSTEP /
        TRUSTREGUP) or (DOGDRIVER / DOGSTEP / TRUSTREGUP)
    Step 4 -- FDJAC (see note below)
    Step 5 -- JAC or FDJAC or BROYUNFAC or (BROYFAC / QRUPDATE)
    Step 7a -- NESTOP

When a step above contains two or more modules or groups of modules separated by the word "or", these are alternative choices selected by the algorithmic option parameters. A given method would use only one choice throughout the algorithm for each such step. Note that step 4 in the iteration list above (FDJAC) is used only when secant approximations to the Jacobian are being used.

When two or more modules are listed above in parentheses separated by slashes (/), the first module in the parenthesized group is called by the driver, the remaining module(s) in the group are called by the first module. For example, NEFN calls FVEC. In addition, the following modules call the following service and function evaluation modules :
   HOOKSTEP calls CHOLDECOMP and LSOLVE
   LINESEARCH calls (NEFN / FVEC)
   TRUSTREGUP calls (NEFN / FVEC)

The following modules also call their following subalgorithms :
   CHOLSOLVE calls LSOLVE and LTSOLVE
   QRSOLVE calls RSOLVE
   QRUPDATE calls JACROTATE

III.3.   GUIDELINE 4

CHOOSING ALGORITHMIC OPTIONS
FOR NONLINEAR EQUATIONS

The following are the algorithmic options available when using this system of algorithms for nonlinear equations, and an indication of how to choose them.

global -- A positive integer designating which global strategy to use at each iteration, as follows
   =1 Line search (Alg. A6.3.1)
   =2 Hookstep trust region (Alg. A6.4.1)
   =3 Dogleg trust region (Alg. A6.4.3)
On some problems, there may be rather large variations in the accuracy and efficiency obtained by interchanging methods 1, 2, and 3 while keeping all other parameters and tolerances unchanged. However, no method seems clearly superior or inferior to the others in general, and each is likely to be best on some problems. The three possibilities are provided for comparison, and to allow a user to select the best method for a particular class of problems.
   Suggested default value : *global* = 1 (because this probably is the easiest method to understand and to code)

analjac -- A Boolean variable that is TRUE if a routine that computes the analytic Jacobian matrix $J(x)$ has been supplied by the user, and is FALSE otherwise. If analytic Jacobians are not supplied, the system will approximate the Jacobian at each iteration by finite differences or secant approximations depending on the value of *cheapF*. The algorithms in this system may perform more efficiently if analytic Jacobians are provided; a routine to evaluate the analytic Jacobian should be supplied if it is readily available.
   Suggested default value : *analjac* = FALSE

cheapF -- A Boolean variable that is TRUE if the objective function $F(x)$ is inexpensive to evaluate and is FALSE otherwise. If *analjac* = TRUE, *cheapF* is disregarded. If *analjac* = FALSE, *cheapF* is used to choose the method of Jacobian approximation: finite difference Jacobians are used if *cheapF* = TRUE, secant approximations to the Jacobian (Broyden's updates) are used if *cheapF* = FALSE. Methods using finite difference Jacobians usually will require fewer iterations than methods using secant approximations on the same problem. Methods using secant approximations usually will require fewer objective function evaluations than methods using finite difference Jacobians (including the function evaluations used for finite differences), especially for problems with medium or large values of n. Thus if the cost of evaluating $F(x)$ is appreciable, *cheapF* should be set FALSE, otherwise *cheapF* should be set TRUE.

Suggested default value : *cheapF* = FALSE

factsec -- A Boolean variable that only is used if secant approximations to the Jacobian are used, i.e. if *analjac* = FALSE and *cheapF* = FALSE, and is disregarded in all other cases. When it is used, the value of *factsec* only affects the algorithmic overhead of the algorithm, the results and sequence of iterates are unaffected. If *factsec* = TRUE, the system uses a factored secant approximation, updating an QR factorization of the Jacobian approximation and requiring $O(n^2)$ operations per iteration; if *factsec* = FALSE, it uses an unfactored approximation, requiring $O(n^3)$ operations per iteration. (See Section 8.3 of the text for details.) A production code should use *factsec* = TRUE because this is more efficient. However, *factsec* = FALSE may be preferred for class projects as this version is easier to understand.

Suggested default value : *factsec* = FALSE for class projects, *factsec* = TRUE otherwise

III.4.   GUIDELINE 5

CHOOSING TOLERANCES FOR NONLINEAR EQUATIONS

The following are the scaling, stopping and other tolerances used for nonlinear equations, and an indication of appropriate values for them.

typx -- see Guideline 2

typF — An n dimensional array whose i^{th} component is a positive scalar specifying the typical magnitude of the ith component function of $F(x)$ at points that are not near a root of $F(x)$. *typF* is used to determine the scale vector S_F and diagonal scaling matrix D_F used throughout the code; $D_F = \text{diag}((S_F)_1, \ldots, (S_F)_n)$ where $S_F[i] = 1/ typF[i]$. It should be chosen so that all the components of $D_F F(x)$ have similar typical magnitudes at points not too near a root, and should be chosen in conjunction with *fvectol* as discussed below. It is important to supply values of *typF* when the magnitudes of the components of $F(x)$ are expected to be very different; in this case, the code may work better with good scale information than with $D_F = I$. If the magnitudes of the components of $F(x)$ are similar, the code will probably work as well with $D_F = I$ as with any other values.

Suggested default value : $typF[i] = 1, i = 1, \cdots, n$

Fdigits -- A positive integer specifying the number of reliable digits returned by the nonlinear equations function FVEC. If the components of $F(x)$ have differing numbers of reliable digits, *Fdigits* should be set to the smallest of these numbers. The remainder of the discussion of *Fdigits* is analogous to the discussion of *fdigits* in Guideline 2.

Suggested default value : *Fdigits* = -1

fvectol -- A positive scalar giving the tolerance at which the scaled function $D_F F(x)$ is considered close enough to zero to terminate the algorithm. The algorithm is halted if the maximum component of $D_F F(x_+)$ is $\leq fvectol$. This is the primary stopping condition for nonlinear equations; the values of $typF$ and $fvectol$ should be chosen so that this test reflects the user's idea of what constitutes a solution to the problem.

Suggested default value : $fvectol = macheps^{1/3}$

steptol -- see Guideline 2

mintol -- A positive scalar used to test whether the algorithm is stuck at a local minimizer of $f(x) = \frac{1}{2}\|D_F F(x)\|_2$ where $F(x) \neq 0$. The algorithm is halted if the maximum component of the scaled gradient of $f(x)$ at x_+ is $\leq mintol$; here $\nabla f(x) = J(x)^T D_F^2 F(x)$ and the scaled value of $\nabla f(x_+)$ is defined as for minimization (see *gradtol*, Guideline 2). *mintol* should be set rather small to be sure this condition isn't invoked inappropriately.

Suggested default value : $mintol = macheps^{2/3}$

maxstep -- see Guideline 2
itnlimit -- see Guideline 2
δ -- see Guideline 2

III.5.   GUIDELINE 6

STORAGE CONSIDERATIONS FOR NONLINEAR EQUATIONS

The storage considerations for nonlinear equations are closely related to those for unconstrained minimization given in Guideline 3; please read Guideline 3 before Guideline 6. The following are additional remarks that apply to the algorithms for nonlinear equations.

1) The entire system of algorithms for nonlinear equations uses three $n \times n$ matrices, J, M, and H. J is used to store the analytic or finite difference Jacobian. When factored secant updates are used, J also is used to hold the matrix $Z = Q^T$ used in Algorithms A3.4.1 and A8.3.2. M is used to hold the (compressed) QR factorization of J; the lower triangle of M also is used to store the matrix L used in the global portion of the algorithm. H is used by the global portion of the algorithm; only the upper triangle including the main diagonal of H is used. If analytic Jacobians, finite difference Jacobians

or unfactored secant approximations are used (i.e. anything but factored secant approximations), H easily can share storage with M if the algorithms are modified slightly as described in item 2 below. This reduces the matrix storage requirement to two $n \times n$ matrices. In addition, when analytic or finite difference Jacobians are used, J also can share storage with M if several additional modifications are made as indicated in item 3 below. This reduces the matrix storage requirement to one $n \times n$ matrix, but this reduction is not possible when secant updates are used. When factored secant approximations to the Jacobian are used, H can share storage with M only if some fairly substantial modifications are made as indicated in item 4 below.

2) To allow H to share storage with M when analytic Jacobians, finite difference Jacobians or unfactored secant approximations are used, Algorithms A5.5.2, A6.4.1-2, and A6.4.5 must be modified as described in item 2 of Guideline 3. In addition, the following modifications must be made to Algorithm A6.5.1 and Driver D6.1.3 :

Algorithm A6.5.1
   i)   Change output parameter $H \in R^{n \times n}$ to output parameter $hdiag \in R^n$
   ii) Change $H[i,j]$ to $M[i,j]$ on lines 4T.1.1, 4T.3.1 and 4E.5.1.2.
   iii) Change $H[1,j]$ to $M[1,j]$ on line 4T.2.
   iv) Change $H[j,i]$ to $M[j,i]$ on line 4T.3.1.
   v)   Change the body of the FOR loop on line 4T.4 to
          $hdiag[i] \leftarrow M[i,i] + \cdots$ (rest is unchanged)
   vi) On line 4T.5, change the parameter H to $hdiag$
   vii) Change $H[i,i]$ to $hdiag[i]$ on line 4E.5.1.1.

Driver D6.1.3
   i)   On lines 10.2T and 10.3b, change the parameter H_c to $hdiag$

3) When analytic or finite difference Jacobians are used, the matrices J and M may share storage simply by combining J and M into one parameter in the call and parameter list of Algorithm A6.5.1, changing $J[i,j]$ to $M[i,j]$ in line 1.1 of Algorithm A6.5.1, and replacing all remaining parameters M in Driver D6.1.3 by J. The only further modification required is to revise steps 4T.1-4 of Algorithm A6.5.1 to form $H \leftarrow R^T R$ instead of $H \leftarrow J^T D_F^2 J$. Since H and M (which contains R in its upper triangle) also may share storage as described in item 2 above, this is a bit tricky. An easy way to accomplish it is to use the lower triangle of M, which is unoccupied at that time, as temporary storage for H. The details of this conversion are left as an exercise.

4) When factored secant approximations to the Jacobian are used, the matrix H may share storage with M only if the following changes are made to the system of algorithms. The details are left as an exercise.
   i)   Factored secant updates are not allowed when $global = 2$.
   ii) Modifications similar to those described in item 2 above are made to Algorithm A6.5.1fac and Driver D6.1.3.
   iii) In steps 3T.1-4 of Algorithm A6.5.1fac, $R^T R$ is formed in the lower triangle of M, and the *CHOLDECOMP* routine called by step 3T.5 is replaced by one that does the decomposition in place in the lower triangle of the matrix. (CHOLDECOMP as written inputs an upper triangular matrix and outputs a lower triangular matrix.)

III.6   ALGORITHM D6.1.4. (NEEXAMPLE)

DRIVER FOR A NONLINEAR EQUATIONS ALGORITHM USING DOGLEG AND FINITE DIFFERENCE JACOBIAN APPROXIMATION

Purpose : Attempt to find a root x_* of $F(x) : R^n \to R^n$ starting from x_0, using a dogleg algorithm with finite difference Jacobians

Input Parameters : $n \in Z$ $(n \geq 1)$, $x_0 \in R^n$, FVEC (the name of a user-supplied function $F: R^n \to R^n$ that is a least once continuously differentiable)
plus optionally : $typx \in R^n$, $typF \in R^n$, $Fdigits \in Z$, $fvectol \in R$, $steptol \in R$, $mintol \in R$, $maxstep \in R$, $itnlimit \in Z$, $\delta \in R$, $printcode \in Z$ (for further information see Algorithm D6.1.3)

Input-Output Parameters : none   (see item 2 under Storage Considerations, Alg. D6.1.3)

Output Parameters : $x_f \in R^n$ (the final approximation to x_*), $termcode \in Z$ (indicates which termination condition caused the algorithm to stop)
plus optionally :
$FV_+ \in R^n$ $(= F(x_f))$, $J_c \in R^{n \times n}$ $(\cong J(x_f))$, other output parameters as desired by the implementor (see Section I.4)

Meaning of Return Codes, Storage Considerations :
same as Algorithm D6.1.3

Algorithm :

(* Initialization section : *)
1. CALL MACHINEPS ($macheps$)
2. CALL NEINCK (parameter list selected by implementor)
    (* see NEINCK description *)
3. IF $termcode < 0$ THEN
    3.1 $x_f \leftarrow x_0$
    3.2 RETURN from Algorithm D6.1.4   (* write appropriate message, if desired *)
4. $itncount \leftarrow 0$
5. CALL NEFN (n, x_0, f_c)
    (* NEFN also calculates $FV_+ = F(x_0)$ and communicates it to the driver via a global parameter *)
6. CALL NESTOP0 $(n, x_0, FV_+, S_F, fvectol, termcode, consecmax)$ (* Alg. A7.2.4 *)
7. IF $termcode > 0$
7T. THEN $x_f \leftarrow x_0$
7E. ELSE (* calculate initial finite difference Jacobian *)
    7E.1 CALL FDJAC $(n, x_0, FV_+, FVEC, S_x, \eta, J_c)$ (* Alg. A5.4.1 *)

$(* \; g_c \; \leftarrow \; J_c^T D_F^2 \, FV_+ \; : \; *)$

7E.2 FOR $i = 1$ TO n DO

$$g_c[i] \leftarrow \sum_{j=1}^{n} J_c[j,i] \; * \; FV_+[j] \; * \; S_F[j]^2$$

7E.3 $FV_c \leftarrow FV_+$

8. $x_c \leftarrow x_0$

9. *restart* $\leftarrow$ TRUE

$(* $ Iteration section : $*)$

10. WHILE *termcode* $= 0$ DO

    10.1 *itncount* $\leftarrow$ *itncount* $+ 1$

    10.2 CALL NEMODEL $(n, \, FV_c, \, J_c, \, g_c, \, S_F, \, S_x, \, macheps, \, global, \, M, \, H_c, \, s_N)$
        $(* $ Alg. A6.5.1 $*)$

    10.3 CALL DOGDRIVER $(n, \, x_c, \, f_c, \, \text{FVEC}, \, g_c, \, M, \, s_N, \, S_x, \, maxstep, \, steptol,$
        $\delta, \, retcode, \, x_+, \, f_+, \, maxtaken)$ $(* $ Alg. A6.4.3 $*)$

    10.4 CALL FDJAC $(n, \, x_+, \, FV_+, \, \text{FVEC}, \, S_x, \, \eta, \, J_c)$ $(* $ Alg. A5.4.1 $*)$

    10.5 $g_c \leftarrow J_c^T * D_F^2 * FV_+$
        $(* $ same as statement 7E.2 $*)$

    10.6 CALL NESTOP $(n, \, x_c, \, x_+, \, FV_+, \, f_+, \, g_c, \, S_x, \, S_F, \, retcode, \, fvectol,$
        $steptol, \, itncount, \, itnlimit, \, maxtaken, \, analjac, \, cheapF, \, mintol,$
        $consecmax, \, termcode)$ $(* $ Alg. A7.2.3 $*)$

    10.7 IF *termcode* > 0

        THEN $x_f \leftarrow x_+$

        ELSE *restart* $\leftarrow$ FALSE

    10.8 $x_c \leftarrow x_+$

    10.9 $f_c \leftarrow f_+$

    10.10 $FV_c \leftarrow FV_+$

    $(* $ END, WHILE loop 10 $*)$

$(* $ write appropriate termination message, if desired $*)$

$(* $ RETURN from Algorithm D6.1.4 $*)$

$(* $ END, Algorithm D6.1.4 $*)$

SECTION IV.   INDIVIDUAL MODULES

IV.1. MODULES SUPPLIED BY THE USER

The following are the specifications for the user-supplied functions that are required or optional when using this system of algorithms for unconstrained minimization or for solving systems of nonlinear equations.

ALGORITHM FN

ROUTINE FOR EVALUATING UNCONSTRAINED MINIMIZATION OBJECTIVE FUNCTION $f(x)$

Purpose : Evaluate the user-supplied function $f(x) : R^n \to R$ at a point x_c.

Input Parameters : $n \in Z$, $x_c \in R^n$
Input-Output Parameters : none
Output Parameters : $f_c \in R$ $(= f(x_c))$

Description :

The user must supply this routine when using this system of algorithms for unconstrained minimization. It may have any name (except the name of any routine in this package); its name is input to Driver D6.1.1. Its input and output parameters must be exactly as shown (or all the calls to FN in the system of algorithms must be changed). The input parameters n and x_c may not be altered by this routine; the routine must assign to f_c the value of the objective function at x_c.

ALGORITHM GRAD

ROUTINE FOR EVALUATING GRADIENT VECTOR $\nabla f(x)$ (Optional)

Purpose : Evaluate the user-supplied gradient $\nabla f(x) : R^n \to R^n$ of the objective function $f(x)$ at a point x_c. $(\nabla f(x)[i] = \partial f(x) / \partial x[i])$

Input Parameters : $n \in Z$, $x_c \in R^n$
Input-Output Parameters : none
Output Parameters : $g \in R^n$ ($= \nabla f(x_c)$)

Description :

The user optionally may supply this routine when using this system of algorithms for unconstrained minimization (see *analgrad* in Guideline 1). It may have any name (except the name of any routine in this package); its name is input to Driver D6.1.1. Its input and output parameters must be exactly as shown (or all the calls to GRAD in the system of algorithms must be changed). The input parameters n and x_c may not be altered by this routine; the routine must assign to g the value of the gradient of the objective function at x_c.

ALGORITHM HESS

ROUTINE FOR EVALUATING HESSIAN MATRIX $\nabla^2 f(x)$ (Optional)

Purpose : Evaluate the user-supplied Hessian $\nabla^2 f(x) : R^n \to R^{n \times n}$ of the objective function $f(x)$ at a point x_c. ($\nabla^2 f(x)[i,j] = \partial^2 f(x) / \partial x[i] \partial x[j]$; it is assumed that $f(x)$ is at least twice continuously differentiable so that $\nabla^2 f(x)$ is symmetric.)

Input Parameters : $n \in Z$, $x_c \in R^n$
Input-Output Parameters : none
Output Parameters : $H \in R^{n \times n}$ symmetric ($= \nabla^2 f(x_c)$)

Storage Considerations :
1) Only the upper triangle of H, including the main diagonal, is used by the remainder of the system of algorithms. Thus, no values need to be inserted in the lower triangle of H, i.e., $H[i,j]$ where $i > j$. If the lower triangle of H is filled, it will be ignored, but no harm will be caused to the remainder of the system of algorithms.
2) There are no changes to this algorithm if one is economizing on matrix storage.
3) Algorithm HESS may require an additional input parameter ndim if these algorithms are implemented in FORTRAN -- see item 5 of Guideline 3.

Description :

The user optionally may supply this routine when using this system of algorithms for unconstrained minimization (see *analhess* in Guideline 1). It may have any name (except the name of any routine in this package); its name is input to Driver D6.1.1. Its input and output parameters must be exactly as shown (or all the calls to HESS in the system of algorithms must be changed). The input parameters n and x_c may not be altered by this routine; the routine must assign to H the value of the Hessian of the objective function at x_c.

ALGORITHM FVEC

ROUTINE FOR EVALUATING NONLINEAR EQUATIONS FUNCTION $F(x)$

Purpose : Evaluate the user-supplied function vector $F(x) : R^n \rightarrow R^n$ at a point x_c.

Input Parameters : $n \in Z$, $x_c \in R^n$
Input-Output Parameters : none
Output Parameters : $F_c \in R^n$ ($= F(x_c)$)

Description :

The user must supply this routine when using this system of algorithms to solve systems of nonlinear equations. It may have any name (except the name of any routine in this package); its name is input to Driver D6.1.3. However if its name is not FVEC, then the word FVEC on line 2 of Algorithm NEFN must be changed to the name of this routine (see Algorithm NEFN for clarification). Its input and output parameters must be exactly as shown (or all the calls to FVEC in the system of algorithms must be changed). The input parameters n and x_c may not be altered by this routine; the routine must assign to F_c the value of the nonlinear equations function vector at x_c.

ALGORITHM JAC

ROUTINE FOR EVALUATING JACOBIAN MATRIX $J(x)$ (Optional)

Purpose : Evaluate the user-supplied Jacobian $J(x) : R^n \rightarrow R^{n \times n}$ of the function vector $F(x)$ at a point x_c. ($J[i,j] = \partial f_i(x) / \partial x[j]$ where $f_i(x)$ is the ith component function of $F(x)$.)

Input Parameters : $n \in Z$, $x_c \in R^n$
Input-Output Parameters : none
Output Parameters : $J \in R^{n \times n}$ ($= J(x_c)$)

Storage Considerations :
1) The full matrix J is used.
2) There are no changes to this algorithm if one is economizing on matrix storage.

Description :

The user optionally may supply this routine when using this system of algorithms for solving systems of nonlinear equations (see *analjac* in Guideline 4). It may have any name (except the name of any routine in this package); its name is input to Driver D6.1.3. Its input and output parameters must be exactly as shown (or all the calls to HESS in the system of algorithms must be changed). The input parameters n and x_c may not be altered by this routine; the routine must assign to J the value of the Jacobian of the nonlinear equations function vector at x_c.

IV.2.   MODULES SUPPLIED BY THE IMPLEMENTOR

ALGORITHM UMINCK

CHECKING INPUT PARAMETERS AND TOLERANCES FOR UNCONSTRAINED MINIMIZATION

Purpose : Check values of algorithmic options, tolerances, and other parameters input by the user to the unconstrained minimization driver, and assign values to algorithmic options and tolerances not input by the user, and to the constants S_x and η.

Input Parameters : $n \in Z$, $macheps \in R$, $x_0 \in R^n$
plus $typx \in R^n$ and $fdigits \in Z$ if input by the user to the unconstrained minimization driver

Optional Input or Output Parameters : $typf \in R$, $gradtol \in R$, $steptol \in R$, $maxstep \in R$, $itnlimit \in Z$, $printcode \in Z$ (see Additional Consideration 1 below); $\delta \in R$ (see Additional Consideration 2 below); $global \in Z$, $analgrad \in$ Boolean, $analhess \in$ Boolean, $cheapf \in$ Boolean, *factsec* ∈Boolean (see Additional Consideration 3 below)

Output Parameters : $S_x \in R^n$ (see Additional Consideration 4 below), $\eta \in R$, *termcode* $\in Z$

Meaning of Return Codes :
termcode $= 0$: all input satisfactory or corrected by this routine
termcode < 0 : fatal input error found, as follows
termcode $= -1$: $n < 1$
termcode < -1 : other termination codes assigned by the implementor of Algorithm UMINCK

Storage Considerations :
1) No additional vector or matrix storage is required.
2) If *typx* is input by the user, it may share storage with S_x, that is, together they may be one input-output parameter to this algorithm.

Additional Considerations :
1) The values of *typf*, *gradtol*, *steptol*, *maxstep*, and *itnlimit* either must be input by the user to the unconstrained minimization driver, or must be assigned by Algorithm UMINCK. The choice is up to the implementor of this system of algorithms. Values input by the user optionally may be checked by Algorithm UMINCK. Values assigned by Algorithm UMINCK either may be the default values given in Guideline 2, or other values selected by the implementor. (x_0 is used to calculate the default value of *maxstep*.) If *printcode* is used as discussed in Section I.4 of this Appendix, it also should be assigned a value by the user or by Algorithm UMINCK.
2) δ must be given a value by the user or by Algorithm UMINCK if *global* = 2 or 3; it is ignored if *global* = 1 or 4. See step 4 below.
3) When implementing the full system of algorithms for unconstrained minimization (Driver D6.1.1), *global*, *analgrad*, *analhess*, *cheapf*, and *factsec* either must have values input to Driver D6.1.1 or must be assigned values by Algorithm UMINCK. The same discussion as in item 1 above applies here. Suggestions for values of these algorithmic options are given in Guideline 1. If one is implementing a subset of the system of algorithms for unconstrained minimization, such as Driver D6.1.2, some or all of these five parameters may not be required.
4) If scaling is not being implemented in the entire system, the output parameter S_x and step 2 below are eliminated.

Description :

1) If $n < 1$ set *termcode* to -1 and terminate Algorithm UMINCK.
2) If *typx* was input by the user, set $S_x[i]$ to $(1/typx[i])$, $i=1, \cdots, n$, otherwise set $S_x[i] = 1, i=1, \cdots, n$.
3) If *fdigits* was input by the user, set η to max{*macheps*, $10^{-fdigits}$}, or to *macheps* if *fdigits* = -1. Otherwise, set η to *macheps* or estimate it, for example by the routine of Hamming [1973] given in Gill, Murray and Wright [1981], Section 8.5.2.3. If $\eta > 0.01$, a negative *termcode* probably should be returned.
4) If *global* = 2 or 3 and no value was input for δ, set δ to -1.
5) Assign values to algorithmic options and tolerances not input by the user, and optionally check those that were input by the user, as discussed in

Additional Considerations 1 and 3 above. Note: if *global* = 2 and *analhess* = FALSE and *cheapf* = FALSE, *factsec* must be set to FALSE. (See Guideline 1.)

ALGORITHM NEINCK

**CHECKING INPUT PARAMETERS AND TOLERANCES
FOR NONLINEAR EQUATIONS**

Purpose : Check values of algorithmic options, tolerances, and other parameters input by the user to the nonlinear equations driver, and assign values to algorithmic options and tolerances not input by the user, and to the constants S_x, S_F, and η.

Input Parameters : $n \in Z$, $macheps \in R$, $x_0 \in R^n$
plus $typx \in R^n$, $typF \in R^n$, and $Fdigits \in Z$ if input by the user to the nonlinear equations driver

Optional Input or Output Parameters : $fvectol \in R$, $steptol \in R$, $mintol \in R$, $maxstep \in R$, $itnlimit \in Z$, $printcode \in Z$ (see Additional Consideration 1 below); $\delta \in R$ (see Additional Consideration 2 below); $global \in Z$, $analjac \in$ Boolean, $cheapF \in$ Boolean, $factsec \in$ Boolean (see Additional Consideration 3 below)

Output Parameters : $S_x \in R^n$, $S_F \in R^n$ (see Additional Consideration 4 below), $\eta \in R$, $termcode \in Z$

Meaning of Return Codes :
   $termcode = 0$: all input satisfactory or corrected by this routine
   $termcode < 0$: fatal input error found, as follows
   $termcode = -1 : n < 1$
   $termcode < -1$: other termination codes assigned by the implementor of Algorithm NEINCK

Storage Considerations :
1) No additional vector or matrix storage is required.
2) If $typx$ or $typF$ are input by the user, they may share storage with S_x and S_F respectively, that is, each pair may be one input-output parameter to this algorithm.

Additional Considerations :
1) The values of $fvectol$, $steptol$, $mintol$, $maxstep$, and $itnlimit$ either must be input by the user to the nonlinear equations driver, or must be assigned by Algorithm NEINCK. The choice is up to the implementor of this system of algorithms. Values input by the user optionally may be checked

by Algorithm NEINCK. Values assigned by Algorithm NEINCK either may be the default values given in Guidelines 2 and 5, or other values selected by the implementor. (x_0 is used to calculate the default value of *maxstep*.) If *printcode* is used as discussed in Section I.4 of this Appendix, it also should be assigned a value by the user or by Algorithm NEINCK.

2) δ must be given a value by the user or by Algorithm NEINCK if *global* = 2 or 3; it is ignored if *global* = 1. See step 5 below.

3) When implementing the full system of algorithms for nonlinear equations (Driver D6.1.3), *global*, *analjac*, *cheapF*, and *factsec* either must have values input to Driver D6.1.3 or must be assigned values by Algorithm NEINCK. The same discussion as in item 1 above applies here. Suggestions for values of these algorithmic options are given in Guideline 4. If one is implementing a subset of the system of algorithms for nonlinear equations, such as Driver D6.1.4, some or all of these four parameters may not be required.

4) If scaling is not being implemented in the entire system, the output parameters S_x and S_F and steps 2 and 3 below are eliminated.

Description :

1) If $n < 1$ set *termcode* to -1 and terminate Algorithm NEINCK.

2) If *typx* was input by the user, set $S_x[i]$ to $(1/typx[i])$, $i=1, \cdots ,n$, otherwise set $S_x[i] = 1, i=1, \cdots ,n$.

3) If *typF* was input by the user, set $S_F[i]$ to $(1/typF[i])$, $i=1, \cdots ,n$, otherwise set $S_F[i] = 1, i=1, \cdots ,n$.

4) If *Fdigits* was input by the user, set η to max$\{macheps, 10^{-Fdigits}\}$, or to *macheps* if *Fdigits* = -1. Otherwise, set η to *macheps* or estimate it, for example by the routine of Hamming [1973] given in Gill, Murray and Wright [1981], Section 8.5.2.3. If $\eta > 0.01$, a negative *termcode* probably should be returned.

5) If *global* = 2 or 3 and no value was input for δ, set δ to -1.

6) Assign values to algorithmic options and tolerances not input by the user, and optionally check those that were input by the user, as discussed in Additional Considerations 1 and 3 above.

IV.3. ALGORITHMIC MODULES

ALGORITHM NEFN

CALCULATING THE SUM OF SQUARES FOR NONLINEAR EQUATIONS

Purpose : Evaluate the nonlinear equations function $F(x)$ at x_+, and calculate $f_+ \leftarrow \frac{1}{2} \|D_F F(x_+)\|_2^2$

Input Parameters : $n \in Z$, $x_+ \in R^n$
Input-Output Parameters : none
Output Parameters : $f_+ \in R$

Storage Considerations :
1) In addition to the above parameters, the input variable $S_F \in R^n$ ($D_F = \text{diag}((S_F)_1, \ldots, (S_F)_n)$) and the output variable $FV_+ \in R^n$ ($= F(x_+)$) *must* be passed as global variables between NEFN and the nonlinear equations driver D6.1.3. (In FORTRAN, this is done by placing S_F and FV_+ in a labeled COMMON block in NEFN and Alg. D6.1.3.) This is the only use of global variables in the entire system of algorithms. It is required so that the parameters of NEFN and the unconstrained minimization objective function FN are the same; this enables the identical global routines to be applicable to both unconstrained minimization and nonlinear equations.

Additional Considerations :
1) If the name of the user-supplied routine that evaluates the nonlinear equations function vector is not FVEC (see Algorithm FVEC in Section IV.1 of this appendix), then the word FVEC on line 2 of this algorithm *must* be changed to the name of this user-supplied routine. This is not necessary if the name of the user-supplied routine for $F(x)$ can be passed as a global variable between Alg. D6.1.3 and NEFN; in FORTRAN this is not possible.

Algorithm :

1. CALL FVEC(n, x_+, FV_+)   (* $FV_+ \leftarrow F(x_+)$ *)
2. $f_+ \leftarrow \frac{1}{2} * \sum_{i=1}^{n} ((S_F[i] * FV_+[i])^2)$
(* RETURN from Algorithm NEFN *)
(* END, Algorithm NEFN *)

ALGORITHM A1.3.1 (MACHINEPS) – CALCULATING MACHINE EPSILON

Purpose : Calculate machine epsilon

Input Parameters : none
Input-Output Parameters : none
Output Parameters : $macheps \in R$

Storage Considerations :
1) No vector or matrix storage is required.

Algorithm :

1. $macheps \leftarrow 1$
2. REPEAT
   $macheps \leftarrow macheps / 2$
   UNTIL $(1 + macheps) = 1$
3. $macheps \leftarrow 2 * macheps$
(* RETURN from Algorithm A1.3.1 *)
(* END, Algorithm A1.3.1 *)

ALGORITHM A3.2.1 (QRDECOMP) – QR DECOMPOSITION

Purpose : Calculate the QR decomposition of a square matrix M using the algorithm in Stewart [1973]. Upon termination of the algorithm, the decomposition is stored in M, $M1$, and $M2$ as described below.

Input Parameters : $n \in Z$
Input-Output Parameters : $M \in R^{n \times n}$ (on output, Q and R are encoded in M, $M1$, and $M2$ as described below)
Output Parameters : $M1 \in R^n$, $M2 \in R^n$, $sing \in$ Boolean

Storage Considerations :
1) No additional vector or matrix storage is required.
2) On output, Q and R are stored as described by Stewart : R is contained in the upper triangle of M except that its main diagonal is contained in $M2$, and $Q^T = Q_{n-1} \cdots Q_1$ where $Q_j = I - (u_j u_j^T / \pi_j)$, $u_j[i] = 0$, $i=1, \cdots, j-1$, $u_j[i] = M[i,j]$, $i=j, \cdots, n$, $\pi_j = M1[j]$.
3) There are no changes to this algorithm if one is economizing on matrix storage as described in Guideline 6.

Description :

The QR decomposition of M is performed using Householder transformations by the algorithm in Stewart [1973]. The decomposition returns $sing$ = TRUE if singularity of M is detected, $sing$ = FALSE otherwise. The decomposition is completed even if singularity is detected.

Algorithm :

1. *sing* ← FALSE
   (* *sing* becomes TRUE if singularity of M is detected during the decomposition *)
2. FOR $k = 1$ TO $n-1$ DO
   2.1 $\eta \leftarrow \max\limits_{k\leq i\leq n}\{|M[i,k]|\}$
   2.2 IF $\eta = 0$
   2.2T THEN   (* matrix is singular *)
       2.2T.1 $M1[k] \leftarrow 0$
       2.2T.2 $M2[k] \leftarrow 0$
       2.2T.3 *sing* ← TRUE
   2.2E ELSE
       (* form Q_k and premultiply M by it *)
       2.2E.1 FOR $i = k$ TO n DO
           $M[i,k] \leftarrow M[i,k] / \eta$
       2.2E.2 $\sigma \leftarrow \mathrm{sign}(M[k,k]) * (\sum\limits_{i=k}^{n} M[i,k]^2)^{\frac{1}{2}}$
       2.2E.3 $M[k,k] \leftarrow M[k,k] + \sigma$
       2.2E.4 $M1[k] \leftarrow \sigma * M[k,k]$
       2.2E.5 $M2[k] \leftarrow -\eta * \sigma$
       2.2E.6 FOR $j = k+1$ TO n DO
           2.2E.6.1 $\tau \leftarrow (\sum\limits_{i=k}^{n} M[i,k] * M[i,j]) / M1[k]$
           2.2E.6.2 FOR $i = k$ TO n DO
               $M[i,j] \leftarrow M[i,j] - \tau * M[i,k]$
3. IF $M[n,n] = 0$ THEN sing ← TRUE
4. $M2[n] \leftarrow M[n,n]$
(* RETURN from Algorithm A3.2.1 *)
(* END, Algorithm A3.2.1 *)

ALGORITHM A3.2.2 (QRSOLVE) – QR SOLVE

Purpose : Solve $(QR)x = b$ for x, where the orthogonal matrix Q and the upper triangular matrix R are stored as described in Algorithm A3.2.1.

Input Parameters : $n \in Z$, $M \in R^{n \times n}$, $M1 \in R^n$, $M2 \in R^n$ (Q and R are encoded in M, $M1$, and $M2$ as described in Algorithm A3.2.1)

Input-Output Parameters : $b \in R^n$ (on output, b is overwritten by the solution x)

Output Parameters : none

Storage Considerations :
1) No additional vector or matrix storage is required.
2) There are no changes to this algorithm if one is economizing on matrix storage as described in Guideline 6.

Additional Considerations :
1) Algorithm A3.2.2a must be a separate routine because it is also called separately by Algorithm A3.3.1 and Algorithm A6.5.1fac.

Description :

1) Multiply b by Q^T. Q^T is stored as $Q_{n-1} \cdots Q_1$, where each Q_j is a House-holder transformation encoded as described in Algorithm A3.2.1. Thus this step consists of premultiplying b by Q_j, $j = 1, \cdots, n-1$.
2) Solve $Rx = Q^T b$.

Algorithm :

$(* \ b \ \leftarrow \ Q^T b \ *)$
1. FOR $j = 1$ TO $n-1$ DO
   $(* \ b \ \leftarrow \ Q_j b \ *)$
   1.1 $\tau \leftarrow (\sum\limits_{i=j}^{n} M[i,j] * b[i]) / M1[j]$
   1.2 FOR $i = j$ TO n DO
      $b[i] \leftarrow b[i] - \tau * M[i,j]$
$(* \ b \ \leftarrow \ R^{-1}b \ *)$
2. CALL RSOLVE($n, M, M2, b$) $(* \ \text{Alg A.3.2.2a} \ *)$
$(* \ \text{RETURN from Algorithm A3.2.2} \ *)$
$(* \ \text{END, Algorithm A3.2.2} \ *)$

ALGORITHM A3.2.2a (RSOLVE) – R SOLVE FOR QR SOLVE

Purpose : Solve $Rx = b$ for x, where the upper triangular matrix R is stored as described in Algorithm A3.2.1.

Input Parameters : $n \in Z$, $M \in R^{n \times n}$, $M2 \in R^n$ ($M2$ contains the diagonal of R, the remainder of R is contained in the upper triangle of M)
Input-Output Parameters : $b \in R^n$ (on output, b is overwritten by the solution x)

Output Parameters : none

Storage Considerations :
1) No additional vector or matrix storage is required.
2) There are no changes to this algorithm if one is economizing on matrix storage as described in Guideline 6.

Algorithm :

1. $b[n] \leftarrow b[n] / M2[n]$
2. FOR $i = n-1$ DOWNTO 1 DO

$$b[i] \leftarrow \frac{b[i] - \sum\limits_{j=i+1}^{n} M[i,j] * b[j]}{M2[i]}$$

(* RETURN from Algorithm A3.2.2a *)
(* END, Algorithm A3.2.2a *)

ALGORITHM A3.2.3 (CHOLSOLVE) – CHOLESKY SOLVE DRIVER

Purpose : Solve $(LL^T)s = -g$ for s

Input Parameters : $n \in Z$, $g \in R^n$, $L \in R^{n \times n}$ lower triangular
Input-Output Parameters : none
Output Parameters : $s \in R^n$

Storage Considerations :
1) No additional vector or matrix storage is required.
2) g is not changed by the solution process.
3) There are no changes to this algorithm if one is economizing on matrix storage as described in Guideline 3.

Additional Considerations :
1) Algorithm A3.2.3a must be a separate routine because it is also called directly by Algorithm A6.4.2 (HOOKSTEP).

Algorithm :

(* Solve $Ly = g$: *)
1. CALL LSOLVE (n, g, L, s) (* Alg. A3.2.3a *)
   (* Solve $L^T s = y$: *)
2. CALL LTSOLVE (n, s, L, s) (* Alg. A3.2.3b *)
3. $s \leftarrow -s$

(* RETURN from Algorithm A3.2.3 *)
(* END, Algorithm A3.2.3 *)

ALGORITHM A3.2.3a (LSOLVE) — L SOLVE

Purpose : Solve $Ly = b$ for y

Input Parameters : $n \in Z$, $b \in R^n$, $L \in R^{n \times n}$ lower triangular
Input-Output Parameters : none
Output Parameters : $y \in R^n$

Storage Considerations :
1) No additional vector or matrix storage is required.
2) b and y are intended to be separate parameters. If the calling sequence causes b and y to correspond to the same vector, the algorithm will work and b will be overwritten.
3) There are no changes to this algorithm if one is economizing on matrix storage as described in Guideline 3.

Algorithm :

1. $y[1] \leftarrow b[1] / L[1,1]$
2. FOR $i = 2$ TO n DO
$$y[i] \leftarrow \frac{b[i] - \sum_{j=1}^{i-1}(L[i,j] * y[j])}{L[i,i]}$$
(* RETURN from Algorithm A3.2.3a *)
(* END, Algorithm A3.2.3a *)

ALGORITHM A3.2.3b (LTSOLVE) − I. TRANSPOSED SOLVE

Purpose : Solve $L^T x = y$ for x

Input Parameters : $n \in Z$, $y \in R^n$, $L \in R^{n \times n}$ lower triangular
Input-Output Parameters : none
Output Parameters : $x \in R^n$

Storage Considerations :
1) No additional vector or matrix storage is required.
2) y and x are intended to be separate parameters. If the calling sequence causes y and x to correspond to the same vector, the algorithm will work and y will be overwritten.
3) There are no changes to this algorithm if one is economizing on matrix storage as described in Guideline 3.

Algorithm :

1. $x[n] \leftarrow y[n] / L[n,n]$
2. FOR $i = n-1$ DOWNTO 1 DO

$$x[i] \leftarrow \frac{y[i] - \sum_{j=i+1}^{n} (L[j,i] * x[j])}{L[i,i]}$$

(* RETURN from Algorithm A3.2.3b *)
(* END, Algorithm A3.2.3b *)

ALGORITHM A3.3.1 (CONDEST)

ESTIMATING THE CONDITION NUMBER OF
AN UPPER TRIANGULAR MATRIX

Purpose : Estimate the l_1 condition number of an upper triangular matrix R, using an algorithm contained in Cline, Moler, Stewart and Wilkinson [1979].

Input Parameters : $n \in Z$, $M \in R^{n \times n}$, $M2 \in R^n$ ($M2$ contains the main diagonal of R, the remainder of R is contained in the upper triangle of M)
Input-Output Parameters : none
Output Parameters : $est \in R$ (an underestimate of the l_1 condition number of R)

Storage Considerations :
1) 3 n-vectors of additional storage are required, for p, pm, and x.
2) There are no changes to this algorithm if one is economizing on matrix storage as described in Guideline 6.

Description :

1) $est \leftarrow \|R\|_1$
2) Solve $R^T x = e$ for x, where $e_i = \pm 1$, $i = 1, \cdots, n$, and the signs of the e_i's are chosen by the algorithm in Cline, Moler, Stewart and Wilkinson, using their equation (3.19).
3) Solve $Ry = x$ for y. (In the implementation below, y overwrites x.)
4) $est \leftarrow est * \|y\|_1 / \|x\|_1$.

Algorithm :

(* steps 1 and 2 set $est \leftarrow \|R\|_1$ *)
1. $est \leftarrow |M2[1]|$
2. FOR $j = 2$ TO n DO

   2.1 $temp \leftarrow |M2[j]| + \sum_{i=1}^{j-1} |M[i,j]|$

   2.2 $est \leftarrow \max\{temp, est\}$

(* steps 3-5 solve $R^T x = e$, selecting e as they proceed *)
3. $x[1] \leftarrow 1 / M2[1]$
4. FOR $i = 2$ TO n DO

   $p[i] \leftarrow M[1,i] * x[1]$
5. FOR $j = 2$ TO n DO

   (* select e_j and calculate $x[j]$ *)

   5.1 $xp \leftarrow (1 - p[j]) / M2[j]$

   5.2 $xm \leftarrow (-1 - p[j]) / M2[j]$

   5.3 $temp \leftarrow |xp|$

   5.4 $tempm \leftarrow |xm|$

   5.5 FOR $i = j+1$ TO n DO

   5.5.1 $pm[i] \leftarrow p[i] + M[j,i] * xm$

   5.5.2 $tempm \leftarrow tempm + (|pm[i]| / |M2[i]|)$

   5.5.3 $p[i] \leftarrow p[i] + M[j,i] * xp$

   5.5.4 $temp \leftarrow temp + (|p[i]| / |M2[i]|)$

   5.6 IF $temp \geq tempm$

   5.6T THEN $x[j] \leftarrow xp$ (* $e_j = 1$ *)

   5.6E ELSE   (* $e_j = -1$ *)

   5.6E.1 $x[j] \leftarrow xm$

   5.6E.2 FOR $i = j+1$ TO n DO

   $p[i] \leftarrow pm[i]$

6. $xnorm \leftarrow \sum_{j=1}^{n} |x[j]|$

7. $est \leftarrow est / xnorm$

8. CALL RSOLVE(n, M, $M2$, x)

    (* call Alg. A3.2.2a to calculate $R^{-1}x$ *)

9. $xnorm \leftarrow \sum\limits_{j=1}^{n} |x[j]|$

10 $est \leftarrow est \ * xnorm$

(* RETURN from Algorithm A3.3.1 *)

(* END, Algorithm A3.3.1 *)

ALGORITHM A3.4.1 (QRUPDATE) – QR FACTORIZATION UPDATE

Purpose : Given the QR factorization of A, calculate the factorization Q_+R_+ of $A_+ = Q(R+uv^T)$ in $O(n)^2$ operations.

Input Parameters : $n \in Z$, $u \in R^n$, $v \in R^n$, $method \in Z$ (=1 for nonlinear equations, 2 for unconstrained minimization)

Input-Output Parameters : $Z \in R^{n \times n}$ (only used when $method = 1$, contains Q^T on input, Q_+^T on output), $M \in R^{n \times n}$ (upper triangle contains R on input, R_+ on output)

Output Parameters : none

Storage Considerations :

1) No additional vector or matrix storage is required.

2) When $method = 2$, Z is not used, and this algorithm requires only one $n \times n$ matrix since the call of Alg. A3.4.1 by Alg. A9.4.2 causes Z and M to refer to the same matrix.

3) The first lower subdiagonal of M, as well as the upper triangle including the main diagonal, are used at intermediate stages of this algorithm.

4) There are no changes to this algorithm if one is economizing on matrix storage as described in Guidelines 3 and 6.

Description :

$R+uv^T$ is premultiplied by $2(n-1)$ Jacobi rotations. The first $n-1$ transform uv^T to the matrix $\|u\|_2 e_1 v^T$ and R to an upper Hessenberg matrix R_H. The last $n-1$ transform the upper Hessenberg matrix $(R_H + \|u\|_2 e_1 v^T)$ to the upper triangular matrix R_+. If $method = 1$, Q^T is premultiplied by the same $2(n-1)$ Jacobi rotations to form Q_+^T. For further details, see section 3.4 of the text.

Algorithm :

0. FOR $i = 2$ TO n DO
    $M[i, i-1] \leftarrow 0$

    (* find the largest k such that $u[k] \neq 0$ *)
1. $k \leftarrow n$
2. WHILE $(u[k] = 0)$ AND $(k > 1)$ DO
    $k \leftarrow k-1$
    (* transform $R+uv^T$ to upper Hessenberg : *)
3. FOR $i = k-1$ DOWNTO 1 DO
    3.1 CALL JACROTATE$(n, i, u[i], -u[i+1], method, Z, M)$
         (* Alg. A3.4.1a *)
    3.2 IF $u[i] = 0$
         THEN $u[i] \leftarrow |u[i+1]|$
         ELSE $u[i] \leftarrow + \sqrt{(u[i])^2 + (u[i+1])^2}$
4. FOR $j = 1$ TO n DO
    $M[1,j] \leftarrow M[1,j] + u[1] * v[j]$
    (* transform upper Hessenberg matrix to upper triangular : *)
5. FOR $i = 1$ TO $k-1$ DO
    CALL JACROTATE$(n, i, M[i,i], -M[i+1,i], method, Z, M)$
(* RETURN from Algorithm A3.4.1 *)
(* END, Algorithm A3.4.1 *)

ALGORITHM A3.4.1a (JACROTATE) – JACOBI ROTATION

Purpose : premultiply M, and if $method = 1$, Z, by the Jacobi rotation matrix $J(i, i+1, a, b)$ as described in section 3.4.

Input Parameters : $n \in Z$, $i \in Z$, $a \in R$, $b \in R$, $method \in Z$ (=1 for nonlinear equations, 2 for unconstrained minimization)
Input-Output Parameters : $Z \in R^{n \times n}$ (only used when $method = 1$), $M \in R^{n \times n}$
Output Parameters : none

Storage Considerations :
same as for Algorithm A3.4.1

Description :

1) $c \leftarrow a / \sqrt{a^2+b^2}$, $s \leftarrow b / \sqrt{a^2+b^2}$
2) New row i of $M \leftarrow c$ * old row i of $M - s$ * old row $i+1$ of M;
    new row $i+1$ of $M \leftarrow s$ * old row i of $M + c$ * old row $i+1$ of M
3) If $method = 1$, perform step 2 on Z

Algorithm :

1. IF $a = 0$
1T. THEN
    1T.1 $c \leftarrow 0$
    1T.2 $s \leftarrow \text{sign}(b)$
1E. ELSE
    1E.1 $den \leftarrow + \sqrt{a^2 + b^2}$
    1E.2 $c \leftarrow a \;/\; den$
    1E.3 $s \leftarrow b \;/\; den$
(* premultiply M by Jacobi rotation : *)
2. FOR $j = i$ TO n DO
    2.1 $y \leftarrow M[i,j]$
    2.2 $w \leftarrow M[i+1,j]$
    2.3 $M[i,j] \leftarrow c \; * y - s \; * w$
    2.4 $M[i+1,j] \leftarrow s \; * y + c \; * w$
3. IF $method = 1$ THEN
    (* premultiply Z by Jacobi rotation : *)
    3.1 FOR $j = 1$ TO n DO
        3.1.1 $y \leftarrow Z[i,j]$
        3.1.2 $w \leftarrow Z[i+1,j]$
        3.1.3 $Z[i,j] \leftarrow c \; * y - s \; * w$
        3.1.4 $Z[i+1,j] \leftarrow s \; * y + c \; * w$
(* RETURN from Algorithm A3.4.1a *)
(* END, Algorithm A3.4.1a *)

ALGORITHM A3.4.2 (QFORM)

FORMING Q FROM THE QR FACTORIZATION

Purpose : Form the orthogonal matrix $Q^T = Q_{n-1} \cdots Q_1$ from the House-holder transformation matrices $Q_1 \cdots Q_{n-1}$ produced by the QR factorization algorithm A3.2.1. (This is only required when the factored form of Broyden's update is used.)

Input Parameters : $n \in Z$, $M \in R^{n \times n}$, $M1 \in R^n$ (Q is contained in M and $M1$ as described in Algorithm A3.2.1)
Input-Output Parameters : none
Output Parameters : $Z \in R^{n \times n}$ (contains Q^T)

Storage Considerations :
1) No additional vector or matrix storage is required.
2) There are no changes to this algorithm if one is economizing on matrix storage as described in Guideline 6.

Description :

1) $Z \leftarrow I$
2) For $k = 1 \cdots n-1$, premultiply Z by Q_k, where $Q_k = I$ if $M1[k]=0$, $Q_k = I - (u_k u_k^T) / \pi_k$ otherwise, $u_k[i]=0$, $i=1, \cdots, k-1$, $u_k[i]=M[i,k]$, $i=k, \cdots, n$, $\pi_k = M1[k]$

Algorithm :

(* $Z \leftarrow I$: *)
1. FOR $i = 1$ TO n DO
    1.1 FOR $j = 1$ TO n DO
        $Z[i,j] \leftarrow 0$
    1.2 $Z[i,i] \leftarrow 1$
2. FOR $k = 1$ TO $n-1$ DO
    2.1 IF $M1[k] \neq 0$ THEN
        (* $Z \leftarrow Q_k * Z$: *)
        2.1.1 FOR $j = 1$ TO n DO
            2.1.1.1 $\tau \leftarrow (\sum_{i=k}^{n} M[i,k] * Z[i,j]) / M1[k]$
            2.1.1.2 FOR $i = k$ TO n DO
                $Z[i,j] \leftarrow Z[i,j] - \tau * M[i,k]$
(* RETURN from Algorithm A3.4.2 *)
(* END, Algorithm A3.4.2 *)

ALGORITHM A5.4.1 (FDJAC)

FINITE DIFFERENCE JACOBIAN APPROXIMATION

Purpose : Calculate a forward difference approximation to $J(x_c)$ (the Jacobian matrix of $F(x)$ at x_c), using values of $F(x)$

Input Parameters : $n \in Z$, $x_c \in R^n$, $F_c \in R^n$ ($= F(x_c)$), FVEC (the name of $F: R^n \rightarrow R^n$), $S_x \in R^n$, $\eta \in R$
Input-Output Parameters : none
Output Parameters : $J \in R^{n \times n} \cong J(x_c)$

Storage Considerations :
1) One n-vector of additional storage is required for Fj.
2) There are no changes to this algorithm if one is economizing on matrix storage as described in Guideline 6.

Additional Considerations :
1) See additional considerations, Alg. A5.6.3

Description :

Column j of $J(x_c)$ is approximated by $(F(x_c+h_je_j)-F(x_c))/h_j$, where e_j is the jth unit vector, and $h_j = \eta^{\frac{1}{2}} * \max\{|x_c[j]|, 1/S_x[j]\} * \text{sign}(x_c[j])$. Here $1/S_x[j]$ is the typical size of $|x_c[j]|$ input by the user, and $\eta = 10^{-DIGITS}$, $DIGITS$ the number of reliable base 10 digits in $F(x)$. The corresponding elements of $J(x_c)$ and J typically will agree in about their first $(DIGITS/2)$ base 10 digits.

Algorithm :

1. $sqrteta \leftarrow \eta^{\frac{1}{2}}$
2. FOR j = 1 to n DO
   (* calculate column j of J *)
   2.1 $stepsizej \leftarrow sqrteta * \max\{|x_c[j]|, 1/S_x[j]\} * \text{sign}(x_c[j])$
       (* to incorporate a different stepsize rule, change line 2.1 *)
   2.2 $tempj \leftarrow x_c[j]$
   2.3 $x_c[j] \leftarrow x_c[j] + stepsizej$
   2.4 $stepsizej \leftarrow x_c[j]-tempj$
       (* line 2.4 reduces finite precision errors slightly; see Section 5.4 *)
   2.5 CALL FVEC(n, x_c, Fj)   (* $Fj \leftarrow F(x_c+stepsizej*e_j)$ *)
   2.6 FOR i = 1 TO n DO
          $J[i,j] \leftarrow (Fj[i] - F_c[i]) / stepsizej$
   2.7 $x_c[j] \leftarrow tempj$ (* reset $x_c[j]$ *)
   (* end of FOR loop 2 *)
(* RETURN from Algorithm A5.4.1 *)
(* END, Algorithm A5.4.1 *)

ALGORITHM A5.5.1 (MODELHESS)

FORMATION OF THE MODEL HESSIAN

Purpose : Find a $\mu \geq 0$ such that $H + \mu I \in R^{n \times n}$ is safely positive definite, where $\mu=0$ if H is already safely positive definite. Then set

$H \leftarrow H + \mu I$, and calculate the Cholesky decomposition LL^T of H.

Input Parameters : $n \in Z$, $S_x \in R^n$, $macheps \in R$
Input-Output Parameters : $H \in R^{n \times n}$ symmetric
Output Parameters : $L \in R^{n \times n}$ lower triangular

Storage Considerations :
1) No additional vector or matrix storage is required.
2) Only the upper triangle of H and the lower triangle of L, including the main diagonals of both, are used. It is assumed for simplicity that each is stored in a full matrix. However, one can economize on matrix storage by modifying the algorithm as explained in Guideline 3.

Description :

1) If H has any negative diagonal elements or the absolute value of the largest off-diagonal element of H is greater than the largest diagonal element of H, $H \leftarrow H + \mu_1 I$, where $\mu_1 > 0$ is chosen so that the new diagonal is all positive, with the ratio of its smallest to largest element $\geq (macheps)^{\frac{1}{2}}$, and the ratio of its largest element to the largest absolute off-diagonal is $\geq 1 + 2 *(macheps)^{\frac{1}{2}}$.
2) A perturbed Cholesky decomposition is performed on H (see Alg. A5.5.2). It results in $H + D = LL^T$, D a non-negative diagonal matrix that is implicitly added to H during the decomposition and contains one or more positive elements if H is not safely positive definite. On output, $maxadd$ contains the maximum element of D.
3) If $maxadd = 0$ (i.e. $D=0$), then $H = LL^T$ is safely positive definite and the algorithm terminates, returning H and L. Otherwise, it calculates the number sdd that must be added to the diagonal of H to make $(H + sdd*I)$ safely strictly diagonally dominant. Since both $(H + maxadd*I)$ and $(H + sdd*I)$ are safely positive definite, it then calculates $\mu_2 = \min\{maxadd, sdd\}$, $H \leftarrow H + \mu_2 I$, calculates the Cholesky decomposition LL^T of H, and returns H and L.

Algorithm :

(* steps 1, 14, and 15 below are omitted if scaling is not being implemented
  *)

(* scale $H \leftarrow D_x^{-1} H D_x^{-1}$, where $D_x = \text{diag}((S_x)_1,...,(S_x)_n)$ *)
1. FOR $i = 1$ TO n DO
    1.1 FOR $j = i$ TO n DO
            $H[i,j] \leftarrow H[i,j] / (S_x[i] * S_x[j])$

(* step 1 in Description : *)
2. $sqrteps \leftarrow (macheps)^{\frac{1}{2}}$
3. $maxdiag \leftarrow \max_{1 \leq i \leq n}\{H[i,i]\}$
4. $mindiag \leftarrow \min_{1 \leq i \leq n}\{H[i,i]\}$

5. $maxposdiag \leftarrow \max\{0, maxdiag\}$
6. IF $mindiag \leq sqrteps * maxposdiag$
6T. THEN   (* μ will contain amount to add to diagonal of H before attempting the Cholesky decomposition *)
    6T.1 $\mu \leftarrow 2 * (maxposdiag - mindiag) * sqrteps - mindiag$
    6T.2 $maxdiag \leftarrow maxdiag + \mu$
6E. ELSE $\mu \leftarrow 0$
7. $maxoff \leftarrow \max_{1 \leq i < j \leq n} |H[i,j]|$
8. IF $maxoff * (1 + 2 * sqrteps) > maxdiag$ THEN
    8.1 $\mu \leftarrow \mu + (maxoff - maxdiag) + 2 * sqrteps * maxoff$
    8.2 $maxdiag \leftarrow maxoff * (1 + 2 * sqrteps)$
9. IF $maxdiag = 0$ THEN   (* $H=0$ *)
    9.1 $\mu \leftarrow 1$
    9.2 $maxdiag \leftarrow 1$
10. IF $\mu > 0$ THEN   (* $H \leftarrow H + \mu I$ *)
    10.1 FOR $i = 1$ TO n DO
        $H[i,i] \leftarrow H[i,i] + \mu$
11. $maxoffl \leftarrow \sqrt{\max\{maxdiag, (maxoff / n)\}}$ (* see Alg. A5.5.2 *)

(* step 2 in Description : call perturbed Cholesky decomposition *)
12. CALL CHOLDECOMP $(n, H, maxoffl, macheps, L, maxadd)$ (* Alg. A5.5.2 *)

(* step 3 in Description : *)
13. IF $maxadd > 0$ THEN
    (* H wasn't positive definite *)
    13.1 $maxev \leftarrow H[1,1]$
    13.2 $minev \leftarrow H[1,1]$
    13.3 FOR $i = 1$ TO n DO
        13.3.1 $offrow \leftarrow \sum_{j=1}^{i-1} |H[j,i]| + \sum_{j=i+1}^{n} |H[i,j]|$
        13.3.2 $maxev \leftarrow \max\{maxev, H[i,i] + offrow\}$
        13.3.3 $minev \leftarrow \min\{minev, H[i,i] - offrow\}$
    13.4 $sdd \leftarrow (maxev - minev) * sqrteps - minev$
    13.5 $sdd \leftarrow \max\{sdd, 0\}$
    13.6 $\mu \leftarrow \min\{maxadd, sdd\}$
    (* $H \leftarrow H + \mu I$: *)
    13.7 FOR $i = 1$ TO n DO
        $H[i,i] \leftarrow H[i,i] + \mu$
(* call Cholesky decomposition of H : *)
    13.8 CALL CHOLDECOMP $(n, H, 0, macheps, L, maxadd)$ (* Alg. A5.5.2 *)

(* unscale $H \leftarrow D_x H D_x$, $L \leftarrow D_x L$ *)
14. FOR $i = 1$ TO n DO

14.1 FOR $j = i$ TO n DO

$$H[i,j] \leftarrow H[i,j] \ast S_x[i] \ast S_x[j]$$

15. FOR $i = 1$ TO n DO

15.1 FOR $j = 1$ TO i DO

$$L[i,j] \leftarrow L[i,j] \ast S_x[i]$$

(* RETURN from Algorithm A5.5.1 *)
(* END, Algorithm A5.5.1 *)

ALGORITHM A5.5.2 (CHOLDECOMP)

PERTURBED CHOLESKY DECOMPOSITION

Purpose : Find the LL^T decomposition of $H+D$, D a non-negative diagonal matrix that is added to H if necessary to allow the decomposition to continue, using an algorithm based on the modified Cholesky decomposition in Gill, Murray and Wright [1981].

Input Parameters : $n \in Z$, $H \in R^{n \times n}$ symmetric, $maxoffl \in R$, $macheps \in R$
Input-Output Parameters : none
Output Parameters : $L \in R^{n \times n}$ lower triangular, $maxadd \in R$ ($= \max\limits_{1 \leq i \leq n}\{D[i,i]\}$)

Storage Considerations :
1) No additional vector or matrix storage is required.
2) Only the upper triangle of H and the lower triangle of L, including the main diagonals of both, are used. It is assumed for simplicity that each is stored in a full matrix. However, one can economize on matrix storage by modifying the algorithm as explained in Guideline 3.

Description :

The normal Cholesky decomposition is attempted. However, if at any point the algorithm determines that H is not positive definite, or if it would set $L[j,j] \leq minl$, or if the value of $L[j,j]$ would cause any element $L[i,j]$, $i>j$, to be greater than $maxoffl$, then a positive number $D[j,j]$ implicitly is added to $H[j,j]$. $D[j,j]$ is determined so that $L[j,j]$ is the maximum of $minl$ and the smallest value that causes $\max\limits_{j<i\leq n}\{|L[i,j]|\} = maxoffl$. This strategy guarantees that $L[i,i] \geq minl$ for all i, and $|L[i,j]| \leq maxoffl$ for all $i>j$. This in turn implies an upper bound on the condition number of $H+D$.

At the same time, the algorithm is constructed so that if $minl =0$ and H is numerically positive definite, then the normal Cholesky decomposition is performed with no loss of efficiency, resulting in $H = LL^T$ and $D=0$. When

Alg. A5.5.2 is called by Alg. A6.4.2, A6.5.1 or A6.5.1fac, the input parameter $maxoffl$ will be zero and so $minl$ will be zero. When Alg. A5.5.2 is called by Alg. A5.5.1, $minl$ will be positive, but D will still equal 0 if H is positive definite and reasonably well conditioned. In general, the algorithm attempts to create D not much larger than necessary.

For further information, including an explanation of the choice of $maxoffl$, see Gill, Murray and Wright [1981].

Algorithm :

1. $minl \leftarrow (macheps)^{1/4} * maxoffl$
2. IF $maxoffl = 0$ THEN
    (* this occurs when Alg. A5.5.2 is called with H known to be positive definite by Alg A6.4.2 or Alg. A6.5.1; these lines are a finite precision arithmetic precaution for this case *)
    2.1 $maxoffl \leftarrow \sqrt{\max_{1 \le i \le n} \{|H[i,i]|\}}$
2a. $minl2 \leftarrow (macheps)^{1/2} * maxoffl$
3. $maxadd \leftarrow 0$
    (* $maxadd$ will contain the maximum amount that is implicitly added to any diagonal element of H in forming $LL^T = H+D$ *)
4. FOR $j = 1$ TO n DO
    (* form column j of L *)
    4.1 $L[j,j] \leftarrow H[j,j] - \sum_{i=1}^{j-1} (L[j,i])^2$
    4.2 $minljj \leftarrow 0$
    4.3 FOR $i = j+1$ TO n DO
         4.3.1 $L[i,j] \leftarrow H[j,i] - \sum_{k=1}^{j-1} (L[i,k] * L[j,k])$
         4.3.2 $minljj \leftarrow \max\{|L[i,j]|, minljj\}$
    4.4 $minljj \leftarrow \max\{\frac{minljj}{maxoffl}, minl\}$
    4.5 IF $L[j,j] > minljj^2$
    4.5T THEN   (* normal Cholesky iteration *)
          $L[j,j] \leftarrow \sqrt{L[j,j]}$
    4.5E ELSE   (* augment $H[j,j]$ *)
          4.5E.1 IF $minljj < minl2$ THEN
                  $minljj \leftarrow minl2$
                  (* only possible when input value of $maxoffl=0$ *)
          4.5E.2 $maxadd \leftarrow \max\{maxadd, minljj^2 - L[j,j]\}$
          4.5E.3 $L[j,j] \leftarrow minljj$
    4.6 FOR $i = j+1$ TO n DO
         $L[i,j] \leftarrow L[i,j] / L[j,j]$
(* RETURN from Algorithm A5.5.2 *)
(* END, Algorithm A5.5.2 *)

ALGORITHM A5.6.1 (FDHESSG)

FINITE DIFFERENCE HESSIAN APPROXIMATION FROM ANALYTIC GRADIENTS

Purpose : Calculate a forward difference approximation to $\nabla^2 f(x_c)$, using analytic values of $\nabla f(x)$

Input Parameters : $n \in Z$, $x_c \in R^n$, $g \in R^n$ ($= \nabla f(x_c)$), GRAD (the name of $\nabla f : R^n \to R^n$), $S_x \in R^n$, $\eta \in R$

Input-Output Parameters : none

Output Parameters : $H \in R^{n \times n}$ symmetric ($\cong \nabla^2 f(x_c)$)

Storage Considerations :
1) No additional vector or matrix storage is required.
2) Only the diagonal and upper triangle of H have correct values upon return from Algorithm A5.6.1, because these are the only portions of H that are referenced by the rest of the system of algorithms. However, the entire matrix H is used at an intermediate stage of this algorithm.
3) There are no changes to this algorithm if one is economizing on matrix storage as described in Guideline 3.

Description :

1) Approximate $\nabla^2 f(x_c)$ by H using Algorithm A5.4.1.
2) $H \leftarrow (H + H^T) / 2$.

Algorithm :

1. CALL FDJAC(n, x_c, g, GRAD, S_x, η, H)
   (* Alg. A5.4.1 *)
2. FOR $i = 1$ TO $n-1$ DO
   2.1 FOR $j = i+1$ TO n DO
         $H[i,j] \leftarrow (H[i,j] + H[j,i]) / 2$
(* RETURN from Algorithm A5.6.1 *)
(* END, Algorithm A5.6.1 *)

ALGORITHM A5.6.2 (FDHESSF)

FINITE DIFFERENCE HESSIAN APPROXIMATION
FROM FUNCTION VALUES

Purpose : Calculate a forward difference approximation to $\nabla^2 f(x_c)$, using values of $f(x)$ only

Input Parameters : $n \in Z$, $x_c \in R^n$, $f_c \in R(= f(x_c))$, FN (the name of $f : R^n \to R$), $S_x \in R^n$, $\eta \in R$
Input-Output Parameters : none
Output Parameters : $H \in R^{n \times n}$ symmetric ($\cong \nabla^2 f(x_c)$)

Storage Considerations :
1) Two n-vectors of additional storage are required, for *stepsize* and *fneighbor*.
2) Only the diagonal and upper triangle of H are filled by Algorithm A5.6.2, as these are the only portions of H that are referenced by the entire system of algorithms.
3) There are no changes to this algorithm if one is economizing on matrix storage as described in Guideline 3.

Description :

$\nabla^2 f(x_c)[i,j]$ is approximated by $H[i,j] = (f(x_c + h_i e_i + h_j e_j) - f(x_c + h_i e_i)$ $- f(x_c + h_j e_j) + f(x_c)) / (h_i * h_j)$, where e_i is the ith unit vector, and $h_i = \eta^{1/3} * \max\{|x_c[i]|, 1/S_x[i]\} * \text{sign}(x_c[i])$. Here $1/S_x[i]$ is the typical size of $|x_c[i]|$ input by the user, and $\eta = 10^{-DIGITS}$, $DIGITS$ the number of reliable base 10 digits in $f(x)$. The corresponding elements of $\nabla^2 f(x_c)$ and H typically will agree in about their first ($DIGITS/3$) base 10 digits.

Algorithm :

1. *cuberteta* $\leftarrow \eta^{1/3}$
2. FOR $i = 1$ TO n DO
   (* calculate *stepsize* $[i]$ and $f(x_c + stepsize[i] * e_i)$ *)
   2.1 *stepsize* $[i] \leftarrow$ *cuberteta* $* \max\{|x_c[i]|, 1/S_x[i]\} * \text{sign}(x_c[i])$
       (* to incorporate a different stepsize rule, change line 2.1 *)
   2.2 *tempi* $\leftarrow x_c[i]$
   2.3 $x_c[i] \leftarrow x_c[i] + stepsize[i]$
   2.4 *stepsize* $[i] \leftarrow x_c[i] - tempi$
       (* line 2.4 reduces finite precision errors slightly; see Section 5.4 *)
   2.5 CALL FN(n, x_c, *fneighbor* $[i]$)
       (* *fneighbor* $[i] \leftarrow f(x_c + stepsize[i] * e_i)$ *)
   2.6 $x_c[i] \leftarrow tempi$ (* reset $x_c[i]$ *)
3. FOR $i = 1$ TO n DO
   (* calculate row i of H *)

3.1 $tempi \leftarrow x_c[i]$

3.2 $x_c[i] \leftarrow x_c[i] + 2 * stepsize[i]$
   (* alternatively, $x_c[i] \leftarrow x_c[i] - stepsize[i]$ *)

3.3 CALL FN(n, x_c, fii)
   (* $fii \leftarrow f(x_c + 2 * stepsize[i] * e_i)$ *)

3.4 $H[i,i] \leftarrow ((f_c - fneighbor[i]) + (fii - fneighbor[i]))$ /
$\qquad\qquad (stepsize[i] * stepsize[i])$

3.5 $x_c[i] \leftarrow tempi + stepsize[i]$

3.6 FOR $j = i+1$ TO n DO
   (* calculate $H[i,j]$ *)

   3.6.1 $tempj \leftarrow x_c[j]$

   3.6.2 $x_c[j] \leftarrow x_c[j] + stepsize[j]$

   3.6.3 CALL FN(n, x_c, fij)
      (* $fij \leftarrow f(x_c + stepsize[i] * e_i + stepsize[j] * e_j)$ *)

   3.6.4 $H[i,j] \leftarrow ((f_c - fneighbor[i]) + (fij - fneighbor[j]))$ /
$\qquad\qquad\quad (stepsize[i] * stepsize[j])$

   3.6.5 $x_c[j] \leftarrow tempj$ (* reset $x_c[j]$ *)

3.7 $x_c[i] \leftarrow tempi$ (* reset $x_c[i]$ *)
   (* end of FOR loop 3 *)
(* RETURN from Algorithm A5.6.2 *)
(* END, Algorithm A5.6.2 *)

ALGORITHM A5.6.3 (FDGRAD)

FORWARD DIFFERENCE GRADIENT APPROXIMATION

Purpose : Calculate a forward difference approximation to $\nabla f(x_c)$, using values of $f(x)$

Input Parameters : $n \in Z$, $x_c \in R^n$, $f_c \in R (= f(x_c))$, FN (the name of $f: R^n \rightarrow R$), $S_x \in R^n$, $\eta \in R$

Input-Output Parameters : none

Output Parameters : $g \in R^n$ ($\cong \nabla f(x_c)$)

Storage Considerations :
1) No additional vector or matrix storage is required.

Additional Considerations :
1) This algorithm is identical to Algorithm A5.4.1, except that the parameters $f_c \in R$ and $g \in R^n$ in Algorithm A5.6.3 are replaced by the parameters

$F_c \in R^n$ and $J \in R^{n \times n}$ in Algorithm A5.4.1, and line 2.6 in Algorithm A5.6.3 is replaced by the corresponding loop in Algorithm A5.4.1.

Description :

$\nabla f(x_c)[j]$ is approximated by $g[j] = (f(x_c + h_j e_j) - f(x_c))/h_j$, where e_j is the jth unit vector, and $h_j = \eta^{1/2} * \max\{|x_c[j]|, 1/S_x[j]\} * \text{sign}(x_c[j])$. Here $1/S_x[j]$ is the typical size of $|x_c[j]|$ input by the user, and $\eta = 10^{-DIGITS}$, *DIGITS* the number of reliable base 10 digits in $f(x)$. The corresponding elements of $\nabla f(x_c)$ and g typically will agree in about their first $(DIGITS/2)$ base 10 digits.

Algorithm :

1. $sqrteta \leftarrow \eta^{1/2}$
2. FOR $j = 1$ to n DO
   (* calculate $g[j]$ *)
   2.1 $stepsizej \leftarrow sqrteta * \max\{|x_c[j]|, 1/S_x[j]\} * \text{sign}(x_c[j])$
      (* to incorporate a different stepsize rule, change line 2.1 *)
   2.2 $tempj \leftarrow x_c[j]$
   2.3 $x_c[j] \leftarrow x_c[j] + stepsizej$
   2.4 $stepsizej \leftarrow x_c[j] - tempj$
      (* line 2.4 reduces finite precision errors slightly; see Section 5.4 *)
   2.5 CALL FN(n, x_c, fj)   (* $fj \leftarrow f(x_c + stepsizej * e_j)$ *)
   2.6 $g[j] \leftarrow (fj - f_c) / stepsizej$
   2.7 $x_c[j] \leftarrow tempj$ (* reset $x_c[j]$ *)
   (* end of FOR loop 2 *)
(* RETURN from Algorithm A5.6.3 *)
(* END, Algorithm A5.6.3 *)

ALGORITHM A5.6.4 (CDGRAD)

CENTRAL DIFFERENCE GRADIENT APPROXIMATION

Purpose : Calculate a central difference approximation to $\nabla f(x_c)$, using values of $f(x)$

Input Parameters : $n \in Z$, $x_c \in R^n$, FN (the name of $f: R^n \to R$), $S_x \in R^n$, $\eta \in R$
Input-Output Parameters : none
Output Parameters : $g \in R^n$ $(\cong \nabla f(x_c))$

Storage Considerations :
1) No additional vector or matrix storage is required.

Additional Considerations :
1) Central difference gradient approximation may be helpful in solving some unconstrained optimization problems, as discussed in section 5.6. Driver D6.1.1 is not set up to use Algorithm A5.6.4, but it can easily be augmented to include an automatic switch from forward to central difference gradient approximation, as is shown below. The switch would be invoked when an algorithm using forward difference gradients fails to achieve sufficient descent at some iteration but has not yet satisfied the gradient termination criterion. In this case, the current iteration would be repeated using a central difference approximation to $\nabla f(x_c)$, and then the algorithm would continue using central difference gradients exclusively until some termination criterion is satisfied.

The revisions to Driver D6.1.1 to incorporate this automatic switch to central differences are:

Change the ELSE clause of statement 6 to:
> 6E. ELSE
>> 6E.1 CALL FDGRAD(n, x_0, f_c, FN, S_x, η, g_c)
>> 6E.2 $fordiff$ ← TRUE

Insert between statements 10.3 and 10.4:
> 10.3½ IF ($global$ =2) OR ($global$ =3) THEN $trustsave$ ← δ

Insert between statements 10.4d and 10.5:
> 10.4½ IF ($retcode$ =1) AND (NOT $analgrad$) AND ($fordiff$ =TRUE) THEN
>> 10.4½.1 CALL CDGRAD(n, x_c, FN, S_x, η, g_c)
>> 10.4½.2 $fordiff$ ← FALSE
>> 10.4½.3 IF ($global$ =2) OR ($global$ =3) THEN δ ← $trustsave$
>> 10.4½.4 IF $global$ =2 THEN
>>> 10.4½.4.1 CALL MODELHESS(n, S_x, $macheps$, H_c, L_c)
>>> 10.4½.4.2 μ ← 0
>> 10.4½.5 GO TO 10.3

Change the ELSE clause of line 10.5.1 to:
> 10.5.1E ELSE
>> 10.5.1E.1 IF $fordiff$ = TRUE
>>> THEN CALL FDGRAD(n, x_+, f_+, FN, S_x, η, g_+)
>>> ELSE CALL CDGRAD(n, x_+, FN, S_x, η, g_+)

Description :

$\nabla f(x_c)[j]$ is approximated by $g[j] = (f(x_c + h_j e_j) - f(x_c - h_j e_j)) / (2^* h_j)$, where e_j is the jth unit vector, and $h_j = \eta^{1/3}$ * max$\{|x_c[j]|, 1/S_x[j]\}$ * sign($x_c[j]$). Here $1/S_x[j]$ is the typical size of $|x_c[j]|$ input by the user, and $\eta = 10^{-DIGITS}$, $DIGITS$ the number of reliable base 10 digits in $f(x)$. The corresponding elements of $\nabla f(x_c)$ and g typically will agree in about their first $(2 \, *DIGITS / 3)$ base 10 digits.

Algorithm :

1. $cuberteta \leftarrow \eta^{1/3}$
2. FOR $j = 1$ to n DO
   (* calculate $g[j]$ *)
   2.1 $stepsizej \leftarrow cuberteta \, * \max\{|x_c[j]|, \, 1/S_x[j]\} \, * \, \text{sign}(x_c[j])$
      (* to incorporate a different stepsize rule, change line 2.1 *)
   2.2 $tempj \leftarrow x_c[j]$
   2.3 $x_c[j] \leftarrow x_c[j] + stepsizej$
   2.4 $stepsizej \leftarrow x_c[j] - tempj$
      (* line 2.4 reduces finite precision errors slightly; see Section 5.4 *)
   2.5 CALL FN(n, x_c, fp) $(* fp \leftarrow f(x_c + stepsizej \, *e_j)$ *)
   2.6 $x_c[j] \leftarrow tempj - stepsizej$
   2.7 CALL FN(n, x_c, fm) $(* fm \leftarrow f(x_c - stepsizej \, *e_j)$ *)
   2.8 $g[j] \leftarrow (fp - fm) / (2 \, * stepsizej)$
   2.9 $x_c[j] \leftarrow tempj$ (* reset $x_c[j]$ *)
   (* end of FOR loop 2 *)
(* RETURN from Algorithm A5.6.4 *)
(* END, Algorithm A5.6.4 *)

ALGORITHM A6.3.1 (LINESEARCH) – LINE SEARCH

Purpose : Given $g^T p < 0$ and $\alpha < \frac{1}{2}$ $(\alpha = 10^{-4}$ is used), find $x_+ = x_c + \lambda p$, $\lambda \in (0,1]$ such that $f(x_+) \leq f(x_c) + \alpha \lambda g^T p$, using a back-tracking line search.

Input Parameters : $n \in Z$, $x_c \in R^n$, $f_c \in R(= f(x_c))$, FN (the name of $f : R^n \rightarrow R$), $g \in R^n (= \nabla f(x_c))$, $p \in R^n$, $S_x \in R^n$, $maxstep \in R$, $steptol \in R$
Input-Output Parameters : none
Output Parameters : $retcode \in Z$, $x_+ \in R^n$, $f_+ \in R(= f(x_+))$, $maxtaken \in$ Boolean

Meaning of Return Codes :
   $retcode = 0$: satisfactory x_+ found
   $retcode = 1$: routine failed to locate satisfactory x_+ sufficiently

distinct from x_c

Storage Considerations :
1) No additional vector or matrix storage is required.

Additional Considerations :
1) In all our calls of this algorithm (in Drivers D6.1.1-3), p is the Newton or secant step. However, this algorithm will work for any p that obeys $g^T p < 0$.

Description :

(Let $x_+(\lambda)$ denote $x_c + \lambda p$.)
1) Set $\lambda = 1$.
2) Decide whether $x_+(\lambda)$ is satisfactory. If so, set $retcode = 0$ and terminate Algorithm A6.3.1. If not:
3) Decide whether steplength is too small. If so, set $retcode = 1$ and terminate Algorithm A6.3.1. If not:
4) Decrease λ by a factor between 0.1 and 0.5, as follows:
   (Let $f_p'(x_c)$ denote the directional derivative of $f(x)$ at x_c in the direction p.)
   a. On the first backtrack: select the new λ such that $x_+(\lambda)$ is the minimizer of the one dimensional quadratic interpolating $f(x_c)$, $f_p'(x_c)$, $f(x_c+p)$, but constrain the new λ to be ≥ 0.1. (It is guaranteed that the new $\lambda < 1/(2(1-\alpha)) = 1/1.9998$; see Section 6.3.2.)
   b. On all subsequent backtracks: select the new λ such that $x_+(\lambda)$ is the local minimizer of the one dimensional cubic interpolating $f(x_c)$, $f_p'(x_c)$, $f(x_+(\lambda))$, $f(x_+(\lambda_{prev}))$, but constrain the new λ to be in $[0.1*$ oldλ, $0.5*$ old$\lambda]$.
5) Return to step 2.

Algorithm :

1. $maxtaken \leftarrow$ FALSE
2. $retcode \leftarrow 2$
3. $\alpha \leftarrow 10^{-4}$
   (* α is a constant used in the step-acceptance test $f(x_+) \leq f(x_c) + \alpha \lambda g^T p$; it can be changed by changing this statement *)
4. $Newtlen \leftarrow \|D_x p\|_2$
   (* $D_x = \text{diag}((S_x)_1,...,(S_x)_n)$ is stored as $S_x \in R^n$. For suggested implementation of expressions involving D_x, see Section I5 of this appendix. *)
5. IF $Newtlen > maxstep$ THEN
   (* Newton step $x_+ = x_c + p$ longer than maximum allowed *)
   5.1 $p \leftarrow p * (maxstep / Newtlen)$
   5.2 $Newtlen \leftarrow maxstep$
6. $initslope \leftarrow g^T p$
7. $rellength \leftarrow \max_{1 \leq i \leq n}\{|p[i]|/(\max\{|x_c[i]|, 1/S_x[i]\})\}$
   (* $rellength$ is relative length of p as calculated in the stopping routine *)

8. *minlambda* ← *steptol / rellength*

    (* *minlambda* is the minimum allowable steplength *)

9. $\lambda \leftarrow 1$

10. REPEAT (* loop to check whether $x_+ = x_c + \lambda p$ is satisfactory, and generate next λ if required *)

    10.1 $x_+ \leftarrow x_c + \lambda p$

    10.2   CALL FN (n, x_+, f_+) (* $f_+ \leftarrow f(x_+)$ *)

    10.3a   IF $f_+ \le f_c + \alpha * \lambda *$ *initslope* THEN

        (* satisfactory x_+ found *)

        10.3a.1   *retcode* ← 0

        10.3a.2   IF $\lambda = 1$ AND (*Newtlen* > 0.99 * *maxstep*) THEN

                maxtaken ← TRUE

        (* Algorithm A6.3.1 RETURNS from here *)

    10.3b   ELSEIF $\lambda <$ *minlambda* THEN

        (* no satisfactory x_+ can be found sufficiently distinct from x_c *)

        10.3b.1   *retcode* ← 1

        10.3b.2 $x_+ \leftarrow x_c$

        (* Algorithm A6.3.1 RETURNS from here *)

    10.3c   ELSE   (* reduce λ *)

        10.3c.1   IF $\lambda = 1$

        10.3c.1T   THEN   (* first backtrack, quadratic fit *)

           10.3c.1T.1 $\lambda temp \leftarrow -initslope / (2 * (f_+ - f_c - initslope))$

        10.3c.1E   ELSE (* all subsequent backtracks, cubic fit *)

           10.3c.1E.1 $\begin{bmatrix} a \\ b \end{bmatrix} \leftarrow \dfrac{1}{(\lambda - \lambda prev)} * \begin{bmatrix} 1/\lambda^2 & -1/(\lambda prev^2) \\ -\lambda prev/\lambda^2 & \lambda/(\lambda prev^2) \end{bmatrix}$

               $* \begin{bmatrix} f_+ - f_c - \lambda * initslope \\ f_+ prev - f_c - \lambda prev * initslope \end{bmatrix}$

           10.3c.1E.2 $disc \leftarrow b^2 - 3 * a * initslope$

           10.3c.1E.3   IF $a = 0$

                THEN (* cubic is a quadratic *)

                   $\lambda temp \leftarrow -initslope / (2 * b)$

                ELSE (* legitimate cubic *)

                   $\lambda temp \leftarrow (-b + (disc)^{1/2}) / (3 * a)$

           10.3c.1E.4   IF $\lambda temp > 0.5 * \lambda$ THEN $\lambda temp \leftarrow 0.5 * \lambda$

        10.3c.2 $\lambda prev \leftarrow \lambda$

        10.3c.3 $f_+ prev \leftarrow f_+$

        10.3c.4   IF $\lambda temp \le 0.1 * \lambda$ THEN $\lambda \leftarrow 0.1 * \lambda$ ELSE $\lambda \leftarrow \lambda temp$

        (* end of alternative 10.3c, "reduce λ", and of multiple branch statement 10.3 *)

10U   UNTIL *retcode* < 2   (* end of REPEAT loop 10 *)

(* RETURN from Algorithm A6.3.1 *)

(* END, Algorithm A6.3.1 *)

ALGORITHM A6.3.1mod (LINESEARCHMOD)

LINE SEARCH WITH DIRECTIONAL DERIVATIVE CONDITION

Purpose : Given $g^T p < 0$, $\alpha < \frac{1}{2}$ ($\alpha = 10^{-4}$ is used) and $\beta \in (\alpha, 1)$ ($\beta = 0.9$ is used), find $x_+ = x_c + \lambda p$, $\lambda > 0$, such that $f(x_+) \le f(x_c) + \alpha \lambda g^T p$ and $\nabla f(x_+)^T p \ge \beta g^T p$.

Input Parameters: those in Algorithm A6.3.1, plus $\eta \in \mathbb{R}$, *analgrad* $\in$ Boolean, GRAD (the name of $\nabla f : R^n \to R^n$ if *analgrad* = TRUE, dummy name otherwise)

Input-Output Parameters : none

Output Parameters : those in Algorithm A6.3.1, plus $g_+ \in R^n$($= \nabla f(x_+)$)

Meaning of Return Codes, Storage Considerations, Additional Considerations :
same as Algorithm A6.3.1

Description :

(Let $x_+(\lambda)$ denote $x_c + \lambda p$, and let the two inequalities $f(x_+) \le f(x_c) + \alpha \lambda g^T p$ and $\nabla f(x_+)^T p \ge \beta g^T p$ be referred to as the "α condition" and the "β condition", respectively.)
1) Set $\lambda = 1$.
2) If $x_+(\lambda)$ satisfies the α and β conditions, set *retcode* = 0 and terminate Algorithm A6.3.1mod. If not:
3) If $x_+(\lambda)$ satisfies the α condition only and $\lambda \ge 1$, set $\lambda \leftarrow 2 * \lambda$ and go to step 2.
4) If $x_+(\lambda)$ satisfies the α condition only and $\lambda < 1$, or $x_+(\lambda)$ doesn't satisfy the α condition and $\lambda > 1$, then:
   a. If $\lambda < 1$, define $\lambda lo = \lambda$, λhi = the last previously attempted value of λ. If $\lambda > 1$, define $\lambda hi = \lambda$, λlo = the last previously attempted value of λ. Note: in both cases, $x_+(\lambda lo)$ satisfies the α condition and not the β condition, $x_+(\lambda hi)$ doesn't satisfy the α condition, and $\lambda lo < \lambda hi$.
   b. Find a $\lambda \in (\lambda lo, \lambda hi)$ for which $x_+(\lambda)$ satisfies the α and β conditions, using successive quadratic interpolations. Then set *retcode* = 0 and terminate Algorithm A6.3.1mod.
5) Otherwise ($x_+(\lambda)$ doesn't satisfy the α condition and $\lambda \le 1$) do the same backtracking step as is described in step 4 of the Description of Algorithm A6.3.1, and go to step 2.

Algorithm :

The same as Algorithm A6.3.1, except that steps 10.3a.1 and 10.3a.2 are replaced by the following:
10.3a.1 IF *analgrad*
   THEN CALL GRAD(n, x_+, g_+)
   ELSE CALL FDGRAD(n, x_+, f_+, FN, S_x, η, g_+)   (* Alg. A5.6.3, forward difference gradient approximation *)

(* or instead CALL CDGRAD(n, x_+, FN, S_x, η, g_+), Alg. A5.6.4, if central difference gradient is desired *)

10.3a.2 $\beta \leftarrow 0.9$

(* to change the constant β in this algorithm, just change this line *)

10.3a.3 $newslope \leftarrow g_+^T p$

10.3a.4 IF $newslope < \beta$ * $initslope$ THEN

    10.3a.4.1 IF $(\lambda = 1)$ AND $(Newtlen < maxstep)$ THEN

        10.3a.4.1.1 $maxlambda \leftarrow maxstep / Newtlen$

        10.3a.4.1.2 REPEAT

            10.3a.4.1.2.1 $\lambda prev \leftarrow \lambda$

            10.3a.4.1.2.2 $f_+prev \leftarrow f_+$

            10.3a.4.1.2.3 $\lambda \leftarrow \min\{2 * \lambda, maxlambda\}$

            10.3a.4.1.2.4 $x_+ \leftarrow x_c + \lambda * p$

            10.3a.4.1.2.5 CALL FN(n, x_+, f_+)   (* $x_+ \leftarrow f(x_+)$ *)

            10.3a.4.1.2.6 IF $f_+ \leq f_c + \alpha * \lambda *$ $initslope$ THEN

                10.3a.4.1.2.6.1 same as line 10.3a.1 above

                10.3a.4.1.2.6.2 $newslope \leftarrow g_+^T p$

            (* end of IFTHEN 10.3a.4.1.2.6 *)

        10.3a.4.1.2U   UNTIL $(f_+ > f_c + \alpha * \lambda * initslope)$ OR $(newslope \geq \beta * initslope)$ OR $(\lambda \geq maxlambda)$

    (* end of IFTHEN 10.3a.4.1 *)

    10.3a.4.2 IF $(\lambda < 1)$ OR $((\lambda > 1)$ AND $(f_+ > f_c + \alpha * \lambda * initslope))$ THEN

        10.3a.4.2.1 $\lambda lo \leftarrow \min\{\lambda, \lambda prev\}$

        10.3a.4.2.2 $\lambda diff \leftarrow |\lambda prev - \lambda|$

        10.3a.4.2.3 IF $\lambda < \lambda prev$

        10.3a.4.2.3T THEN

            10.3a.4.2.3T.1 $flo \leftarrow f_+$

            10.3a.4.2.3T.2 $fhi \leftarrow f_+prev$

        10.3a.4.2.3E ELSE

            10.3a.4.2.3E.1 $flo \leftarrow f_+prev$

            10.3a.4.2.3E.2 $fhi \leftarrow f_+$

        10.3a.4.2.4 REPEAT

$$10.3a.4.2.4.1 \quad \lambda incr \leftarrow \frac{- newslope * (\lambda diff)^2}{2 * (fhi - (flo + newslope * \lambda diff))}$$

            10.3a.4.2.4.2 IF $\lambda incr < 0.2 * \lambda diff$ THEN

                $\lambda incr \leftarrow 0.2 * \lambda diff$

            10.3a.4.2.4.3 $\lambda \leftarrow \lambda lo + \lambda incr$

            10.3a.4.2.4.4 $x_+ \leftarrow x_c + \lambda * p$

            10.3a.4.2.4.5 CALL FN(n, x_+, f_+)   (* $x_+ \leftarrow f(x_+)$ *)

            10.3a.4.2.4.6 IF $f_+ > f_c + \alpha * \lambda * initslope$

            10.3a.4.2.4.6T THEN

                10.3a.4.2.4.6T.1 $\lambda diff \leftarrow \lambda incr$

                10.3a.4.2.4.6T.2 $fhi \leftarrow f_+$

            10.3a.4.2.4.6E ELSE

10.3a.4.2.4.6E.1 same as line 10.3a.1 above

10.3a.4.2.4.6E.2 $newslope \leftarrow g_+^T p$

10.3a.4.2.4.6E.3 IF $newslope < \beta * initslope$ THEN

    10.3a.4.2.4.6E.3.1 $\lambda lo \leftarrow \lambda$

    10.3a.4.2.4.6E.3.2 $\lambda diff \leftarrow \lambda diff - \lambda incr$

    10.3a.4.2.4.6E.3.3 $flo \leftarrow f_+$

(* end of IFTHEN 10.3a.4.2.4.6E.3 and IFTHENELSE 10.3a.4.2.4.6 *)

10.3a.4.2.4U   UNTIL   ($newslope \geq \beta * initslope$)   OR

    ($\lambda diff < minlambda$)

10.3a.4.2.5 IF $newslope < \beta * initslope$ THEN

    (* couldn't satisfy β condition *)

    10.3a.4.2.5.1 $f_+ \leftarrow flo$

    10.3a.4.2.5.2 $x_+ \leftarrow x_c + \lambda lo * p$

(* end of IFTHEN 10.3a.4.2 and IFTHEN 10.3a.4 *)

10.3a.5 $retcode \leftarrow 0$

10.3a.6 IF ($\lambda * Newtlen > 0.99 * maxstep$) THEN

    $maxtaken \leftarrow$ TRUE

(* END, additions in Algorithm A6.3.1mod *)

ALGORITHM A6.4.1 (HOOKDRIVER)

LOCALLY CONSTRAINED OPTIMAL ("HOOK") DRIVER

Purpose : Find an $x_+ = x_c - (H + \mu D_x^2)^{-1} g$, $\mu \geq 0$, such that $f(x_+) \leq f(x_c) + \alpha g^T(x_+ - x_c)$ ($\alpha = 10^{-4}$ is used) and scaled steplength $\in [.75\delta, 1.5\delta]$, starting with the input δ but increasing or decreasing δ if necessary. ($D_x = \text{diag}((S_x)_1, ..., (S_x)_n)$.) Also, produce starting trust region δ for the next iteration.

Input Parameters : $n \in Z$, $x_c \in R^n$, $f_c \in R (= f(x_c))$, FN (the name of $f : R^n \rightarrow R$), $g \in R^n (= \nabla f(x_c))$, $L \in R^{n \times n}$ lower triangular, $H \in R^{n \times n}$ symmetric $(= LL^T)$, $s_N \in R^n (= -H^{-1}g)$, $S_x \in R^n$, $maxstep \in R$, $steptol \in R$, $itncount \in Z$, $macheps \in R$

Input-Output Parameters : $\delta \in R$, $\mu \in R$, $\delta prev \in R$, $\varphi \in R$, $\varphi' \in R$ (the last four will not have values the first time this algorithm is called; the last two will have current values on output only if $\mu \neq 0$)

Output Parameters : $retcode \in Z$, $x_+ \in R^n$, $f_+ \in R (= f(x_+))$, $maxtaken \in$ Boolean

Meaning of Return Codes :

    $retcode$ = 0 : satisfactory x_+ found

    $retcode$ = 1 : routine failed to locate satisfactory x_+ sufficiently distinct from x_c

Storage Considerations :

1) Two n-vectors of additional storage are required, for s and x_+prev.

2) Only the upper triangle of H and the lower triangle of L, including the main diagonals of both, are used. It is assumed for simplicity that each is stored in a full matrix. However, one can economize on matrix storage by modifying the algorithm as explained in Guideline 3.

Description :

1) Call Algorithm A6.4.2 to find a step $s = -(H+\mu D_x^2)^{-1}g$, $\mu \geq 0$, such that $\|D_x s\|_2 \in [0.75\delta, 1.5\delta]$, or set $s = s_N$ if $\|D_x s_N\|_2 \leq 1.5\delta$.

2) Call Algorithm A6.4.5 to decide whether to accept x_+ as the next iterate, and to calculate the new value of δ. If $retcode$ = 2 or 3, return to step 1; if $retcode$ = 0 or 1, terminate algorithm A6.4.1.

Algorithm :

1. $retcode \leftarrow 4$

    (* signifies initial call of Algorithm A6.4.2 *)

2. $firsthook \leftarrow$ TRUE.

3. $Newtlen \leftarrow \|D_x s_N\|_2$

    (* $D_x = \text{diag}((S_x)_1,...,(S_x)_n)$ is stored as $S_x \in R^n$. For suggested implementation of expressions involving D_x, see Section I.5 of this appendix. *)

4. IF $(itncount = 1)$ OR $(\delta = -1)$ THEN

    4.1 $\mu \leftarrow 0$ (* needed by Alg. A6.4.2 *)

    4.2 IF $\delta = -1$ THEN

        (* no initial trust region was provided by user (or nonlinear equations secant method restart); set δ to length of scaled Cauchy step *)

        4.2.1 $\alpha \leftarrow \|D_x^{-1}g\|_2^2$

        (* 4.2.2-3 implement $\beta \leftarrow \|L^T D_x^{-2}g\|_2^2$ *)

        4.2.2 $\beta \leftarrow 0$

        4.2.3 FOR $i = 1$ TO n DO

            4.2.3.1 $temp \leftarrow \sum_{j=i}^{n}(L[j,i] * g[j]/(S_x[j] * S_x[j]))$

            4.2.3.2 $\beta \leftarrow \beta + temp * temp$

        4.2.4 $\delta \leftarrow \alpha * \alpha^{\frac{1}{2}}/\beta$

        4.2.5 IF $\delta > maxstep$ THEN $\delta \leftarrow maxstep$

5. REPEAT (* calculate and check a new step *)

    5.1 CALL HOOKSTEP $(n, g, L, H, s_N, S_x, Newtlen, macheps, \delta prev, \delta, \mu,$

        $\varphi, \varphi', firsthook, \varphi'init, s, Newttaken)$

        (* Find new step by hook step Algorithm A6.4.2 *)

5.2 $\delta prev \leftarrow \delta$

5.3 CALL TRUSTREGUP $(n, x_c, f_c, \text{FN}, g, L, s, S_x, Newttaken, maxstep,$
steptol, 1, $H, \delta, retcode, x_+prev, f_+prev, x_+, f_+, maxtaken)$
(* check new point and update trust radius, Alg. A6.4.5 *)

5U UNTIL $retcode < 2$

(* RETURN from Algorithm A6.4.1 *)
(* END, Algorithm A6.4.1 *)

ALGORITHM A6.4.2 (HOOKSTEP)

LOCALLY CONSTRAINED OPTIMAL ("HOOK") STEP

Purpose : Find an approximate solution to

$$\min_{s \in R^n} g^T s + \tfrac{1}{2} s^T H s$$

subject to $\|D_x s\|_2 \leq \delta$
by   selecting   an $s = -(H + \mu D_x^2)^{-1} g$, $\mu \geq 0$,   such   that
$\|D_x s\|_2 \in [0.75\delta, 1.5\delta]$,   or $s = H^{-1} g$ if $Newtlen \leq 1.5\delta$.
$(D_x = \text{diag}((S_x)_1, ..., (S_x)_n).)$

Input   Parameters : $n \in Z$, $g \in R^n$, $L \in R^{n \times n}$ lower   triangular,
$H \in R^{n \times n}$ symmetric $(=LL^T)$, $s_N \in R^n$ $(= -H^{-1}g)$, $S_x \in R^n$,
$Newtlen \in R$ $(=\|D_x s_N\|_2)$, $macheps \in R$, $\delta prev \in R$

Input-Output Parameters : $\delta \in R$, $\mu \in R$, $\varphi \in R$, $\varphi' \in R$, $firsthook \in$Boolean,
$\varphi'init \in R$ (φ, φ' will have new values on output only if $\mu \neq 0$; $\varphi'init$ will
have a new value on output only if $firsthook$ has become false during
this call.)

Output Parameters : $s \in R^n$, $Newttaken \in$Boolean

Storage Considerations :
1) One additional n-vector of storage is required, for $tempvec$.
2) Only the upper triangle of H and the lower triangle of L, including the
main diagonals of both, are used. It is assumed for simplicity that each is
stored in a full matrix. However, one can economize on matrix storage by
modifying the algorithm as explained in Guideline 3.

Description :

1. If $1.5\delta \geq Newtlen$, set $s = s_N$, $\delta = \min \{\delta, Newtlen\}$, and terminate Algorithm A6.4.2. If not:
2. Calculate the starting value of μ from the final value of μ at the previous

iteration, and lower and upper bounds μlow and μup on the exact solution μ_*, as explained in Section 6.4.1 of the text.

3. If $\mu \notin [\mu low, \mu up]$, set $\mu = \max\{(\mu low * \mu up)^{\frac{1}{2}}, 10^{-3}\mu up\}$

4. Calculate $s = s(\mu) = -(H + \mu D_x^2)^{-1}g$. Also calculate $\varphi = \varphi(\mu) = \|D_x s(\mu)\|_2 - \lambda$, and $\varphi' = \varphi'(\mu)$.

5. If $\|D_x s\|_2 \in [0.75\delta, 1.5\delta]$, terminate Algorithm A6.4.2. If not:

6. Calculate new values of μ, μlow, and μup as explained in Section 6.4.1, and return to step 3.

Algorithm :

1. $hi \leftarrow 1.5$

2. $lo \leftarrow 0.75$

    (*hi and lo are constants used in the step acceptance test $\|D_x s\|_2 \in [lo * \delta, hi * \delta]$; they can be changed by changing statements 1 and 2 *)

3. IF $Newtlen \leq hi * \delta$

3T. THEN (* s is Newton step *)

   3T.1 $Newttaken \leftarrow$ TRUE

   3T.2 $s \leftarrow s_N$

   3T.3 $\mu \leftarrow 0$

   3T.4 $\delta \leftarrow \min\{\delta, Newtlen\}$

   (* Algorithm A6.4.2 RETURNS from here *)

3E. ELSE (* find μ such that $\|D_x(H + \mu D_x^2)^{-1}g\|_2 \in [lo * \delta, hi * \delta\ ]$*)

   3E.1 $Newttaken \leftarrow$ FALSE

   (* compute starting value of μ, if previous step wasn't Newton step *)

   3E.2 IF $\mu > 0$ THEN

$$\mu \leftarrow \mu - \left[\frac{\varphi + \delta prev}{\delta}\right] * \left[\frac{(\delta prev - \delta) + \varphi}{\varphi'}\right]$$

   (* the term $(\delta prev - \delta)$ should be parenthesized as $\delta prev = \delta$ is possible *)

   (* steps 3E.3-6 compute lower and upper bounds on μ_* *)

   3E.3 $\varphi \leftarrow Newtlen - \delta$

   3E.4 IF $firsthook$ THEN (* calculate $\varphi'init$ *)

      3E.4.1 $firsthook \leftarrow$ FALSE

      (* steps 3E.4.2-3 result in $tempvec \leftarrow L^{-1}D_x^2 s_N$ *)

      3E.4.2 $tempvec \leftarrow D_x^2 s_N$

      (* $D_x = \text{diag}((S_x)_1,\ldots,(S_x)_n)$ is stored as $S_x \in R^n$. For suggested implementation of expressions involving D_x, see Section I.5 of this appendix. *)

      3E.4.3 CALL LSOLVE $(n, tempvec, L, tempvec)$

      (* solve $L * tempvec = D_x^2 s_N$ by Alg. A3.2.3a *)

      3E.4.4 $\varphi'init \leftarrow -\|tempvec\|_2^2\ /\ Newtlen$

   3E.5 $\mu low \leftarrow -\varphi/\varphi'init$

   3E.6 $\mu up \leftarrow \|D_x^{-1}g\|_2\ /\ \delta$

3E.7 *done* ← FALSE

3E.8 REPEAT

(* test value of μ, generate next μ if necessary *)

3E.8.1 IF $((\mu < \mu low)$ OR $(\mu > \mu up))$ THEN

$\mu \leftarrow \max\{(\mu low * \mu up)^{\frac{1}{2}}, 10^{-3}\mu up\}$

(* $H \leftarrow H + \mu D_x^2$: *)

3E.8.2 FOR $i = 1$ TO n DO

$H[i,i] \leftarrow H[i,i] + \mu * S_x[i] * S_x[i]$

(* calculate the LL^T factorization of new H, using Alg. A5.5.2 : *)

3E.8.3 CALL CHOLDECOMP $(n, H, 0, macheps, L, maxadd)$

(* solve $(LL^T)s = -g$, using Alg. A3.2.3 : *)

3E.8.4 CALL CHOLSOLVE (n, g, L, s)

(* reset H to original value: $H \leftarrow H - \mu * D_x^2$ *)

3E.8.5 FOR $i = 1$ TO n DO

$H[i,i] \leftarrow H[i,i] - \mu * S_x[i] * S_x[i]$

3E.8.6 *steplen* ← $\|D_x s\|_2$

3E.8.7 φ ← *steplen* $- \delta$

(* steps 3E.8.8-9 result in *tempvec* ← $L^{-1}D_x^2 s$ *)

3E.8.8 *tempvec* ← $D_x^2 s$

3E.8.9 CALL LSOLVE $(n, tempvec, L, tempvec)$

(* solve $L * tempvec = D_x^2 s_N$ by Alg. A3.2.3a *)

3E.8.10 φ' ← $- \|tempvec\|_2^2$ / *steplen*

3E.8.11   IF $((steplen \geq lo * \delta)$ AND $(steplen \leq hi * \delta))$ OR

$(\mu up - \mu low \leq 0)$

3E.8.11T THEN (* s is acceptable step *)

3E.8.11T.1 *done* ← TRUE

(* Algorithm A6.4.2 RETURNS from here *)

3E.8.11E ELSE (* s not acceptable, calculate new μ and bounds

μlow and μup *)

3E.8.11E.1 μlow ← $\max\{\mu low, \mu - (\varphi / \varphi')\}$

3E.8.11E.2 IF $\varphi < 0$ THEN $\mu up \leftarrow \mu$

3E.8.11E.3 $\mu \leftarrow \mu - \left[\dfrac{steplen}{\delta}\right] * \left[\dfrac{\varphi}{\varphi'}\right]$

(* end of IFTHENELSE 3E.8.11 *)

3E.8U UNTIL *done* = TRUE

(* end of IFTHENELSE 3 *)

(* RETURN from Algorithm A6.4.2 *)

(* END, Algorithm A6.4.2 *)

ALGORITHM A6.4.3 (DOGDRIVER)

DOUBLE DOGLEG DRIVER

Purpose : Find an x_+ on the double dogleg curve such that $f(x_+) \le f(x_c) + \alpha g^T(x_+ - x_c)$ ($\alpha = 10^{-4}$ is used) and scaled steplength = δ, starting with the input δ but increasing or decreasing δ if necessary. Also, produce starting trust region δ for the next iteration.

Input Parameters : $n \in Z$, $x_c \in R^n$, $f_c \in R(= f(x_c))$, FN (the name of $f : R^n \to R$), $g \in R^n (= \nabla f(x_c))$, $L \in R^{n \times n}$ lower triangular, $s_N \in R^n (= -(LL^T)^{-1}g)$, $S_x \in R^n$, $maxstep \in R$, $steptol \in R$

Input-Output Parameters : $\delta \in R$

Output Parameters : $retcode \in Z$, $x_+ \in R^n$, $f_+ \in R(= f(x_+))$, $maxtaken \in$ Boolean

Meaning of Return Codes :
$retcode = 0$: satisfactory x_+ found
$retcode = 1$: routine failed to locate satisfactory x_+ sufficiently distinct from x_c

Storage Considerations :
1) Four n-vectors of additional storage are required, for s, $\hat{s}_{SD}$, $\hat{v}$, and x_+prev.
2) There are no changes to this algorithm if one is economizing on matrix storage as described in Guidelines 3 and 6.

Description :

1) Call Algorithm A6.4.4 to find the step s on the double dogleg curve such that $\|D_x s\|_2 = \delta$, or set $s = s_N$ if $\|D_x s_N\|_2 \le \delta$. ($D_x = $ diag $((S_x)_1, ..., (S_x)_n)$).
2) Call Algorithm A6.4.5 to decide whether to accept x_+ as the next iterate, and to calculate the new value of δ. If $retcode = 2$ or 3, return to step 1; if $retcode = 0$ or 1, terminate algorithm A6.4.3.

Algorithm :

1. $retcode \leftarrow 4$
   (* signifies initial call of Algorithm A6.4.4 *)
2. $firstdog \leftarrow$ TRUE.
3. $Newtlen \leftarrow \|D_x s_N\|_2$
   (* $D_x = $ diag $((S_x)_1, ..., (S_x)_n)$ is stored as $S_x \in R^n$. For suggested implementation of expressions involving D_x, see Section I.5 of this appendix. *)
4. REPEAT (* calculate and check a new step *)
   4.1 CALL DOGSTEP $(n, g, L, s_N, S_x, Newtlen, maxstep, \delta, firstdog, Cauchylen, \eta, \hat{s}_{SD}, \hat{v}, s, Newttaken)$

(* Find new step by double dogleg Algorithm A6.4.4 *)

4.2 CALL TRUSTREGUP $(n, x_c, f_c,$ FN, $g, L, s, S_x, Newttaken, maxstep,$
$steptol, 2, L, \delta, retcode, x_+prev, f_+prev, x_+, f_+, maxtaken)$

(* check new point and update trust radius, Alg. A6.4.5 *)

(* second L in above parameter list is a dummy parameter *)

4U UNTIL $retcode < 2$

(* RETURN from Algorithm A6.4.3 *)

(* END, Algorithm A6.4.3 *)

ALGORITHM A6.4.4 (DOGSTEP) – DOUBLE DOGLEG STEP

Purpose : Find an approximate solution to

$$\underset{s \in R^n}{\text{minimize}} \; g^T s + \tfrac{1}{2} s^T LL^T s$$

subject to $\|D_x s\|_2 \le \delta$

by selecting the s on the double dogleg curve described below such that $\|D_x s\|_2 = \delta$, or $s = s_N$ if $Newtlen \le \delta$. ($D_x = \text{diag}((S_x)_1,...,(S_x)_n).$)

Input Parameters : $n \in Z$, $g \in R^n$, $L \in R^{n \times n}$ lower triangular, $s_N \in R^n$ ($= -(LL^T)^{-1}g$), $S_x \in R^n$, $Newtlen \in R$ ($= \|D_x s_N\|_2$), $maxstep \in R$

Input-Output Parameters : $\delta \in R$, $firstdog \in$ Boolean, $Cauchylen \in R$, $\eta \in R$, $\hat{s}_{SD} \in R^n$, $\hat{v} \in R^n$ (the last four parameters will have current values, on input or output, only if $firstdog$ = FALSE at that point)

Output Parameters : $s \in R^n$, $Newttaken \in$ Boolean

Storage Considerations :

1) No additional vector or matrix storage is required.

2) There are no changes to this algorithm if one is economizing on matrix storage as described in Guidelines 3 and 6.

Description :

1) If $Newtlen \le \delta$, set $s = s_N$, $\delta = Newtlen$, and terminate Algorithm A6.4.4. If not:

2) Calculate the double dogleg curve if it hasn't already been calculated during this call of Algorithm A6.4.3. The double dogleg curve is the piecewise linear curve connecting x_c, $x_c + s_{SD}$, $x_c + \eta s_N$ and $x_c + s_N$, where s_{SD} is the Cauchy step to the minimizer of the quadratic model in the scaled steepest descent direction, and $\eta \le 1$. (See Section 6.4.2 of the text for further details.)

3) Calculate the unique point $x_c + s$ on the double dogleg curve such that

$\|D_x s\|_2 = \delta.$

Algorithm :

1. IF *Newtlen* $\leq \delta$
1T. THEN (* s is Newton step *)
    1T.1 *Newttaken* ← TRUE
    1T.2 $s \leftarrow s_N$
    1T.3 $\delta \leftarrow$ *Newtlen*
    (* Algorithm A6.4.4 RETURNS from here *)
1E. ELSE
    (* Newton step too long, s on double dogleg curve *)
    1E.1 *Newttaken* ← FALSE
    1E.2 IF *firstdog* THEN
        (* calculate double dogleg curve *)
        1E.2.1 *firstdog* ← FALSE
        1E.2.2 $\alpha \leftarrow \|D_x^{-1}g\|_2^2$
            (* $D_x = \mathrm{diag}\,((S_x)_1,...,(S_x)_n)$ is stored as $S_x \in R^n$. For suggested implementation of expressions involving D_x, see Section I.5 of this appendix. *)
        (* 1E.2.3-4 implement $\beta \leftarrow \|L^T D_x^{-2}g\|_2^2$ *)
        1E.2.3 $\beta \leftarrow 0$
        1E.2.4 FOR $i = 1$ TO n DO
            1E.2.4.1 $temp \leftarrow \sum_{j=i}^{n}(L[j,i] * g[j] \,/\, (S_x[j] * S_x[j]))$
            1E.2.4.2 $\beta \leftarrow \beta + temp * temp$
        1E.2.5 $\hat{s}_{SD} \leftarrow -(\alpha/\beta)D_x^{-1}g$
            (* $\hat{s}_{SD}$ is Cauchy step in scaled metric; s_{SD} in Description is $D_x^{-1}\hat{s}_{SD}$ *)
        1E.2.6 *Cauchylen* $\leftarrow \alpha * \alpha^{\frac{1}{2}} / \beta$ (* $= \|\hat{s}_{SD}\|_2$ *)
        1E.2.7 $\eta \leftarrow 0.2 + (0.8 * \alpha^2 / (\beta * |g^T s_N|))$
            (* $\eta \leq 1$; see Section 6.4.2 *)
        1E.2.8 $\hat{v} \leftarrow \eta D_x s_N - \hat{s}_{SD}$
            (* $\hat{v}$ is $(\eta s_N - s_{SD})$ in the scaled metric *)
        1E.2.9 IF $\delta = -1$ THEN $\delta \leftarrow \min\{$*Cauchylen*, *maxstep*$\}$
            (* first iteration, and no initial trust region was provided by the user *)
        (* end of 1E.2, calculate double dogleg curve *)
    1E.3a IF $\eta *$ *Newtlen* $\leq \delta$ THEN
        (* take partial step in Newton direction *)
        1E.3a.1 $s \leftarrow (\delta/$ *Newtlen*$) * s_N$
        (* Algorithm A6.4.4 RETURNS from here *)

1E.3b ELSEIF $Cauchylen \geq \delta$ THEN
(* take step in steepest descent direction *)

1E.3b.1 $s \leftarrow (\delta / \ Cauchylen) * D_x^{-1}\hat{s}_{SD}$
(* Algorithm A6.4.4 RETURNS form here *)

1E.3c ELSE
(* calculate convex combination of s_{SD} and ηs_N $(= D_x^{-1}\hat{s}_{SD} + \lambda D_x^{-1}\hat{v})$ that has scaled length $= \delta$ *)

1E.3c.1 $temp \leftarrow \hat{v}^T \hat{s}_{SD}$

1E.3c.2 $tempv \leftarrow \hat{v}^T \hat{v}$

1E.3c.3 $\lambda \leftarrow (- temp + \sqrt{temp^2 - tempv * (Cauchylen^2 - \delta^2)}\ )$ $/$
$tempv$

1E.3c.4 $s \leftarrow D_x^{-1} (\hat{s}_{SD} + \lambda \hat{v})$

(* end of ELSE 1E.3c, and of multiple branch 1E.3, and of IFTHENELSE 1 *)

(* RETURN from Algorithm A6.4.4 *)

(* END, Algorithm A6.4.4 *)

ALGORITHM A6.4.5 (TRUSTREGUP)

UPDATING THE MODEL TRUST REGION

Purpose : Given a step s produced by the hookstep or dogstep algorithm, decide whether $x_+ = x_c + s$ should be accepted as the next iterate. If so, choose the initial trust region for the next iteration. If not, decrease or increase the trust region for the current iteration.

Input Parameters : $n \in Z$, $x_c \in R^n$, $f_c \in R(= f(x_c))$, FN (the name of $f : R^n \to R$), $g \in R^n$, $L \in R^{n \times n}$ lower triangular, $s \in R^n$, $S_x \in R^n$, $Newttaken \in$ Boolean, $maxstep \in R$, $steptol \in R$, $steptype \in R$ ($= 1$ for hookstep, 2 for dogstep), $H \in R^{n \times n}$ symmetric (only used if $steptype = 1$, contains model Hessian in this case)

Input-Output Parameters : $\delta \in R$, $retcode \in Z$, $x_+ prev \in R^n$, $f_+ prev \in R$ (the last two parameters will have current values on input, or new values on output, only if $retcode = 3$ at that point)

Output Parameters : $x_+ \in R^n$, $f_+ \in R (= f(x_+))$, *maxtaken* $\in$ Boolean

Meaning of Return Codes :
   retcode = 0 : x_+ accepted as next iterate, δ is trust region for next
      iteration
   retcode = 1 : x_+ unsatisfactory but global step is terminated because
      the relative length of $(x_+ - x_c)$ is within the stopping algorithm's
      steplength stopping tolerance, indicating in most cases that no
      further progress is possible -- see *termcode* = 3 in Algorithms
      A7.2.1 and A7.2.3
   retcode = 2 : $f(x_+)$ too large, current iteration to be continued with a
      new, reduced δ
   retcode = 3 : $f(x_+)$ sufficiently small, but the chance of taking a
      longer successful step seems sufficiently good that the current
      iteration is to be continued with a new, doubled δ

Storage Considerations :
1) No additional vector or matrix storage is required, except the n-vector
F_+prev for nonlinear equations only.
2) Only the upper triangle of H and the lower triangle of L, including the
main diagonals of both, are used. It is assumed for simplicity that each is
stored in a full matrix. However, one can economize on matrix storage by
modifying the algorithm as explained in Guideline 3.
3) When *steptype* = 2, H is not used, and this algorithm requires only one
$n \times n$ matrix since the call of Alg. A6.4.5 by Alg. A6.4.3 causes H and L to
refer to the same matrix.
4) For nonlinear equations, the global variable FV_+ is used.

Description :

1) Calculate $x_+ = x_c + s$, $f_+ = f(x_+)$, $\Delta f = f(x_+) - f(x_c)$
2) If *retcode* = 3 and $(f_+ \geq f_+prev$ or $\Delta f > 10^{-4}g^T s)$, reset x_+, f_+, to
x_+prev, f_+prev, $\delta = \delta/2$, *retcode* = 0 and terminate Algorithm A6.4.5. If
not:
3) If $\Delta f > 10^{-4}g^T s$, step is unacceptable. If relative steplength is too small,
set *retcode* = 1 and terminate Algorithm A6.4.5. If not, calculate the λ for
which $x_c + \lambda s$ is the minimizer of the one dimensional quadratic interpolating $f(x_c)$, $f(x_c + s)$, and the directional derivative of f at x_c in the direction
s; set $\delta = \lambda \|s\|$, except constrain the new δ to be between 0.1 and 0.5 of the
old δ. Terminate Algorithm A6.4.5.
4) $\Delta f \leq 10^{-4}g^T s$, so the step is acceptable. Calculate $\Delta fpred =$ the value of
$f(x_+) - f(x_c)$ predicted by the quadratic model $(= g^T s + \frac{1}{2}s^T LL^T s)$. Then
do one of:
   a. If Δf and $\Delta fpred$ agree to within relative error 0.1 or negative curvature is indicated, and a longer step is possible, and δ hasn't been
   decreased during this iteration, set $\delta = 2*\delta$, *retcode* = 3, $x_+prev = x_+$,
   $f_+prev = f_+$, and terminate Algorithm A6.4.5. If not:
   b. Set *retcode* = 0. If $\Delta f > 0.1*\Delta fpred$ set $\delta = \delta/2$, otherwise if
   $\Delta f < 0.75*\Delta fpred$ set $\delta = 2*\delta$, otherwise don't change δ. Terminate
   Algorithm A6.4.5.

Algorithm :

1. $maxtaken \leftarrow$ FALSE
2. $\alpha \leftarrow 10^{-4}$
   (* α is a constant used in the step-acceptance test $f(x_+) \leq f(x_c) + \alpha g^T(x_+ - x_c)$; it can be changed by changing this statement *)
3. $steplen \leftarrow \|D_x s\|_2$
   (* $D_x = \text{diag}((S_x)_1, ..., (S_x)_n)$ is stored as $S_x \in R^n$. For suggested implementation of expressions involving D_x, see Section 1.5 of this appendix. *)
4. $x_+ \leftarrow x_c + s$
5. CALL FN $(n, x+, f+)$ (* $f_+ \leftarrow f(x_+)$ *)
6. $\Delta f \leftarrow f_+ - f_c$
7. $initslope \leftarrow g^T s$
8. IF $retcode \neq 3$ THEN $f_+prev \leftarrow 0$
   (* may be necessary to prevent run-time error in following statement *)
9a. IF $(retcode = 3)$ AND $((f_+ \geq f_+prev)$ OR $(\Delta f > \alpha * initslope))$ THEN
    (* reset x_+ to x_+prev and terminate global step *)
    9a.1 $retcode \leftarrow 0$
    9a.2 $x_+ \leftarrow x_+prev$
    9a.3 $f_+ \leftarrow f_+prev$
    9a.4 $\delta \leftarrow \delta/2$
    9a.5 $FV_+ \leftarrow F_+prev$ (* Nonlinear Equations only *)
    (* Algorithm A6.4.5 RETURNS from here *)
9b. ELSEIF $\Delta f \geq \alpha * initslope$ THEN
    (* $f(x_+)$ too large *)
    9b.1 $rellength \leftarrow \max_{1 \leq i \leq n} \left\{ \dfrac{|s[i]|}{\max\{|x_+[i]|, 1/S_x[i]\}} \right\}$
    9b.2 IF $rellength < steptol$
    9b.2T THEN
          (* $x_+ - x_c$ too small, terminate global step *)
          9b.2T.1 $retcode \leftarrow 1$
          9b.2T.2 $x_+ \leftarrow x_c$
          (* Algorithm A6.4.5 RETURNS from here *)
    9b.2E ELSE
          (* reduce δ, continue global step *)
          9b.2E.1 $retcode \leftarrow 2$
          9b.2E.2 $\delta temp \leftarrow \dfrac{-initslope * steplen}{(2 * (\Delta f - initslope))}$
          9b.2E.3a IF $\delta temp < 0.1 * \delta$
                   THEN $\delta \leftarrow 0.1 * \delta$
          9b.2E.3b ELSEIF $\delta temp > 0.5 * \delta$
                   THEN $\delta \leftarrow 0.5 * \delta$
          9b.2E.3c ELSE $\delta \leftarrow \delta temp$

        (* Algorithm A6.4.5 RETURNS from here *)

9c ELSE (* $f(x_+)$ sufficiently small *)

    (* 9c.1-2 set $\Delta fpred \leftarrow g^T s + \frac{1}{2} s^T Hs$ *)

    9c.1 $\Delta fpred \leftarrow initslope$

    9c.2 IF $steptype = 1$

    9c.2T THEN   (* hookstep, calculate $s^T Hs$ *)

        9c.2T.1 FOR $i = 1$ TO n DO

$$9c.2T.1.1 \quad temp \leftarrow \tfrac{1}{2} * H[i,i] * s[i]^2 + \sum_{j=i+1}^{n} H[i,j] * s[i] * s[j]$$

        9c.2T.1.2 $\Delta fpred \leftarrow \Delta fpred + temp$

    9c.2E ELSE   (* dogleg step, calculate $s^T LL^T s$ instead *)

        9c.2E.1 FOR $i = 1$ TO n DO

$$9c.2E.1.1 \quad temp \leftarrow \sum_{j=i}^{n}(L[j,i] * s[j])$$

        9c.2E.1.2 $\Delta fpred \leftarrow \Delta fpred + (temp * temp / 2)$

    9c.3   IF $retcode \neq 2$ AND $((|\Delta fpred - \Delta f| \leq 0.1 * |\Delta f|)$ OR

        $(\Delta f \leq initslope))$ AND $(Newttaken = \text{FALSE})$ AND

        $(\delta \leq 0.99 * maxstep)$

    9c.3T THEN

        (* double δ and continue global step *)

        9c.3T.1 $retcode \leftarrow 3$

        9c.3T.2 $x_+prev \leftarrow x_+$

        9c.3T.3 $f_+prev \leftarrow f_+$

        9c.3T.4 $\delta \leftarrow \min\{2 * \delta, maxstep\}$

        9c.3T.5 $F_+prev \leftarrow FV_+$ (* Nonlinear Equations only *)

        (* Algorithm A6.4.5 RETURNS from here *)

    9c.3E ELSE

        (* accept x_+ as new iterate, choose new δ *)

        9c.3E.1 $retcode \leftarrow 0$

        9c.3E.2 IF $steplen > 0.99 * maxstep$ THEN

                $maxtaken \leftarrow \text{TRUE}$

        9c.3E.3a IF $\Delta f \geq 0.1 * \Delta fpred$ THEN $\delta \leftarrow \delta / 2$

                (* decrease δ for next iteration *)

        9c.3E.3b ELSEIF $\Delta f \leq 0.75 * \Delta fpred$ THEN

                $\delta \leftarrow \min\{2 * \delta, maxstep\}$

                (* increase δ for next iteration *)

        (* 9c.3E.3c ELSE δ is unchanged *)

    (* end of ELSE 9c.3E, and of ELSE 9c, and of multiple branch 9 *)

(* RETURN from Algorithm A6.4.5 *)

(* END, Algorithm A6.4.5 *)

ALGORITHM A6.5.1 (NEMODEL)

FORMATION OF THE AFFINE MODEL
FOR NONLINEAR EQUATIONS

Purpose : Factor the model Jacobian, calculate the Newton step, and provide the necessary information for the global algorithm. If the model Jacobian is singular or ill-conditioned, modify it before calculating the Newton step, as described in Section 6.5 in the text.

Input Parameters : $n \in Z$, $F_c \in R^n$ ($= F(x_c)$), $J \in R^{n \times n}$ ($= J(x_c)$ or an approximation to it), $g \in R^n$ ($= J^T D_F^2 F_c$), $S_F \in R^n$, $S_x \in R^n$, $macheps \in R$, $global \in Z$

Input-Output Parameters : none

Output Parameters : $M \in R^{n \times n}$, $H \in R^{n \times n}$, $s_N \in R^n$

Storage Considerations :
1) Two n-vectors of additional storage are required, for $M1$ and $M2$.
2) Only the upper triangle and main diagonal of H is used. H may share storage with M, if some minor modifications are made to this algorithm (and others) as explained in Guidelines 6 and 3.
3) M is used in the first part of this algorithm to contain the QR factorization of $D_F J$ as described in Alg. A3.2.1. On output, if $global$ = 2 or 3, its lower triangle and main diagonal contain the Cholesky factor L of the matrix H described in step 3 and 4 below. If $global$ = 2, this same matrix H is contained in the upper triangle and main diagonal of the parameter H on output. (H may be used within the algorithn regardless of the value of $global$.)

Description :

1) Calculate the QR decomposition of $D_F J$. ($D_F = \text{diag}((S_F)_1, \ldots, (S_F)_n)$.)
2) Estimate the l_1 condition number of RD_x^{-1}. ($D_x = \text{diag}((S_x)_1, \ldots, (S_x)_n)$.)
3) If RD_x^{-1} is singular or ill-conditioned, set $H = J^T D_F^2 J + \sqrt{n * macheps} * \|J^T D_F^2 J\|_1 * D_x^2$ (see Section 6.5), calculate the Cholesky factorization of H, set $s_N \leftarrow -H^{-1}g$ where $g = J^T D_F^2 F_c$, and terminate Algorithm A6.5.1.
4) If RD_x^{-1} is well conditioned, calculate $s_N \leftarrow -J^{-1}F_c$. If $global$ = 2 or 3, store R^T in the lower triangle of M. If $global$ = 2, set $H = R^T R$. Terminate Algorithm A6.5.1.

Algorithm :

(* $M \leftarrow D_F J$; $D_F = \text{diag}((S_F)_1, \ldots, (S_F)_n)$ *)
1. FOR i = 1 TO n DO
    1.1 FOR j = 1 TO n DO
        $M[i,j] \leftarrow S_F[i] * J[i,j]$
2. CALL QRDECOMP(n, M, $M1$, $M2$, $sing$)
    (* call Alg. A3.2.1 to calculate the QR factorization of M *)

3. IF NOT *sing*

3T. THEN   (* estimate $K_1(RD_x^{-1})$ *)

(* $R \leftarrow RD_x^{-1}$ *)

3T.1 FOR $j = 1$ TO n DO

3T.1.1 FOR $i = 1$ TO $j-1$ DO

$M[i,j] \leftarrow M[i,j] \,/\, S_x[j]$

3T.1.2 $M2[j] \leftarrow M2[j] \,/\, S_x[j]$

3T.2 CALL CONDEST(n, M, $M2$, *est*)

(* call Alg. A3.3.1 to estimate the condition number of R *)

3E. ELSE *est* $\leftarrow 0$

4. IF (*sing*) OR (*est* $> 1/ \sqrt{macheps}$)

4T. THEN   (* perturb Jacobian as described in Section 6.5 *)

(* $H \leftarrow J^T D_F^2 J$ *)

4T.1 FOR $i = 1$ TO n DO

4T.1.1 FOR $j = i$ TO n DO

$$H[i,j] \leftarrow \sum_{k=1}^{n} J[k,i] * J[k,j] * S_F[k]^2$$

(* steps 4T.2-3 calculate $Hnorm \leftarrow \|D_x^{-1} H D_x^{-1}\|_1$ *)

4T.2 $Hnorm \leftarrow (1/ S_x[1]) * \sum_{j=1}^{n} (|H[1,j]| \,/\, S_x[j])$

4T.3 FOR $i = 2$ TO n DO

4T.3.1 $temp \leftarrow (1/ S_x[i]) *$

$$\{\sum_{j=1}^{i} (|H[j,i]| \,/\, S_x[j]) + \sum_{j=i+1}^{n} (|H[i,j]| \,/\, S_x[j])\}$$

4T.3.2 $Hnorm \leftarrow \max\{temp, Hnorm\}$

(* $H \leftarrow H + \sqrt{n} * macheps * Hnorm * D_x^2$ *)

4T.4 FOR $i = 1$ TO n DO

$H[i,i] \leftarrow H[i,i] + \sqrt{n * macheps} * Hnorm * S_x[i]^2$

(* steps 4T.5-6 calculate $s_N \leftarrow H^{-1}g$ *)

4T.5 CALL CHOLDECOMP(n, H, 0, *macheps*, M, *maxadd*)   (* Alg. A5.5.2 *)

4T.6 CALL CHOLSOLVE(n, g, M, s_N)   (* Alg. A3.2.3 *)

4E. ELSE   (* calculate normal Newton step *)

(* reset $R \leftarrow RD_x$ *)

4E.1 FOR $j = 1$ TO n DO

4E.1.1 FOR $i = 1$ TO $j-1$ DO

$M[i,j] \leftarrow M[i,j] * S_x[j]$

4E.1.2 $M2[j] \leftarrow M2[j] * S_x[j]$

4E.2 $s_N \leftarrow -D_F * F_c$

(* $D_F = \text{diag}((S_F)_1, \ldots, (S_F)_n)$ *)

4E.3 CALL QRSOLVE(n, M, $M1$, $M2$, s_N)

(* call Alg. A3.2.2 to calculate $M^{-1}s_N$ *)

4E.4 IF (*global* = 2) OR (*global* = 3) THEN

(* lower triangle of $M \leftarrow R^T$ *)

4E.4.1 FOR $i = 1$ TO n DO

    4E.4.1.1 $M[i,i] \leftarrow M2[i]$

    4E.4.1.2 FOR $j = 1$ TO $i-1$ DO

        $M[i,j] \leftarrow M[j,i]$

4E.5 IF $global = 2$ THEN

  $(* \ H \leftarrow LL^T \ *)$

  4E.5.1 FOR $i = 1$ TO n DO

    4E.5.1.1 $H[i,i] \leftarrow \sum_{k=1}^{i} (M[i,k])^2$

    4E.5.1.2 FOR $j = i+1$ TO n DO

        $H[i,j] \leftarrow \sum_{k=1}^{i} (M[i,k] * M[j,k])$

$(*$ RETURN from Algorithm A6.5.1 $*)$
$(*$ END, Algorithm A6.5.1 $*)$

ALGORITHM A6.5.1fac (NEMODELFAC)

FORMATION OF THE AFFINE MODEL FOR NONLINEAR EQUATIONS WITH FACTORED SECANT UPDATES

Purpose : A modification of Algorithm A6.5.1 that is used with factored secant updates. This algorithm differs from Algorithm A6.5.1 mainly in the storage of the QR factorization of the Jacobian approximation.

Input Parameters : $n \in Z$, $F_c \in R^n$ $(= F(x_c))$, $g \in R^n$, $S_F \in R^n$, $S_x \in R^n$, $macheps \in R$, $global \in Z$, $restart \in$ Boolean

Input-Output Parameters : $M \in R^{n \times n}$, $M2 \in R^n$, $J \in R^{n \times n}$

Output Parameters : $H \in R^{n \times n}$, $s_N \in R^n$

Storage Considerations :
1) One n-vector of additional storage is required, for $M1$.
2) Only the upper triangle and main diagonal of H are used. H may share storage with M, if some minor modifications are made to this algorithm as explained in Guidelines 6 and 3.
3) If this is a restart J contains $J(x_c)$ on input and Q^T on output, where $QR = D_F J(x_c)$. Otherwise, J contains Q^T on input, where QR is the factored secant approximation to $D_F J(x_c)$, and is unchanged by this algorithm. M contains R in its upper triangle on output (sometimes excluding its main diagonal), and also on input (including the main diagonal) unless this is a restart. The diagonal of R is stored in $M2$ on output, and also on input unless this is a restart. If $global = 2$ or 3, the main diagonal and lower triangle of M will contain on output the same matrix L as is output in M by

Algorithm A6.5.1. If *global* = 2, H contains the same matrix on output as is output in H by Algorithm A6.5.1.

Description :

1) If this is a restart, calculate the QR decomposition of $D_F J$, placing Q^T into J and R into M. $(D_F = \text{diag } ((S_F)_1, \ldots, (S_F)_n).)$
2) Estimate the l_1 condition number of RD_x^{-1}. $(D_x = \text{diag } ((S_x)_{1,\ldots,}(S_x)_n).)$
3) If RD_x^{-1} is singular or ill-conditioned, set $H = R^T D_F^2 R + \sqrt{n * macheps} * \|R^T D_F^2 R\|_1 * D_x^2$ (see Section 6.5), calculate the Cholesky factorization of H, set $s_N \leftarrow -H^{-1} g$, where $g = R^T Q^T D_F F_c$, and terminate Algorithm A6.5.1fac.
4) If RD_x^{-1} is well conditioned, calculate $s_N \leftarrow -R^{-1} Q^T D_F F_c$. If *global* = 2 or 3, store R^T in the lower triangle of M. If *global* = 2, set $H \leftarrow R^T R$. Terminate Algorithm A6.5.1fac.

Algorithm :

1. IF *restart*
1T. THEN   (* calculate new QR factorization *)
　　(* $M \leftarrow D_F J(x_c)$; $D_F = \text{diag } ((S_F)_1, \ldots, (S_F)_n)$ *)
　　1T.1 FOR $i = 1$ TO n DO
　　　　1T.1.1 FOR $j = 1$ TO n DO
　　　　　　$M[i,j] \leftarrow S_F[i] * J[i,j]$
　　1T.2 CALL QRDECOMP(n, M, $M1$, $M2$, *sing*)
　　　　(* call Alg. A3.2.1 to calculate the QR factorization of M *)
　　1T.3 CALL QFORM(n, M, $M1$, J)
　　　　(* call Alg. A3.4.2 to calculate Q^T and store it in J *)
　　1T.4 FOR $i = 1$ TO n DO
　　　　$M[i,i] \leftarrow M2[i]$
1E. ELSE   (* check if R is singular *)
　　1E.1 *sing* $\leftarrow$ FALSE
　　1E.2 FOR $i = 1$ TO n DO
　　　　1E.2.1 IF $M[i,i] = 0$ THEN *sing* $\leftarrow$ TRUE
2. IF NOT *sing*
2T. THEN   (* estimate $K_1(RD_x^{-1})$ *)
　　(* $R \leftarrow RD_x^{-1}$ *)
　　2T.1 FOR $j = 1$ TO n DO
　　　　2T.1.1 FOR $i = 1$ TO $j-1$ DO
　　　　　　$M[i,j] \leftarrow M[i,j] / S_x[j]$
　　　　2T.1.2 $M2[j] \leftarrow M2[j] / S_x[j]$
　　2T.2 CALL CONDEST(n, M, $M2$, *est*)
　　　　(* call Alg. A3.3.1 to estimate the condition number of R *)
　　(* reset $R \leftarrow RD_x$ *)
　　2T.3 FOR $j = 1$ TO n DO
　　　　2T.3.1 FOR $i = 1$ TO $j-1$ DO

$$M[i,j] \leftarrow M[i,j] * S_x[j]$$
2T.3.2 $M2[j] \leftarrow M2[j] * S_x[j]$
2E. ELSE $est \leftarrow 0$
3. IF $(sing)$ OR $(est > 1/ \sqrt{macheps}\,)$
3T. THEN (* perturb Jacobian as described in Section 6.5 *)
  (* $H \leftarrow R^T R$ *)
  3T.1 FOR $i = 1$ TO n DO
    3T.1.1 FOR $j = i$ TO n DO
$$H[i,j] \leftarrow \sum_{k=1}^{i} M[k,i] * M[k,j]$$
  (* steps 3T.2-3 calculate $Hnorm \leftarrow \|D_x^{-1}HD_x^{-1}\|_1$ *)
  3T.2 $Hnorm \leftarrow (1/ S_x[1]) * \sum_{j=1}^{n}(|H[1,j]| / S_x[j])$
  3T.3 FOR $i = 2$ TO n DO
    3T.3.1 $temp \leftarrow (1/ S_x[i]) *$
$$\{\sum_{j=1}^{i}(|H[j,i]| / S_x[j]) + \sum_{j=i+1}^{n}(|H[i,j]| / S_x[j])\}$$
    3T.3.2 $Hnorm \leftarrow \max\{temp, Hnorm\}$
  (* $H \leftarrow H + \sqrt{n * macheps} * Hnorm * D_x^2$ *)
  3T.4 FOR $i = 1$ TO n DO
$$H[i,i] \leftarrow H[i,i] + \sqrt{n * macheps} * Hnorm * S_x[i]^2$$
  (* steps 3T.5-6 calculate $s_N \leftarrow H^{-1}g$ *)
  3T.5 CALL CHOLDECOMP$(n, H, 0, macheps, M, maxadd)$ (* Alg. A5.5.2
  *)
  3T.6 CALL CHOLSOLVE(n, g, M, s_N) (* Alg. A3.2.3 *)
3E. ELSE (* calculate normal Newton step *)
  (* $s_N \leftarrow -JD_F F_c$; recall $J = Q^T$ *)
  3E.1 FOR $i = 1$ TO n DO
$$s_N[i] \leftarrow - \sum_{j=1}^{n}(J[i,j] * S_F[j] * F_c[j])$$
  3E.2 CALL RSOLVE$(n, M, M2, s_N)$
    (* call Alg. A3.2.2a to calculate $R^{-1}s_N$ *)
  3E.3 IF $(global = 2)$ OR $(global = 3)$ THEN
    (* lower triangle of $M \leftarrow R^T$ *)
    3E.3.1 FOR $i = 2$ TO n DO
      3E.3.1.1 FOR $j = 1$ TO $i-1$ DO
$$M[i,j] \leftarrow M[j,i]$$
  3E.4 IF $global = 2$ THEN
    (* $H \leftarrow LL^T$ *)
    3E.4.1 FOR $i = 1$ TO n DO
      3E.4.1.1 FOR $j = i$ TO n DO
$$H[i,j] \leftarrow \sum_{k=1}^{i}(M[i,k] * M[j,k])$$
(* RETURN from Algorithm A6.5.1fac *)

(* END, Algorithm A6.5.1fac *)

ALGORITHM A7.2.1 (UMSTOP)

STOPPING CONDITIONS FOR UNCONSTRAINED MINIMIZATION

Purpose : Decide whether to terminate unconstrained minimization algorithm D6.1.1, for any of the reasons listed below

Input Parameters : $n \in Z$, $x_c \in R^n$, $x_+ \in R^n$, $f \in R (= f(x_+))$, $g \in R^n (= \nabla f(x_+))$,
$S_x \in R^n$, $typf \in R$, $retcode \in Z$, $gradtol \in R$, $steptol \in R$, $itncount \in Z$,
$itnlimit \in Z$, $maxtaken \in$ Boolean
Input-Output Parameters : $consecmax \in Z$
Output Parameters : $termcode \in Z$

Meaning of Return Codes :
$termcode = 0$: no termination criterion satisfied
$termcode > 0$: some termination criterion satisfied, as follows:
$termcode = 1$: norm of scaled gradient less than $gradtol$; x_+ probably is
an approximate local minimizer of $f(x)$ (unless $gradtol$ is too large)
$termcode = 2$: scaled distance between last two steps less than $steptol$;
x_+ may be an approximate local minimizer of $f(x)$, but it is also possible that the algorithm is making very slow progress and is not near
a minimizer, or that $steptol$ is too large
$termcode = 3$: last global step failed to locate a lower point than x_c;
either x_c is an approximate local minimizer and no more accuracy is
possible, or an inaccurately coded analytic gradient is being used, or
the finite difference gradient approximation is too inacurate, or $step$-
tol is too large
$termcode = 4$: iteration limit exceeded
$termcode = 5$: five consecutive steps of length $maxstep$ have been
taken; either $f(x)$ is unbounded below, or $f(x)$ has a finite asymp-
tote in some direction, or $maxstep$ is too small

Storage Considerations :
1) No additional vector or matrix storage is required.

Description :

$termcode$ is set to 1, 2, 3, 4, or 5 if the corresponding condition described
above is satisfied; if none of these conditions are satisfied, $termcode \leftarrow 0$.
Also, $consecmax$ is augmented by one if $maxtaken =$ TRUE, otherwise consecmax is set to zero.

Algorithm :

1. $termcode \leftarrow 0$
2a. IF $retcode = 1$ THEN $termcode \leftarrow 3$
2b. ELSEIF $\displaystyle\max_{1 \le i \le n}\left\{ |g[i]| * \frac{\max\{|x_+[i]|, 1/S_x[i]\}}{\max\{|f|, typf\}} \right\} \le gradtol$ THEN

$termcode \leftarrow 1$

   (* this is maximum component of scaled gradient; see Section 7.2.
   WARNING: if typf is too large, algorithm may terminate prematurely
   *)

2c. ELSEIF $\displaystyle\max_{1 \le i \le n}\left\{ \frac{|x_+[i] - x_c[i]|}{\max\{|x_+[i]|, 1/S_x[i]\}} \right\} \le steptol$ THEN $termcode \leftarrow 2$

   (* this is maximum component of scaled step; see Section 7.2 *)
2d. ELSEIF $itncount \ge itnlimit$ THEN $termcode \leftarrow 4$
2e. ELSEIF $maxtaken = $ TRUE THEN
   (* step of length $maxstep$ was taken; print this message if desired *)
   2e.1 $consecmax \leftarrow consecmax + 1$
   2e.2 IF $consecmax = 5$ THEN $termcode \leftarrow 5$
   (* limit of five maxsteps is completely arbitrary and can be changed by
   changing this statement *)
2f. ELSE $consecmax \leftarrow 0$
(* end of multiple branch 2 *)
(* if desired, output appropriate termination message from here if
$termcode > 0$ *)
(* RETURN from Algorithm A7.2.1 *)
(* END, Algorithm, A7.2.1 *)

ALGORITHM A7.2.2 (UMSTOP0)

STOPPING CONDITIONS FOR UNCONSTRAINED MINIMIZATION AT ITERATION ZERO

Purpose : Decide whether to terminate Algorithm D6.1.1 at iteration zero
   because x_0 is an approximate critical point of $f(x)$

Input Parameters : $n \in Z$, $x_0 \in R^n$, $f \in R$ $(=f(x_0))$, $g \in R^n$ $(=\nabla f(x_0))$, $S_x \in R^n$,
   $typf \in R$, $gradtol \in R$
Input-Output Parameters : none
Output Parameters : $termcode \in Z$, $consecmax \in Z$

Meaning of Return Codes :
    $termcode = 0 : x_0$ is not an approximate critical point of $f(x)$
    $termcode = 1 : x_0$ is an approximate critical point of $f(x)$

Storage Considerations :
1) No additional vector or matrix storage is required.

Additional Considerations :
1) If $termcode = 1$, it is possible that x_0 is an approximate local maximizer or saddle point of $f(x)$. This can be determined by calculating whether $\nabla^2 f(x_0)$ is positive definite. Alternatively, Algorithm D6.1.1 can be restarted from some point in the vicinity of x_0. Then it will most likely converge back to x_0 only if x_0 is an approximate local minimizer of $f(x)$.

Description :

$termcode$ is set to 1 if the absolute value of the maximum component of the scaled gradient is less than $10^{-3} * gradtol$. (A stricter gradient tolerance is used at iteration zero.) Otherwise, $termcode \leftarrow 0$.

Algorithm :

1. $consecmax \leftarrow 0$
    (* $consecmax$ will be used by Algorithm A7.2.1 to check for divergence; it contains the number of consecutive past steps whose scaled length was equal to $maxstep$ *)

2. IF $\max\limits_{1 \leq i \leq n}\left\{|g[i]| * \dfrac{\max\{|x_0[i]|, 1/S_x[i]\}}{\max\{|f|, typf\}}\right\} \leq 10^{-3} * gradtol$
    THEN $termcode \leftarrow 1$
        (* if desired, output appropriate termination message *)
    ELSE $termcode \leftarrow 0$
(* RETURN from Algorithm A7.2.2 *)
(* END, Algorithm A7.2.2 *)

ALGORITHM A7.2.3 (NESTOP)

STOPPING CONDITIONS FOR NONLINEAR EQUATIONS

Purpose : Decide whether to terminate nonlinear equations algorithm D6.1.3, for any of the reasons listed below

Input   **Parameters**   : $n \in Z$, $x_c \in R^n$, $x_+ \in R^n$, $F \in R^n (=F(x_+))$,

$Fnorm \in R\,(=\!\frac{1}{2}\|D_F F(x_+)\|_2^2), \quad g \in R^n\,(=J_+^T D_F^2 F_+), \quad S_x \in R^n, \quad S_F \in R^n,$
$retcode \in Z, \quad fvectol \in R, \quad steptol \in R, \quad itncount \in Z, \quad itnlimit \in Z,$
$maxtaken \in \text{Boolean}, \; analjac \in \text{Boolean}, \; cheapF \in \text{Boolean}, \; mintol \in R$

Input-Output Parameters : $consecmax \in Z$

Output Parameters : $termcode \in Z$

Meaning of Return Codes :

$termcode = 0$: no termination criterion satisfied

$termcode > 0$: some termination criterion satisfied, as follows:

$termcode = 1$: norm of scaled function value is less than $fvectol$; x_+ probably is an approximate root of $F(x)$ (unless $fvectol$ is too large)

$termcode = 2$: scaled distance between last two steps less than $steptol$; x_+ may be an approximate root of $F(x)$, but it is also possible that the algorithm is making very slow progress and is not near a root, or that $steptol$ is too large

$termcode = 3$: last global step failed to decrease $\|F(x)\|_2$ sufficiently; either x_c is close to a root of $F(x)$ and no more accuracy is possible, or an incorrectly coded analytic Jacobian is being used, or the secant approximation to the Jacobian is inaccurate, or $steptol$ is too large

$termcode = 4$: iteration limit exceeded

$termcode = 5$: five consecutive steps of length $maxstep$ have been taken; either $\|F(x)\|_2$ asymptotes from above to a finite value in some direction, or $maxstep$ is too small

$termcode = 6$: x_c seems to be an approximate local minimizer of $\|F(x)\|_2$ that is not a root of $F(x)$ (or $mintol$ is too small); to find a root of $F(x)$, the algorithm must be restarted from a different region

Storage Considerations :
1) No additional vector or matrix storage is required.

Description :

$termcode$ is set to 1, 2, 3, 4, 5 or 6 if the corresponding condition described above is satisfied; if none of these conditions are satisfied, $termcode \leftarrow 0$. Also, $consecmax$ is augmented by one if $maxtaken = \text{TRUE}$, otherwise consecmax is set to zero.

Algorithm :

1. $termcode \leftarrow 0$

2a. IF $retcode = 1$ THEN $termcode \leftarrow 3$

2b. ELSEIF $\max_{1 \leq i \leq n}\{S_F[i] * |F[i]|\} \leq fvectol$ THEN $termcode \leftarrow 1$

(* this is maximum component of scaled function value; see Section 7.2. *)

2c. ELSEIF $\max_{1 \leq i \leq n}\left\{\dfrac{|x_+[i] - x_c[i]|}{\max\{|x_+[i]|, 1/S_x[i]\}}\right\} \leq steptol$ THEN $termcode \leftarrow 2$

(* this is maximum component of scaled step; see Section 7.2 *)

2d. ELSEIF *itncount* $\geq$ *itnlimit* THEN *termcode* $\leftarrow$ 4

2e. ELSEIF *maxtaken* = TRUE THEN

(* step of length *maxstep* was taken; print this message if desired *)

2e.1 *consecmax* $\leftarrow$ *consecmax* + 1

2e.2 IF *consecmax* = 5 THEN *termcode* $\leftarrow$ 5

(* limit of five maxsteps is completely arbitrary and can be changed by changing this statement *)

2f. ELSE

2f.1 *consecmax* $\leftarrow$ 0

2f.2 IF *analjac* OR *cheapf* THEN

2f.21 IF $\max\limits_{1 \leq i \leq n} \left\{ \dfrac{|g[i]| * \max\{|x_+[i]|, 1/S_x[i]\}}{\max\{Fnorm, n/2\}} \right\} \leq mintol$ THEN

termcode $\leftarrow$ 6

(* test for local minimizer of $\|F(x)\|_2$ omitted when using secant methods because g may not be accurate in this case *)

(* end of multiple branch 2 *)

(* if desired, output appropriate termination message from here if *termcode* > 0 *)

(* RETURN from Algorithm A7.2.3 *)

(* END, Algorithm, A7.2.3 *)

ALGORITHM A7.2.4 (NESTOP0)

STOPPING CONDITIONS FOR NONLINEAR EQUATIONS AT ITERATION ZERO

Purpose : Decide whether to terminate Algorithm D6.1.3 at iteration zero because x_0 is an approximate root of $F(x)$

Input Parameters : $n \in Z$, $x_0 \in R^n$, $F \in R^n$ $(=F(x_0))$, $S_F \in R^n$, $fvectol \in R$
Input-Output Parameters : none
Output Parameters : *termcode* $\in Z$, *consecmax* $\in Z$

Meaning of Return Codes :

termcode = 0 : x_0 is not an approximate root of $F(x)$
termcode = 1 : x_0 is an approximate root of $F(x)$

Storage Considerations :
1) No additional vector or matrix storage is required.

Description :

termcode is set to 1 if the absolute value of the maximum component of the scaled function is less than $10^{-2} * fvectol$. (A stricter function tolerance is used at iteration zero.) Otherwise, *termcode* $\leftarrow$ 0.

Algorithm :

1. *consecmax* $\leftarrow$ 0
   (* *consecmax* will be used by Algorithm A7.2.3 to check for divergence; it contains the number of consecutive past steps whose scaled length was equal to *maxstep* *)
2. IF $\max_{1 \leq i \leq n}\{S_F[i] * |F[i]|\} \leq 10^{-2} * fvectol$
       THEN *termcode* $\leftarrow$ 1
           (* if desired, output appropriate termination message *)
       ELSE *termcode* $\leftarrow$ 0
(* RETURN from Algorithm A7.2.4 *)
(* END, Algorithm A7.2.4 *)

ALGORITHM A8.3.1 (BROYUNFAC)

BROYDEN'S UPDATE, UNFACTORED FORM

Purpose : Update A by the formula

$$A \leftarrow A + \frac{(y - As)(D_x^2 s)^T}{s^T D_x^2 s}$$

where $s = x_+ - x_c$, $y = F_+ - F_c$, skipping all or part of the update under the conditions described below

Input   Parameters   : $n \in Z$, $x_c \in R^n$, $x_+ \in R^n$, $F_c \in R^n$ ($= F(x_c)$),
$F_+ \in R^n$ ($= F(x_+)$), $\eta \in R$, $S_x \in R^n$ ($D_x = $ diag $((S_x)_1,...,(S_x)_n)$)
Input-Output Parameters : $A \in R^{n \times n}$
Output Parameters : none

Storage Considerations :
1) One n-vector of additional storage is required, for s .
2) There are no changes to this algorithm if one is economizing on matrix

storage as described in Guideline 6.

Description :

For each row i, the update to row i is made unless $|(y - As)[i]|$ is less than the estimated noise in $y[i]$, which is $\eta * (|F_+[i]| + |F_c[i]|)$.

Algorithm :

1. $s \leftarrow x_+ - x_c$
2. $denom \leftarrow \|D_x s\|_2^2$
   (* $D_x = \text{diag} ((S_x)_1,...,(S_x)_n)$ is stored as $S_x \in R^n$. For suggested implementation of expressions involving D_x, see Section I.5 of this appendix. *)
3. FOR $i = 1$ TO n DO
   (* $tempi \leftarrow (y - As)[i]$ *)
   3.1 $tempi \leftarrow F_+[i] - F_c[i] - \sum_{j=1}^{n} A[i,j] * s[j]$
   3.2 IF $|tempi| \geq \eta * (|F_+[i]| + |F_c[i]|)$ THEN
       (* update row i *)
       3.2.1 $tempi \leftarrow tempi / denom$
       3.2.2 FOR $j = 1$ TO n DO
             $A[i,j] \leftarrow A[i,j] + tempi * s[j] * S_x[j]^2$
(* RETURN from Algorithm A8.3.1 *)
(* END, Algorithm A8.3.1 *)

ALGORITHM A8.3.2 (BROYFAC)

BROYDEN'S UPDATE, FACTORED FORM

Purpose : Update the QR factorization of A, an approximation to $D_F J(x_c)$, into the factorization $Q_+ R_+$ of

$$A + \frac{(D_F y - As)(D_x^2 s)^T}{s^T D_x^2 s},$$

an approximation to $D_F J(x_+)$, where $s = x_+ - x_c$, $y = F_+ - F_c$, skipping all or part of the update under the conditions described below.

Input   Parameters   : $n \in Z$, $x_c \in R^n$, $x_+ \in R^n$, $F_c \in R^n$ ($= F(x_c)$),
$F_+ \in R^n$ ($= F(x_+)$), $\eta \in R$, $S_x \in R^n$ ($D_x = \text{diag} ((S_x)_1,...,(S_x)_n)$), $S_F \in R^n$

$(D_F = \text{diag}\,((S_F)_1, \ldots, (S_F)_n))$

Input-Output Parameters : $Z \in R^{n \times n}$ (contains Q^T on input, Q_+^T on output), $M \in R^{n \times n}$ (the upper triangle and main diagonal of M contain R on input, R_+ on output), $M2 \in R^n$ (contains the main diagonal of R on input, of R_+ on output)

Output Parameters : none

Storage Considerations :
1) Three n-vectors of additional storage are required, for s, t and w.
2) The first lower subdiagonal of M, as well as the upper triangle including the main diagonal, are used at intermediate stages of this algorithm.
3) There are no changes to this algorithm if one is economizing on matrix storage as described in Guideline 6.

Description :

1) Move $M2$ to the diagonal of M.
2) $s \leftarrow x_+ - x_c$.
3) Calculate $t = Rs$.
4) Calculate $w = D_F y - As = D_F y - Z^T t$. If $|w[i]|$ is less than the estimated noise in $D_F y[i]$, measured by $\eta * D_F[i] * (|F_+[i]| + |F_c[i]|)$, then $w[i]$ is set to zero.
5) If $w = 0$, skip the update. Otherwise :
   5a) Set $t = Q^T w = Zw$, $s = D_x^2 s / s^T D_x^2 s$.
   5b) Call Alg. A3.4.1 to update the QR factorization of A to the factorization $Q_+ R_+$ of $Q(R + ts^T)$.
   5c) Move the new diagonal of M to $M2$.

Algorithm :

1. FOR $i = 1$ TO n DO
   $M[i,i] \leftarrow M2[i]$
2. $s \leftarrow x_+ - x_c$
3. *skipupdate* $\leftarrow$ TRUE
   $(* \, t \leftarrow Rs \, : \, *)$
4. FOR $i = 1$ TO n DO
   $t[i] \leftarrow \sum_{j=i}^{n} M[i,j] * s[j]$
5. FOR $i = 1$ TO n DO
   $(* \, w[i] \leftarrow (D_F y - Z^T t)[i] \, : \, *)$
   5.1 $w[i] \leftarrow S_F[i] * F_+[i] - S_F[i] * F_c[i] - \sum_{j=1}^{n} Z[j,i] * t[j]$
   5.2 IF $w[i] > \eta * S_F[i] * (|F_+[i]| + |F_c[i]|)$
      THEN *skipupdate* $\leftarrow$ FALSE
      ELSE $w[i] \leftarrow 0$
6. IF *skipupdate* = FALSE THEN
   $(* \, t \leftarrow Z * w \, : \, *)$

6.1 FOR $i = 1$ TO n DO

$$t[i] \leftarrow \sum_{j=1}^{n} Z[i,j] * w[j]$$

6.2 $denom \leftarrow \|D_x s\|_2^2$

(* $D_x = \text{diag}((S_x)_1,...,(S_x)_n)$ is stored as $S_x \in R^n$. For suggested implementation of expressions involving D_x, see Section I.5 of this appendix. *)

6.3 $s \leftarrow D_x^2 * s / denom$

6.4 CALL QRUPDATE(n, t, s, 1, Z, M)   (* Alg. A3.4.1 *)

6.5 FOR $i = 1$ TO n DO

$$M2[i] \leftarrow M[i,i]$$

(* RETURN from Algorithm A8.3.2 *)

(* END, Algorithm A8.3.2 *)

ALGORITHM A9.4.1 (BFGSUNFAC)

POSITIVE DEFINITE SECANT UPDATE (BFGS), UNFACTORED FORM

Purpose : Update H by the BFGS formula

$$H \leftarrow H - \frac{Hss^T H}{s^T Hs} + \frac{yy^T}{y^T s}$$

where $s = x_+ - x_c$, $y = g_+ - g_c$, skipping the update under the conditions stated below

Input Parameters : $n \in Z$, $x_c \in R^n$, $x_+ \in R^n$, $g_c \in R^n$, $g_+ \in R^n$, $macheps \in R$, $\eta \in R$, $analgrad \in$ Boolean (= TRUE if analytic gradient is being used, FALSE otherwise)

Input-Output Parameters : $H \in R^{n \times n}$ symmetric

Output Parameters : none

Storage Considerations :

1) Three n-vectors of additional storage are required, for s, y, and t.

2) Only the upper triangle of H, including the main diagonal, is used. (These are the only portions of H that are referenced by the rest of the system of algorithms.)

3) If one is conserving matrix storage, this algorithm is modified slightly as explained in Guideline 3.

Description :

The above update is made unless either : i) $y^T s < (macheps)^{\frac{1}{2}} \|s\|_2\|y\|_2$, or
ii) for every i, $|(y-Hs)[i]|$ is less than the estimated noise in $y[i]$. The
estimated noise in $y[i]$ is calculated by $tol * (|g_c[i]| + |g_+[i]|)$ where
$tol = \eta$ if analytic gradients are being used, $tol = \eta^{\frac{1}{2}}$ if finite difference gradients are being used.

Algorithm :

1. $s \leftarrow x_+ - x_c$
2. $y \leftarrow g_+ - g_c$
3. $temp1 \leftarrow y^T s$
4. IF $temp1 \geq (macheps)^{\frac{1}{2}} \|s\|_2\|y\|_2$ THEN
 (* ELSE update is skipped and algorithm terminates *)
 4.1 IF $analgrad$ = TRUE
   THEN $tol \leftarrow \eta$
   ELSE $tol \leftarrow \eta^{\frac{1}{2}}$
 4.2 $skipupdate \leftarrow$ TRUE
 4.3 FOR $i = 1$ TO n DO
  (* $t[i] \leftarrow Hs[i]$: *)
  4.3.1 $t[i] \leftarrow \sum_{j=1}^{i} H[j,i] * s[j] + \sum_{j=i+1}^{n} H[i,j] * s[j]$
  4.3.2 IF $|y[i] - t[i]| \geq tol * \max\{|g_c[i]|, |g_+[i]|\}$ THEN
   $skipupdate \leftarrow$ FALSE
 4.4 IF $skipupdate$ = FALSE THEN
  (* do update; ELSE update is skipped and algorithm terminates *)
  4.4.1 $temp2 \leftarrow s^T t$ (* $= s^T Hs$ *)
  4.4.2 FOR $i=1$ TO n DO
   4.4.2.1 FOR $j=i$ TO n DO
    $H[i,j] \leftarrow H[i,j] + \dfrac{y[i] * y[j]}{temp1} - \dfrac{t[i] * t[j]}{temp2}$
(* end of IFTHEN 4.4, and of IFTHEN 4 *)
(* RETURN from Algorithm A9.4.1 *)
(* END, Algorithm A9.4.1 *)

ALGORITHM A9.4.2 (BFGSFAC)

POSITIVE DEFINITE SECANT UPDATE (BFGS),
FACTORED FORM

Purpose : Update the LL^T factorization of H into the factorization $L_+L_+^T$ of

$$H - \frac{Hss^TH}{s^THs} + \frac{yy^T}{y^Ts} = JJ^T,$$

$$J = L + \frac{(y-\alpha Hs)(\alpha L^Ts)^T}{y^Ts}, \quad \alpha = (\frac{y^Ts}{s^THs})^{\frac{1}{2}}$$

where $s = x_+ - x_c$, $y = g_+ - g_c$, by using Algorithm A3.4.1 to calculate the QR factorization of J^T in $O(n^2)$ operations. The update is skipped under conditions stated below.

Input Parameters : $n \in Z$, $x_c \in R^n$, $x_+ \in R^n$, $g_c \in R^n$, $g_+ \in R^n$, $macheps \in R$, $\eta \in R$, $analgrad \in$ Boolean (= TRUE if analytic gradient is being used, FALSE otherwise)

Input-Output Parameters : $M \in R^{n \times n}$ (the lower triangle and main diagonal of M contain L on input, L_+ on output)

Output Parameters : none

Storage Considerations :
1) Four n-vectors of additional storage are required, for s, y, t, and u. This can be reduced to two vectors simply by replacing t by s and u by y in all instances in the algorithm below. (I.e., t and u are only used to make the algorithm easier to follow.)
2) The entire matrix M is used at intermediate stages of this algorithm, although only the lower triangle and main diagonal are needed for input to and output from the algorithm.
3) There are no changes to this algorithm if one is economizing on matrix storage as described in Guideline 3.

Description :

1) If $y^Ts < (macheps)^{\frac{1}{2}} \|s\|_2\|y\|_2$, or for every i, $|(y-LL^Ts)[i]|$ is less than the estimated noise in $y[i]$, the update is skipped. The estimated noise in $y[i]$ is calculated by $tol * (|g_c[i]| + |g_+[i]|)$ where $tol = \eta$ if analytic gradients are being used, $tol = \eta^{\frac{1}{2}}$ if finite difference gradients are being used.
2) If the update is to be performed, $u = L^Ts/\sqrt{y^Ts * s^TLL^Ts}$ and $t = y - \alpha LL^Ts$ are calculated, and L^T is copied into the upper triangle of M. Then Algorithm A3.4.1 is called, returning in the upper triangle of M the matrix R_+ for which $R_+^TR_+ = (L + tu^T)(L^T + ut^T)$. R_+^T is then copied into the lower triangle of M and returned as the new value L_+.

Algorithm :

1. $s \leftarrow x_+ - x_c$
2. $y \leftarrow g_+ - g_c$
3. $temp1 \leftarrow y^Ts$
4. IF $temp1 \geq (macheps)^{\frac{1}{2}} \|s\|_2\|y\|_2$ THEN
   (* ELSE update is skipped and algorithm terminates *)

(* $t \leftarrow L^T s$: *)

4.1 FOR $i = 1$ TO n DO

$$t[i] \leftarrow \sum_{j=i}^{n} M[j,i] * s[j]$$

4.2 $temp2 \leftarrow t^T t$ (* $= s^T L L^T s$ *)

4.3 $\alpha \leftarrow \sqrt{temp1 / temp2}$ (* $= \sqrt{y^T s / s^T L L^T s}$ *)

4.4 IF $analgrad$ = TRUE

THEN $tol \leftarrow \eta$

ELSE $tol \leftarrow \eta^{\frac{1}{2}}$

4.5 $skipupdate \leftarrow$ TRUE

4.6 FOR $i = 1$ TO n DO

(* $temp3 \leftarrow (L L^T s)[i]$: *)

4.6.1 $temp3 \leftarrow \sum_{j=1}^{i} M[i,j] * t[j]$

4.6.2 IF $|y[i] - temp3| \geq tol * \max\{|g_c[i]|, |g_+[i]|\}$ THEN

$skipupdate \leftarrow$ FALSE

4.6.3 $u[i] \leftarrow y[i] - \alpha * temp3$

4.7 IF $skipupdate$ = FALSE THEN

(* do update; ELSE update is skipped and algorithm terminates *)

4.7.1 $temp3 \leftarrow 1 / \sqrt{temp1 * temp2}$

(* $= 1 / \sqrt{y^T s} * s^T L L^T s$ *)

4.7.2 FOR $i = 1$ TO n DO

$t[i] \leftarrow temp3 * t[i]$

(* copy L^T into upper triangle of M : *)

4.7.3 FOR $i = 2$ TO n DO

4.7.3.1 FOR $j = 1$ TO $i-1$ DO

$M[j,i] \leftarrow M[i,j]$

(* Call Alg. A3.4.1 to calculate the QR factorization of $L^T + u t^T$: *)

4.7.4 CALL QRUPDATE$(n, t, u, 2, M, M)$

(* second L is a dummy parameter *)

(* copy transpose of upper triangle of M into L : *)

4.7.5 FOR $i = 2$ TO n DO

4.7.5.1 FOR $j = 1$ TO $i-1$ DO

$M[i,j] \leftarrow M[j,i]$

(* end of IFTHEN 4.4, and of IFTHEN 4 *)

(* RETURN from Algorithm A9.4.2 *)

(* END, Algorithm A9.4.2 *)

ALGORITHM A9.4.3 (INITHESSUNFAC)

INITIAL HESSIAN FOR SECANT UPDATES
IN UNFACTORED FORM

Purpose : Set $H = \max\{|f(x_0)|, typf\} * D_x^2$, where $D_x = \text{diag}\,((S_x)_{1,...,}(S_x)_n)$

Input Parameters : $n \in Z$, $f \in R$ $(=f(x_0))$, $typf \in R$, $S_x \in R^n$
Input-Output Parameters : none
Output Parameters : $H \in R^{n \times n}$ symmetric

Storage Considerations :
1) No additional vector or matrix storage is required.
2) Only the upper triangle of H, including the main diagonal, is used.
(These are the only portions of H that are referenced by the rest of the system of algorithms.)
3) There are no changes to this algorithm if one is economizing on matrix storage as described in Guideline 3.

Algorithm :

1. $temp \leftarrow \max\{|f|, typf\}$
2. FOR $i = 1$ TO n DO
   2.1 $H[i,i] \leftarrow temp * S_x[i] * S_x[i]$
   2.2 FOR $j = i+1$ TO n DO
      $H[i,j] \leftarrow 0$
(* RETURN from Algorithm A9.4.3 *)
(* END, Algorithm A9.4.3 *)

ALGORITHM A9.4.4 (INITHESSFAC)

INITIAL HESSIAN FOR SECANT UPDATES
IN FACTORED FORM

Purpose: Set $L = \sqrt{\max\{|f(x_0)|, typf\}} * D_x$, where $D_x = \text{diag}\,((S_x)_{1,...,}(S_x)_n)$

Input Parameters : $n \in Z$, $f \in R$ $(=f(x_0))$, $typf \in R$, $S_x \in R^n$
Input-Output Parameters : none
Output Parameters : $L \in R^{n \times n}$ lower triangular

Storage Considerations :
1) No additional vector or matrix storage is required.
2) Only the lower triangle of L, including the main diagonal, is used.
3) There are no changes to this algorithm if one is economizing on matrix storage as described in Guideline 3.

Algorithm :

1. $temp \leftarrow \sqrt{\max\{|f|, typf\}}$
2. FOR i = 1 TO n DO
    2.1 $L[i, i] \leftarrow temp * S_x[i]$
    2.2 FOR j = 1 TO $i - 1$ DO
            $L[i,j] \leftarrow 0$
(* RETURN from Algorithm A9.4.4 *)
(* END, Algorithm A9.4.4 *)

Appendix B

Test Problems[*]

Test problems are discussed in Section 7.3 and the interested reader should read that section before this one. Moré, Garbow, and Hillstrom [1981] provide a set of approximately 15 test problems each for unconstrained minimization, systems of nonlinear equations, and nonlinear least squares. Many of these problems can be run with various values of n. A standard starting point x_0 is given for each problem, and most are intended to be started from $10 * x_0$ and $100 * x_0$ as well. Thus the number of possible problems provided by Moré, Garbow, and Hillstrom actually is quite large. Below we give a small subset of these problems that might be used in testing a class project or in very preliminary testing of a new method. Anyone interested in comprehensive testing is urged to read Moré, Garbow, and Hillstrom and to use the larger set of test problems contained therein. Hiebert [1982] discusses a comparison of codes for solving systems of nonlinear equations using the test problems in Moré, Garbow and Hillstrom, and includes suggestions for modifying these problems into problems with poor scaling or problems where the objective function is noisy.

Problems 1-4 below are standard test problems for both nonlinear equations and unconstrained minimization. In each case, a set of n single-valued functions of n unknowns, $f_i(x)$, $i=1, \cdots ,n$, is given. The nonlinear equations function vector is

$$F(x) = \begin{bmatrix} f_1(x) \\ \cdots \\ f_n(x) \end{bmatrix}$$

while the unconstrained minimization objective function is

* by Robert B. Schnabel

$$f(x) = \sum_{i=1}^{n} f_i(x)^2$$

Problems 1-3 may be run with various values of n. The standard dimensions for problems 1 and 2 are $n=2$ and $n=4$ respectively, and these should be tried first. A reasonable first choice for problem 3 is $n=10$. Problem 5 is a difficult and often used test problem for unconstrained minimization only.

We copy Moré, Garbow, and Hillstrom and list for each problem: a) the dimension n; b) tne nonlinear function(s); c) the starting point x_0; d) the root and minimizer x_* if it is available in closed form. The minimum value of $f(x)$ is zero in all cases. All these problems may be started from x_0, $10*x_0$, or $100*x_0$. For the original source of each problem, see Moré, Garbow, and Hillstrom.

1) *Extended Rosenbrock Function*
   a) n = any positive multiple of 2
   b) for $i=1, \cdots, n/2$:
      $$f_{2i-1}(x) = 10(x_{2i} - x_{2i-1}^2)$$
      $$f_{2i}(x) = 1 - x_{2i-1}$$
   c) $x_0 = (-1.2, 1, \cdots, -1.2, 1)$
   d) $x_* = (1, 1, \cdots, 1, 1)$

2) *Extended Powell Singular Function*
   a) n = any positive multiple of 4
   b) for $i=1, \cdots, n/4$:
      $$f_{4i-3}(x) = x_{4i-3} + 10x_{4i-2}$$
      $$f_{4i-2}(x) = \sqrt{5}(x_{4i-1} - x_{4i})$$
      $$f_{4i-1}(x) = (x_{4i-2} - 2x_{4i-1})^2$$
      $$f_{4i}(x) = \sqrt{10}(x_{4i-3} - x_{4i})^2$$
   c) $x_0 = (3, -1, 0, 1, \cdots, 3, -1, 0, 1)$
   d) $x_* = (0, 0, 0, 0, \cdots, 0, 0, 0, 0)$
      note: both $F'(x_*)$ and $\nabla^2 f(x_*)$ are singular

3) *Trigonometric Function*
   a) n = any positive integer
   b) for $i=1, \cdots, n$:
      $$f_i(x) = n - \sum_{j=1}^{n}(\cos x_j + i(1 - \cos x_i) - \sin x_i)$$
   c) $x_0 = (1/n, 1/n, \cdots, 1/n)$

4) *Helical Valley Function*
   a) n = 3
   b) $f_1(x) = 10(x_3 - 10\Theta(x_1, x_2))$
      $$f_2(x) = 10((x_1^2 + x_2^2)^{\frac{1}{2}} - 1)$$
      $$f_3(x) = x_3$$
   where
      $$\Theta(x_1, x_2) = (1/2\pi) \arctan(x_2/x_1) \quad \text{if } x_1 > 0$$

$\Theta(x_1, x_2) = (1/2\pi) \arctan(x_2/x_1) + 0.5$ if $x_1 < 0$

c) $x_0 = (-1, 0, 0)$

d) $x_* = (1, 0, 0)$

5) *Wood Function*

a) $n = 4$

b) $f(x) = 100(x_1^2 - x_2)^2 + (1 - x_1)^2$ $+ 90(x_3^2 - x_4)^2 + (1 - x_3)^2$
   $+ 10.1((1 - x_2)^2 + (1 - x_4)^2) + 19.8(1 - x_2)(1 - x_4)$

c) $x_0 = (-3, -1, -3, -1)$

d) $x_* = (1, 1, 1, 1)$

References

Aasen, J. O. (1971), "On the reduction of a symmetric matrix to tridiagonal form," *BIT* **11**, 233–242.

Aho, A. V., J. E. Hopcroft, and J. D. Ullman (1974). *The Design and Analysis of Computer Algorithms*, Addison-Wesley, Reading, Mass.

Allgower, E., and K. Georg (1980), "Simplicial and continuation methods for approximating fixed points and solutions to systems of equations," *SIAM Review* **22**, 28–85.

Armijo, L., (1966), "Minimization of functions having Lipschitz-continuous first partial derivatives," *Pacific J. Math.* **16**, 1–3.

Avriel, M. (1976), *Nonlinear Programming: Analysis and Methods*, Prentice-Hall, Englewood Cliffs, N.J.

Bard, Y. (1970), *Nonlinear Parameter Estimation*, Academic Press, New York.

Barnes, J. (1965), "An algorithm for solving nonlinear equations based on the secant method," *Comput. J.* **8**, 66–72.

Bartels, R., and A. Conn. (1982), "An approach to nonlinear l_1 data fitting," in *Numerical Analysis, Cocoyoc* 1981, ed. by J. P. Hennart, Springer Verlag. Lecture Notes in Math. 909, 48–58.

Bates, D. M., and D. G. Watts (1980), "Relative curvature measures of nonlinearity," *J. Roy. Statist. Soc. Ser. B 42*, 1–25, 235.

Beale, E. M. L. (1977), "Integer programming," in *The State of the Art in Numerical Analysis*, D. Jacobs, ed., Academic Press, London, 409–448.

Beaton, A. E., and J. W. Tukey (1974), "The fitting of power series, meaning polynomials, illustrated on hand-spectroscopic data," *Technometrics* **16**, 147–192.

Boggs, P. T., and J. E. Dennis, Jr. (1976), "A stability analysis for perturbed nonlinear iterative methods," *Math. Comp.* **30**, 1–17.

Brent, R. P. (1973), *Algorithms For Minimization Without Derivatives*, Prentice-Hall, Englewood Cliffs, N.J.

Brodlie, K. W. (1977), "Unconstrained minimization," in *The State of the Art in Numerical Analysis*, D. Jacobs, ed., Academic Press, London, 229–268.

Broyden, C. G. (1965), "A class of methods for solving nonlinear simultaneous equations," *Math. Comp.* **19**, 577–593.

Broyden, C. G. (1969), "A new double-rank minimization algorithm," *AMS Notices* **16**, 670.

Broyden, C. G. (1970), "The covergence of a class of double-rank minimization algorithms," Parts I and II, *J.I.M.A.* **6**, 76–90, 222–236.

Broyden, C. G. (1971), "The convergence of an algorithm for solving sparse nonlinear systems," *Math. Comp.* **25**, 285–294.

Broyden, C. G., J. E. Dennis, Jr., and J. J. Moré (1973), "On the local and superlinear covergence of quasi-Newton methods," *J.I.M.A.* **12**, 223–246.

Bryan, C. A. (1968), "Approximate solutions to nonlinear integral equations," *SIAM J. Numer. Anal.* **5**, 151–155.

Buckley, A. G. (1978), "A combined conjugate gradient quasi-Newton minimization algorithm," *Math. Prog.* **15**, 200–210.

Bunch, J. R., and B. N. Parlett (1971), "Direct methods for solving symmetric indefinite systems of linear equations," *SIAM J. Numer. Anal.* **8**, 639–655.

Cline, A. K., C. B. Moler, G. W. Stewart, and J. H. Wilkinson (1979), "An estimate for the condition number of a matrix," *SIAM J. Numer. Anal.* **16**, 368–375.

Coleman, T. F., and J. J. Moré (1983), "Estimation of sparse Jacobian matrices and graph coloring problems," *SIAM J. Numer. Anal.* **20**, 187–209.

Coleman, T. F., and J. J. Moré (1982), "Estimation of sparse Hessian matrices and graph coloring problems," Argonne National Labs, Math–C. S. Div. TM-4.

Conte, S. D., and C. de Boor (1980), *Elementary Numerical Analysis: An Algorithmic Approach*, 3d ed., McGraw-Hill, New York.

Curtis, A., M. J. D. Powell, and J. K. Reid (1974), "On the estimation of sparse Jacobian matrices," *J.I.M.A.* **13**, 117–120.

Dahlquist, G., A. Björck, and N. Anderson (1974), *Numerical Methods*, Prentice-Hall, Englewood Cliffs, N.J.

Davidon, W. C. (1959), "Variable metric methods for minimization," Argonne National Labs Report ANL-5990.

Davidon, W. C. (1975), "Optimally conditioned optimization algorithms without line searches," *Math. Prog.* **9**, 1–30.

Dembo, R. S., S. C. Eisenstat, and T. Steihaug (1982). "Inexact Newton methods," *SIAM J. Numer. Anal.* **19**, 400–408.

Dennis, J. E., Jr. (1971), "Toward a unified convergence theory for Newton-like methods," in *Nonlinear Functional Analysis and Applications*, L. B. Rall, ed., Academic Press, New York, 425–472.

Dennis, J. E., Jr. (1977), "Nonlinear least squares and equations," in *The State of the Art in Numerical Analysis*, D. Jacobs, ed., Academic Press, London, 269–312.

Dennis, J. E., Jr. (1978), "A brief introduction to quasi-Newton methods," in Numerical Analysis, G. H. Golub and J. Oliger, eds., AMS, Providence, R.I., 19–52.

Dennis, J. E., Jr., D. M. Gay, and R. E. Welsch (1981a), "An adaptive nonlinear least-squares algorithm," *TOMS* **7**, 348–368.

Dennis, J. E., Jr., D. M. Gay, and R. E. Welsch (1981b), "Algorithm 573 NL2SOL—An adaptive nonlinear least-squares algorithm [E4]," *TOMS* 7, 369–383.

Dennis, J. E., Jr., and E. S. Marwil (1982), "Direct secant updates of matrix factorizations," *Math. Comp.* 38, 459–474.

Dennis, J. E., Jr., and H. H. W. Mei (1979), "Two new unconstrained optimization algorithms which use function and gradient values," *J. Optim. Theory Appl.* 28, 453–482.

Dennis, J. E., Jr., and J. J. Moré (1974), "A characterization of superlinear convergence and its application to quasi-Newton methods," *Math. Comp.* 28, 549–560.

Dennis, J. E., Jr., and J. J. Moré (1977), "Quasi-Newton methods, motivation and theory," *SIAM Review* 19, 46–89.

Dennis, J. E., Jr., and R. B. Schnabel (1979), "Least change secant updates for quasi-Newton methods," *SIAM Review* 21, 443–459.

Dennis, J. E., Jr., and R. A. Tapia (1976), "Supplementary terminology for nonlinear iterative methods," *SIGNUM Newsletter* 11:4, 4–6.

Dennis, J. E., Jr., and H. F. Walker (1981), "Convergence theorems for least-change secant update methods," *SIAM J. Numer. Anal.* 18, 949–987, 19, 443.

Dennis, J. E., Jr., and H. F. Walker (1983), "Sparse secant update methods for problems with almost sparse Jacobians," in preparation.

Dixon, L. C. W. (1972a), "Quasi-Newton family generate identical points," Parts I and II, *Math. Prog.* 2, 383–387, 2nd *Math. Prog.* 3, 345–358.

Dixon, L. C. W. (1972b), "The choice of step length, a crucial factor in the performance of variable metric algorithms," in *Numerical Methods for Non-linear Optimization*, F. Lootsma, ed., Academic Press, New York, 149–170.

Dixon, L. C. W., and G. P. Szegö (1975, 1978), *Towards Global Optimization*, Vols. 1, 2, North-Holland, Amsterdam.

Dongarra, J. J., J. R. Bunch, C. B. Moler, and G. W. Stewart (1979), *LINPACK Users Guide*, SIAM Publications, Philadelphia.

Fletcher, R. (1970), "A new approach to variable metric algorithms," *Comput. J.* 13, 317–322.

Fletcher, R. (1980), *Practical Methods of Optimization*, Vol. 1, *Unconstrained Optimization*, John Wiley & Sons, New York.

Fletcher, R., and M. J. D. Powell (1963), "A rapidly convergent descent method for minimization," *Comput. J.* 6, 163–168.

Ford, B. (1978), "Parameters for the environment for transportable numerical software," *TOMS* 4, 100–103.

Fosdick, L. ed. (1979), *Performance Evaluation of Numerical Software*, North-Holland, Amsterdam.

Frank, P., and R. B. Schnabel (1982), "Calculation of the initial Hessian approximation in secant algorithms," in preparation.

Gander, W. (1978), "On the linear least squares problem with a quadratic constraint," Stanford Univ. Computer Science Tech. Rept. STAN-CS-78-697, Stanford, Calif.

Gander, W. (1981), "Least squares with a quadratic constraint," *Numer. Math.* 36, 291–307. [This is an abbreviated version of Gander (1978).]

Garfinkel, R. S., and G. L. Nemhauser (1972), *Integer Programming*, John Wiley & Sons, New York.

Gay, D. M. (1979), "Some convergence properties of Broyden's method," *SIAM J. Numer. Anal..* **16**, 623–630.

Gay, D. M. (1981), "Computing optimal locally constrained steps," *SIAM J. Sci. Stat. Comp.* **2**, 186–197.

Gay, D. M., and R. B. Schnabel (1978), "Solving systems of nonlinear equations by Broyden's method with projected updates," in *Nonlinear Programming 3*, O. Mangasarian, R. Meyer, and S. Robinson, eds., Academic Press, New York, 245–281.

Gill, P. E., G. H. Golub, W. Murray, and M. A. Saunders (1974), "Methods for modifying matrix factorizations," *Math. Comp.* **28**, 505–535.

Gill, P. E., and W. Murray (1972), "Quasi-Newton methods for unconstrained optimization," *J.I.M.A.* **9**, 91–108.

Gill, P. E., W. Murray and M. H. Wright (1981), *Practical Optimization*, Academic Press, London.

Goldfarb, D. (1976), "Factorized variable metric methods for unconstrained optimization," *Math. Comp.* **30**, 796–811.

Goldfarb, D. (1970), "A family of variable metric methods derived by variational means," *Math. Comp.* **24**, 23–26.

Goldfeldt, S. M., R. E. Quandt, and H. F. Trotter (1966), "Maximization by quadratic hill-climbing," *Econometrica* **34**, 541–551.

Goldstein, A. A. (1967), *Constructive Real Analysis*, Harper & Row, New York.

Golub, G. H., and V. Pereyra (1973), "The differentiation of pseudo-inverse and non-linear least squares problems whose variables separate," *SIAM J. Numer. Anal.* **10**, 413–432.

Golub, G. H., and C. Van Loan (1983), *Matrix Computations*, the Johns Hopkins University Press.

Greenstadt, J. L. (1970), "Variations on variable-metric methods," *Math. Comp.* **24**, 1–22.

Griewank, A. O., (1982), "A short proof of the Dennis-Schnabel theorem," *B.I.T.* **22**, 252–256.

Griewank, A. O., and Ph. L. Toint (1982a), "Partitioned variable metric updates for large sparse optimization problems," *Numer. Math.* **39**, 119–37.

Griewank, A. O., and Ph. L. Toint (1982b), "Local convergence analysis for partitioned quasi-Newton updates in the Broyden class," to appear in *Numer. Math.*

Griewank, A. O., and Ph. L. Toint (1982c), "On the unconstrained optimization of partially separable functions," in *Nonlinear Optimization*, M. J. D. Powell, ed., Academic Press, London.

Hamming, R. W. (1973), *Numerical Methods for Scientists and Engineers*, 2 ed., McGraw-Hill, New York.

Hebden, M. D. (1973), "An algorithm for minimization using exact second derivatives," Rept. TP515, A.E.R.E., Harwell, England.

Hestenes, M. R. (1980), *Conjugate-Direction Methods In Optimization*, Springer Verlag, New York.

Hiebert, K. L. (1982), "An evaluation of mathematical software that solves systems of nonlinear equations," *TOMS* **8**, 5–20.

Huber, P. J. (1973), "Robust regression: asymptotics, conjectures, and Monte Carlo," *Anals of Statistics* **1**, 799–821.

Huber, P. J. (1981), *Robust Statistics*, John Wiley & Sons, New York.

Johnson, G. W., and N. H. Austria (1983), "A quasi-Newton method employing direct secant updates of matrix factorizations," *SIAM J. Numer. Anal.* **20**, 315–325.

Kantorovich, L. V. (1948), "Functional analysis and applied mathematics," *Uspehi Mat. Nauk.* **3**, 89–185; transl. by C. Benster as N.B.S. Rept. 1509, Washington, D. C., 1952.

Kaufman, L. C. (1975), "A variable projection method for solving separable nonlinear least squares problems," *BIT* **15**, 49–57.

Lawson, C. L., R. J. Hanson, D. R. Kincaid, and F. T. Krogh (1979), "Basic linear algebra subprograms for Fortran usage," *ACM TOMS* **5**, 308–323.

Levenberg, K. (1944), "A method for the solution of certain problems in least squares," *Quart. Appl. Math.* **2**, 164–168.

Marquardt, D. (1963), "An algorithm for least-squares estimation of nonlinear parameters," *SIAM J. Appl. Math.* **11**, 431–441.

Marwil, E. S. (1978), "Exploiting sparsity in Newton-type methods," Cornell Applied Math. Ph.D. Thesis.

Marwil, E. S. (1979), "Convergence results for Schubert's method for solving sparse nonlinear equations," *SIAM J. Numer. Anal.* **16**, 588–604.

Moré, J. J. (1977), "The Levenberg-Marquardt algorithm: implementation and theory," in *Numerical Analysis*, G. A. Watson, ed., Lecture Notes in Math. 630, Springer Verlag, Berlin, 105–116.

Moré, J. J., B. S. Garbow, and K. E. Hillstrom (1980), "User guide for MINPACK-1," Argonne National Labs Report ANL-80-74.

Moré, J. J., B. S. Garbow, and K. E. Hillstrom (1981a), "Testing unconstrained optimization software," *TOMS* **7**, 17–41.

Moré, J. J., B. S. Garbow, and K. E. Hillstrom (1981b), "Fortran subroutines for testing unconstrained optimization software," *TOMS* **7**, 136–140.

Moré, J. J., and D. C. Sorensen (1979), "On the use of directions of negative curvature in a modified Newton method," *Math. Prog.* **16**, 1–20.

Murray, W. (1972), *Numerical Methods for Unconstrained Optimization*, Academic Press, London.

Murray, W., and M. L. Overton (1980), "A projected Lagrangian algorithm for nonlinear minimax optimization," *SIAM J. Sci. Statist. Comput.* **1**, 345–370.

Murray, W., and M. L. Overton (1981), "A projected Lagrangian algorithm for nonlinear l_1 optimization," *SIAM J. Sci. Statist. Comput.* **2**, 207–224.

Nelder, J. A., and R. Mead (1965), "A simplex method for function minimization," *Comput. J.* **7**, 308–313.

Oren, S. S., (1974), "On the selection of parameters in self-scaling variable metric algorithms," *Math. Prog.* **7**, 351–367.

Ortega, J. M., and W. C. Rheinboldt (1970), *Iterative Solution of Nonlinear Equations in Several Variables*, Academic Press, New York.

Osborne, M. R. (1976), "Nonlinear least squares—the Levenberg algorithm revisited," *J. Austral. Math. Soc.* **19** (Series B), 343–357.

Powell, M. J. D. (1970a), "A hybrid method for nonlinear equations," in *Numerical Methods for Nonlinear Algebraic Equations*, P. Rabinowitz, ed., Gordon and Breach, London, 87–114.

Powell, M. J. D. (1970b), "A new algorithm for unconstrained optimization," in *Nonlinear Programming*, J. B. Rosen, O. L. Mangasarian, and K. Ritter, eds., Academic Press, New York, 31–65.

Powell, M. J. D. (1975), "Convergence properties of a class of minimization algorithms," in *Nonlinear Programming* 2, O. Mangasarian, R. Meyer, and S. Robinson, eds. Academic Press, New York, 1–27.

Powell, M. J. D. (1976), "Some global convergence properties of a variable metric algorithm without exact line searches," in *Nonlinear Programming*, R. Cottle and C. Lemke, eds., AMS, Providence, R. I., 53–72.

Powell, M. J. D. (1981), "A note on quasi-Newton formulae for sparse second derivative matrices," *Math. Prog.* **20**, 144–151.

Powell, M. J. D., and Ph. L. Toint (1979), "On the estimation of sparse Hessian matrices," *SIAM J. Numer. Anal.* **16**, 1060–1074.

Pratt, J. W. (1977), "When to stop a quasi-Newton search for a maximum likelihood estimate," Harvard School of Business WP 77-16.

Reid, J. K. (1973), "Least squares solution of sparse systems of non-linear equations by a modified Marquardt algorithm," in *Proc. NATO Conf. at Cambridge, July 1972*, North-Holland, Amsterdam, 437–445.

Reinsch, C. (1971), "Smoothing by spline functions, II," *Numer. Math.* **16**, 451–454.

Schnabel, R. B. (1977), "Analysing and improving quasi-Newton methods for unconstrained optimization," Ph.D. Thesis, Cornell Computer Science TR-77-320.

Schnabel, R. B. (1982), "Convergence of quasi-Newton updates to correct derivative values," in preparation.

Schnabel, R. B., B. E. Weiss, and J. E. Koontz (1982), "A modular system of algorithms for unconstrained minimization," Univ. Colorado Computer Science, TR CU-CS-240-82.

Schubert, L. K. (1970), "Modification of a quasi-Newton method for nonlinear equations with a sparse Jacobian," *Math. Comp.* **24**, 27–30.

Shanno, D. F. (1970), "Conditioning of quasi-Newton methods for function minimization," *Math. Comp.* **24**, 647–657.

Shanno, D. F. (1978), "Conjugate-gradient methods with inexact searches," *Math. of Oper. Res.* **3**, 244–256.

Shanno, D. F. (1980), "On the variable metric methods for sparse Hessians," *Math. Comp.* **34**, 499–514.

Shanno, D. F., and K. H. Phua (1978a), "Matrix conditioning and nonlinear optimization," *Math. Prog.* **14**, 145–160.

Shanno, D. F., and K. H. Phua (1978b), "Numerical comparison of several variable metric algorithms," *J. Optim. Theory Appl.* **25**, 507–518.

Shultz, G. A., R. B. Schnabel, and R. H. Byrd (1982), "A family of trust region based algorithms for unconstrained minimization with strong global convergence properties," Univ. Colorado Computer Science TR CU-CS-216-82.

Sorensen, D. C. (1977), "Updating the symmetric indefinite factorization with applications in a modified Newton's method," Ph.D. Thesis, U. C. San Diego, Argonne National Labs Report ANL-77-49.

Sorensen, D. C. (1982), "Newton's method with a model trust region modification," *SIAM J. Numer. Anal.* **19**, 409–426.

Steihaug, T. (1981), "Quasi-Newton methods for large scale nonlinear problems," Ph.D. Thesis, Yale University.

Stewart, G. W., III (1967), "A modification of Davidon's method to accept difference approximations of derivatives," *J. ACM* **14**, 72–83.

Stewart, G. W., III (1973), *Introduction to Matrix Computations*, Academic Press, New York.

Strang, G. (1976), *Linear Algebra and Its Applications*, Academic Press, New York.

Toint, Ph. L. (1977), "On sparse and symmetric matrix updating subject to a linear equation," *Math. Comp.* **31**, 954–961.

Toint, Ph. L. (1978), "Some numerical results using a sparse matrix updating formula in unconstrained optimization," *Math. Comp.* **32**, 839–851.

Toint, Ph. L. (1981), "A sparse quasi-Newton update derived variationally with a non-diagonally weighted Frobenius norm," *Math. Comp.* **37**, 425–434.

Vandergraft, J. S. (1978), *Introduction to Numerical Computations*, Academic Press, New York.

Wilkinson, J. H. (1963), *Rounding Errors in Algebraic Processes*, Prentice-Hall, Englewood Cliffs, N.J.

Wilkinson, J. H. (1965), *The Algebraic Eigenvalue Problem*, Oxford University Press, London.

Wolfe, P. (1969), "Convergence conditions for ascent methods," *SIAM Review* **11**, 226–235.

Wolfe, P. (1971), "Convergence conditions for ascent methods. II: Some corrections," *SIAM Review* **13**, 185–188.

Zirilli, F. (1982), "The solution of nonlinear systems of equations by second order systems of o.d.e. and linearly implicit A-stable techniques," *SIAM J. Numer. Anal.* **19**, 800–815.

Author Index

Subject Index